SCHOOL EFF

CONTEXTS OF LEARNING
Classrooms, Schools and Society

Managing Editors:

Bert Creemers, *GION, Groningen, The Netherlands*
David Reynolds, *School of Education, University of Newcastle upon Tyne, England*
Sam Stringfield, *Center for the Social Organization of Schools, Johns Hopkins University*

SCHOOL EFFECTIVENESS
COMING OF AGE IN THE TWENTY-FIRST CENTURY

PAM SAMMONS
Reader in Education
Institute of Education
University of London

PUBLISHERS

| LISSE | ABINGDON | EXTON (PA) | TOKYO |

Library of Congress Cataloging-in-Publication Data

Sammons, Pam
   School effectiveness / Pam Sammons.
         p.   cm.
   Includes biographical references and index.
   ISBN 9026515499 (hbk.) —ISBN 9026515502 (pbk.)
   1. Educational evaluation—Great Britain. 2. Educational accountability—Great Britain. 3. School improvement programs—-Great Britain—Evaluation. 4. Education—Standards—Great Britain.
   I. Title. II. Series
LB2822.75.S24 1999
379.1'58'—DC21   99-33302
                                                                              CIP

Cover design: Ivar Hamelink
Typesetting: Red Barn Publishing, Skeagh, Skibbereen, Co. Cork, Ireland
Printed in The Netherlands by Krips b.v., Meppel

Copyright © 1999 Swets & Zeitlinger B.V., Lisse, The Netherlands

All rights reserved. No part of this publication may be reproduced, stored in a retrieval system, or transmitted in any form or by any means, electronic, mechanical, photocopying, recording, or otherwise, without the prior written permission of the publishers.

ISBN 90 265 15499 Hardback
ISBN 90 265 15502 Paperback

Contents

Acknowledgements	vii
Introduction	ix

Part 1: The Impact of Intake: Measuring School Effectiveness

Chapter 1: The Influence of Background on Educational Outcomes	1
Chapter 2: Contextualising School Performance	21
Chapter 3: Issues in School Effectiveness Research	71
Chapter 4: Differential School Effectiveness	101
Chapter 5: Continuity of School Effects	129

Part 2: Understanding School Effectiveness

Chapter 6: The Question of Education Quality	149
Chapter 7: Effective Teaching – Findings from the 'School Matters' Research	173
Chapter 8: Key Characteristics of Effective Schools	183
Chapter 9: Theory and School Effectiveness Research	227

Part 3: Using School Effectiveness Research: Implications for Raising Standards and School Improvement

Chapter 10: The Practitioners' Perspective	253
Chapter 11: Using Value Added Research	271
Chapter 12: Evaluating School Improvement	291
Chapter 13: Beyond the Millennium: Future Directions for School Effectiveness Research	339
References	361
Subject index	389

For David, Bethan and Megan

Acknowledgements

Portions of the following chapters were adapted from previously published material, listed below, and are reprinted with permission. Chapter 1: Sammons, P. et al. (1983) 'Educational Priority Indices: A new perspective' *British Educational Research Journal 9, (1): 27–40.* Chapter 2: Sammons, P. et al. (1994) 'Assessing School Effectiveness: Developing measures to put school performance in context' *extract from the full report for OFSTED.* Chapter 3: Sammons, P. (1996) 'Complexities in the judgement of school effectiveness' *Educational Research and Evaluation 2, (2): 113–149.* Chapter 4: Sammons, P. et al. (1993) 'Differential School Effectiveness: Results from a reanalysis of the Inner London Education Authority's Junior School Project Data' *British Educational Research Journal 19, (4): 381–405.* Chapter 5: Sammons, P. et al. (1995) 'Continuity of School Effects: A longitudinal analysis of primary and secondary school effects on GCSE performance. *School Effectiveness and School Improvement 6, (4): 285-3-7.* Chapter 6: Sammons, P. (1994) 'Findings from School Effectiveness Research: Some implications for improving the quality of schools. *Improving Education: Promoting Quality in Schools 4, 32–51.* P. Ribbens and E. Surridge (eds.) Cassells. Chapter 7: Mortimore, P. et al. (1987) For Effective Classroom Practice' *Forum 30, (1): 8–11.* Chapter 8: Sammons, P., Hillman, J. & Mortimore, P. (1995) 'Key Characteristics of Effective Schools: A review of school effectiveness research' *OFSTED.* Hamilton, D. (1996) 'Peddling Feel-good Fictions' *Forum, 38, (2): 54–56.* Sammons P., Mortimore, P. & Hillman J., (1996) 'Key Charicteristics of Effective Schools: a response to Peddling Feel-good Fictions' *Forum, 38, (3): 88–90.* Chapter 9: Sammons, P., Thomas, S. & Mortimore, P. (1997) 'Towards a model of academic effectiveness in schools' *Forging Links: Effective Schools & Effective Departments Chapter 8.* London: Paul Chapman. Chapter 10: Sammons, P. et al. (1998) 'Practitioners' Views of Effectiveness' *Improving Schools 1, (1): 33–40.* Chapter 11: Sammons, P. & Smees, R. (1998) 'Measuring Pupil Progress at KS1: Using baseline assessment to investigate value added' *School Leadership and Management.* Chapter 12: Taggart, B. & Sammons, P. (1998) 'Evaluating the impact of the Raising Standards Initiative in Belfast' *Paper presented at the annual meeting of the American Educational Research Association, San Diego.*

I am very grateful to Bert Creemers and David Reynolds, the series editors, who encouraged me to write this volume and provided valuable advice on its structure and content. I also wish to record my appreciation to the many colleagues involved in the different studies featured here for their important contributions to my thinking and to the staff and pupils in schools which cooperated with the various research projects over the years.

Thanks are also due to Peter Mortimore and all my colleagues in the International School Effectiveness and Improvement Centre at the London Institute of Education with whom I have debated ideas and findings over the last six years, and who share my excitement at the continuing challenge of linking research and practice.

Lastly, I would like to express particular thanks to Ros Marsh and Adrian Walker for their patience and care in collating the articles and references and to Ros Marsh for the major task of word processing the final manuscript.

Introduction

Has school effectiveness research come of age? After 18 years personally engaged in investigating the topic, this seems to me an important question both for practitioners and policy makers seeking to make use of its findings as well as for researchers seeking a better understanding of the impact of their work. It is difficult to pin point the 'start' of school effectiveness research exactly since many different sub-disciplines have studied schools and classrooms from a variety of perspectives (Creemers, 1994a). Nonetheless, in the US and UK the chief catalyst seems to have been the publication of work by Coleman (1966) and Jencks et al. (1972). These researchers suggested that, although schooling was undoubtedly important, the *particular* school attended by a student had little influence on their educational outcomes in comparison with the influence of factors such as IQ, race, and socio-economic status (SES). The sociological determinism evident in mainstream educational research in the UK which focused on structural inequalities, in particular, meant that studies of school effectiveness were initiated by 'outsiders' working in different traditions, and the early results of seminal studies such as *Fifteen Thousand Hours* (Rutter et al., 1979) were regarded with hostility by many (Reynolds, 1997). As we enter the new millennium it is right to look back over the last quarter century or so of school effectiveness research and reflect on its growth and development, the problems encountered and overcome, continuing challenges and directions for the future.

My first encounter with school effectiveness research came as a post graduate student studying a very different field – vocational further education and the factors influencing young people's choices at the end of compulsory schooling. In 1979 I read *Fifteen Thousand Hours* with fascination, although at that stage I was primarily interested in the concept of neighbourhood or area effects on students' achievements and choices rather than with school influences *per se*, a reflection of my geographical background. The school effectiveness seed was sown, however, and in less than 18 months I was fortunate to be offered the chance to participate in the larger follow up *School Matters* (Mortimore et al. 1988a) research study which focused on primary schools and was led by one of the Rutter team members, Peter Mortimore, then Director of Research and Statistics in the ILEA.

When asked to write a volume for the Contexts of Learning series, I was doubtful about making any claims to a special contribution to the development of the school effectiveness field. Much of my research experience has involved working with teams of academics and practitioners, a fruitful and often provocative combination, frequently engendering much heat as well as light, as we debated – often passionately – the aims, methods, results and their interpretation of our investigations. The results of my research thus reflect this collaboration. Indeed, I would argue that successful studies require just such team work and contributions from a variety of perspectives and backgrounds. It is quite notable that many involved in leading school effectiveness studies were originally drawn from other disciplines, rather than main stream educational research as Reynolds, (1997) observed.

From my early work as a research officer with the former Inner London Education Authority's Research and Statistics Branch, I was fortunate to have worked with two of the leading authorities in the school effectiveness field, Peter Mortimore and later Desmond Nuttall. They both provided unfailing encouragement and exciting opportunities to engage in major research projects which provided a valuable training ground and expanded my interest in the academic study of schools and their influence on students. The *School Matters* research proved to be particularly influential due to the long time scale over which an age group of children were followed – nine years in total.

Over the last 18 years the development of my research career has both reflected and helped to shape the growth in research and in policy and practitioner interest in school effectiveness and its potential as a catalyst for school improvement. From the time of my first work in the area in 1981, the field has grown from a topic of marginal interest outside the main stream of educational research, to one of the major levers of current government policy in the UK which has set up a new Standards and Effectiveness Unit at the Department for Education and Employment (DfEE), with the explicit intention of leading its school improvement agenda. This strategy provides a bold though controversial attempt to use research results to influence practice. The extent to which the research message can be simplified and made accessible to a wider audience, and the difficulties in translating research findings into practice, remain considerable challenges to all in the field. The limitations as well as the strengths of school effectiveness approaches need to be recognised and acknowledged, so that unrealistic expectations of the ability to bring about radical changes in students' educational performance are not raised, and yet without dampening concern to explore and document how schools can make a difference, and avoiding a return to the sociological determinism which has often played down the school's influence, and led to a culture of low expectations for particular groups of students in some schools.

Controversy, of course, is not new to school effectiveness research. Indeed it has been part and parcel of the field from its early days in both the US and UK as noted earlier. For example, Peter Mortimore has provided a fascinating account of the reception accorded both the *Fifteen Thousand Hours* secondary school study and the later *School Matters* primary school research with which I was involved as research co-ordinator for five years (see Mortimore, 1998). The challenge to the sociological determinism (Mortimore, 1995a,b; Reynolds, 1995) of educational research in the 1970s, and the methodological debates surrounding the most appropriate techniques required to unpack the influences of students' backgrounds from those of their schools in seeking to understand the range in their educational outcomes, have remained continuing sources of tension. Even greater argument has centred around philosophical issues pertaining to the question of how best to approach the study of schools and teachers, and to explore their influences on students.

The question of values in education and their influences on students
Questions about values in education, the purposes of schooling, the quality of students' educational experiences, and of what constitutes a 'good school' rightly remain the subject of much argument and are unlikely to be resolved (White & Barber, 1997). Views rightly differ amongst practitioners, parents, and students as well as amongst policymakers. The chief contribution of school effectiveness research, which I believe will continue to be a focus of future research studies, is to seek to disentangle the complex links between the student's 'dowry' (the mix of abilities, prior attainments, and personal and family attributes) which any student brings to school, from those of their educational experiences at school, and the way these jointly influence their later attainment, progress, and development. Critics of school effectiveness have argued that, if the teacher–student learning relationship is 'right', then the educational outcomes will take care of themselves (Elliott, 1996). Against this the need to gauge learning (which cannot be observed) by measuring its outcomes in some way, and to investigate how these are influenced by teachers' classroom practices and by wider features of school process over several years, has been argued by proponents of school effectiveness research (Sammons & Reynolds, 1997). Indeed, the very term 'right' is in my view essentially problematic, since different groups of practitioners, parents, and students may quite justifiably have very different views. Fitness for purpose surely needs to be explored before we can judge what is 'right'. How can we assess what is 'right' without studying the impact of different approaches to classroom practice on students?

Rather than being viewed as a panacea for all educational ills (real or imagined), I believe school effectiveness is most appropriately seen and used as a method of increasing our understanding of school and classroom processes, and the way these can influence students' educational outcomes. Such research provides much needed empirical evidence which should assist in the essential process of the evaluation and critique of classroom practice and educational policy (Sammons, Thomas & Mortimore 1997; Reynolds, 1997; Gray, 1998). It should not be treated prescriptively and, of course, cannot of itself engender improvement. School effectiveness at its best, however, seeks to empower practitioners by this increased understanding of the way school and classroom organisation and teaching behaviour can influence both student attainment and progress *and* social and affective outcomes valued by society. In this way it can stimulate reflection, self-evaluation and review (Mortimore & Sammons, 1997), all of which are essential to the development of teachers' professional practice, as well as for instructional development.

The question of the coming of age of school effectiveness research thus requires us to balance the positive contribution made by the growing field of increasing knowledge about school and teacher influences on students, against the dangers of prescriptive attempts to decontextualise and over-simplify results. School effectiveness cannot provide 'quick fixes' for schools in difficulty (Reynolds, 1996; Stoll & Myers, 1997). The research base has provided an important and powerful critique of raw league table approaches to school accountability (a topic discussed in more depth in subsequent chapters). In the UK and in Australia the misleading nature of raw league table publication of schools' assessment and public examination results has become increasingly recognised and the need for alternative approaches accepted (see Gray, Goldstein & Jesson, 1996; Hill & Rowe, 1996). The ideal of 'like with like' comparisons and the concept of assessing the value added by schools to student outcomes by measuring progress over time, show the direct impact of school effectiveness research on public debate. The methodological advances of school effectiveness research in recent years, particularly through the development of multilevel approaches (Goldstein, 1995), has enabled a better understanding of the relative importance of the contribution of the school to student outcomes, and highlighted issues such as the need to take note of the statistical uncertainty attached to estimates of individual school effects. Such studies reveal the benefits of using multiple measures, and of longitudinal designs which study outcomes over several years. Nonetheless, there remains a danger of potential misuse of the research for purposes of accountability only, rather than as one way of assisting schools in the process of self-evaluation and review, and in encouraging practitioners to discuss and evaluate different approaches to school and classroom organisation and practice (Sammons, 1996; Goldstein, 1997).

If school effectiveness research is beginning to come of age as I believe from my personal experience of the field over 18 years, it is perhaps time to reflect once again on its origins which focused on equity and excellence for all (and concern in particular with the under-achievement of poor and ethnic minority students in disadvantaged urban contexts). It is also necessary to document the achievements and problems encountered during its formative years. Drawing on particular research studies with which I have been involved, this volume seeks to chart the sometimes muddy waters of school effectiveness research, highlight the limitations of the field and dangers of over simplification, and re-evaluate its contribution and legacy. From this analysis a number of conclusions are drawn about the requirements for further development of the field if it is to continue to have a positive impact on educational policy and practice as we move into the new millennium.

This book is divided into three main sections: "Measuring School Effectiveness – accountability and judging school performance"; "Understanding School Effectiveness – how do school and classroom processes influence students"; and "Using School Effectiveness Research – implications for raising standards and school improvement". Each section uses particular articles or chapters drawing on research conducted in a variety of contexts and for different audiences to identify and explore key issues and tease out underlying relationships. My choice of specific pieces is personal, but I have tried to make a choice which covers main themes of continuing relevance in the field and not merely an autobiographical or sequential approach. Inevitably, each piece of research has its own particular historical context which I have tried to summarise. There can sometimes be a desire to update or alter pieces in the light of later experience in any retrospective account written with the benefits of hindsight. However, I have tried to avoid this potential source of bias by seeking to use original pieces where possible in their entirety. Where selections have been made or my ideas changed I have addressed this by means of the introduction or sometimes a postscript to the chapter. With the opportunity for mature reflection, of course, I have been able to critique some of my original research contributions! Writing this book has provided a very valuable opportunity for me to evaluate their impact on academic debate or indeed sometimes on policy. Although this volume is written very much as a personal account and not all might agree with my interpretations, many of the pieces are jointly authored and reflect the important contribution of teams with which I have had the opportunity to work over the years and from whom I have learnt a great deal. Many other school effectiveness researchers have also shaped my ideas over the last two decades – particularly Peter Coleman, Bert Creemers, Peter Daly, Harvey Goldstein, John Gray, Peter Hill, David Hopkins, David Reynolds, and Jaap Scheerens, all of whom have commented on different studies and often suggested new avenues of investigation or alternative interpretations.

Research is rarely a journey into the completely unknown. Nonetheless, the route often encompasses unexplored byways and strange surprises. My experience of school effectiveness research has provided methodological as well as theoretical challenges, while its practical application is not always within the control of the academic. The impact of media coverage, in particular, is two edged and has not always proved positive. The 'holy grail' of raising standards and reducing social exclusion (currently very much at the forefront of educational policy development in the UK, but also a focus of many other national systems), is perhaps likely to remain a chimera, and it would be foolhardy to expect school effectiveness research to overturn deep seated inequalities in educational opportunity by itself. However, I believe it still has an important role to play in increasing our understanding of the complex relationships between influences on students' educational experiences and their subsequent outcomes. Equity and excellence for all may remain unattained goals but are undoubtedly worthy aspirations and attempting to reduce inequalities by improving opportunities and by fostering the *positive* impact of schools must surely remain at the forefront of the school effectiveness research agenda.

In recent years a number of writers such as West and Hopkins (1996) have drawn attention to the need to reconceptualise school effectiveness and improvement, pointing to the need for a clearer appreciation of the strengths and weaknesses of the two traditions, but also for a paradigm shift which involves "*a fundamental revision of existing views*" (West & Hopkins, 1996, p.3). Their analysis draws attention to measurement deficiencies in the SER tradition and an over-emphasis on the potential of managerialist solutions to the problems of so called ineffective schools. In their critique of school improvement, they note the tendency for many school improvement projects to focus on teachers' perspectives and concerns and on the potential of staff development, while frequently avoiding the question of impact on student learning and outcomes. They also note a tendency to seek to apply 'off the shelf' recipes rather than focus on the history and context of the particular school. West and Hopkins (1996) propose a more comprehensive model for the effective school which focuses on four domains:

- student achievements
- student experiences
- teacher and school development
- community involvement

and argue that a reconceptualisation of school effectiveness and school improvement is both timely and philosophically appropriate.

My own experiences working in the field for more than 18 years support this contention. I hope this volume, despite its limitations as a personal collection of work, helps to provide some indication of the achievements and limitations of SER. Over recent years my research has increasingly involved engagement with the issues of school improvement, particularly in my work as a member of International School Effectiveness & Improvement Centre (ISEIC) at the London Institute. Questions of policy and practitioner relevance and the development of joint ways of working feature highly in our concerns (matters discussed more fully in Chapter 13). The opportunity to re-examine particular pieces of work over the years has proved illuminating in helping to uncover unexpected patterns and connections in the field. Using the three basic divisions of *measuring*, *understanding* and *using* school effectiveness research, I hope that the material in this volume helps to identify key issues in SER and provide pointers to future directions in the field. My reflections suggest that SER may indeed be coming of age, making both intellectual and practical advances and through these an increasing contribution to the vital task of promoting the quality of schooling as we enter the 21st century.

Part 1

The Impact of Intake: Measuring School Effectiveness

How can we try to measure the influence of schools, and by implication of teachers, on their students? This deceptively simple question lies at the heart of school effectiveness research. Methodological controversy over this issue dogged the early years of the field, while question of values are intimately tied up with decisions about what is measured.

In many ways school effectiveness research reflects wider debates within the social and educational research communities about the merits and limitations of empirical research. The way we view the world affects our approach to academic enquiry. The ontological and epistemological assumptions underlying school effectiveness research place it in the positivist tradition. Although some studies have employed qualitative approaches, commonly via case studies of particular institutions, the more dominant mode is typified by large scale quantitative studies involving the longitudinal follow up of fairly large numbers of students and schools and by a search for generalisations rather than particularities. The use of quantitative methods, however, does not mean that school effectiveness research is necessarily deterministic or mechanistic in nature. Indeed, it stresses the probabilistic nature of the findings and highlights the need to measure change over time and the impact of context. The use of empirical approaches does, however, seek to obtain valid and reliable (replicable) measures of key variables whether they be of students' attainment or background, or of aspects of school processes such as school organisation or policies and classroom practice. This does not mean that the subjective views of those involved (students, parents or teachers) must be ignored, however. On the contrary, these are vital keys which help to illuminate our understanding of the experience of schools and the way in which school culture can develop and influence both staff and students.

Different sections of this book will address the methodological and philosophical approaches which underpin school effectiveness research. Some of the criticisms levelled at school effectiveness stem from disagreements about the attempt to focus on students' educational outcomes as the ultimate measure of effectiveness. School effectiveness researchers do not accept the assumptions of those from the action research tradition that the educational process is paramount. Rather, as Reynolds (1997) argues

> For us, our "touchstone criteria" to be applied to all educational matters concern whether children learn more or less because of the policy or practice (p. 97).

For example, Elliott (1996) claimed that teachers should focus on the quality of the teaching-learning process rather than on educational outcomes since, if the former is right, the students themselves will take care of the latter. Elliott also disparaged the concept of effectiveness (as defined by the school effectiveness tradition's emphasis on educational outcomes) in comparison with ideas of a 'good' school. In my view, however, such arguments are at best ambiguous (Sammons & Reynolds, 1997). How can we know when the teaching-learning process is 'right'? Who makes the judgement – teachers? – pupils? – parents? – inspectors? Learning, by its very nature is not observable and can only be gauged indirectly by measuring the outcomes of student learning in some way. Assessment, both informal and formal, for a variety of purposes – formative, diagnostic and summative – is hence an essential part of every teacher's repertoire.

> Rather than attempting to define "good", and thus by implication "bad" schools, school effectiveness research focuses deliberately on the narrower concept of effectiveness which concerns the achievement of educational goals measured by student progress ... promoting progress lies at the heart of the educational process and is accepted by teachers and students as well as parents and policy makers as one purpose, and probably the *fundamental purpose* of all schools (Sammons, Thomas & Mortimore, 1997, p. 8).

Measurement issues (questions of reliability and of validity outcomes) thus remain of fundamental concern to school effectiveness researchers. We need to examine both *what* is measured, the choices of student outcomes used to gauge progress and thus help define effectiveness, and the way such outcomes relate to the aims, goals and curriculum of schools, and *how* it is measured.

Of equal importance as the outcomes chosen is the issue of how we can try to measure school effectiveness. The question of how we can properly make 'like with like' comparisons of schools, given the marked differences in intake and the educational challenges faced by teachers in some schools, is crucial. Such intake differences often reflect schools' geographical locations as well as informal or formal selection processes which affect recruitment and the 'dowry' of previous educational experiences, attainments etc. which children bring to school.

There is considerable evidence that the characteristics of student intakes show a powerful link with educational outcomes. The legacy of the Coleman et al. (1966) and Jencks et al. (1972) research was to stimulate better methods to try to disentangle the influences of student background (including prior attainments) and of schools on educational attainment and progress. Measurement issues surrounding what factors should be included as controls in studies of school effectiveness remain hotly disputed. It is one matter to demonstrate a statistical relationship between say socio-economic status and student attainment in reading, for example, quite another to interpret this relationship. Questions of equity and social justice, and concerns over the possible self-fulfilling prophesy of a culture of low expectations for disadvantaged groups of students mean that school effectiveness research remains deeply controversial. The development of better methodological approaches for the analysis of relationships between student background, school processes, and educational outcomes, particularly multilevel modelling techniques (for example, Goldstein, 1987, 1995), although vital for the proper estimation of the relative size of school effects, does not remove the need for debate about measurements and their interpretation.

The first part of this book focuses on the measurement of school effectiveness and draws on both primary and secondary school research with which I have been involved to illustrate some of the key features of school effectiveness studies and complexities in the judgement of school effectiveness. The need for caution in interpretation and use of the statistical results is highlighted.

1

The Influence of Background on Educational Outcomes

Introduction

Classical educational sociology emphasised the impact of social class and inequality as determinants of students' (usually males') educational and later occupational outcomes, and was concerned with the reproduction of the social order. In the UK in particular, the selection and sorting function of schools operating via a 'high stakes' public examination system which controls access to higher education has received considerable attention. Longitudinal studies have highlighted the impact of students' social and family background, and developed concepts such as the cycle of disadvantage to account for inter-generational continuities in educational under-achievement (see Davie, Butler & Goldstein, 1972; Hutchinson, Prosser & Wedge, 1979; or Mortimore & Blackstone, 1982). The recognition of the need to widen educational opportunities evident during the 1960s by policy makers saw schools as one mechanism to help alleviate social disadvantage. Recognition of the existence of geographical concentrations of disadvantaged students in particular areas often, but not only, in inner cities, led to the development of educational priority areas (EPA) after the publication of the 1967 Plowden Report. This report drew attention to the problem of 'social handicaps', reinforcing 'educational handicaps', and advocated positive discrimination of resources to schools. The positive discrimination policy, although officially recognising links between social disadvantage and educational outcomes, and seeing schools as a potentially powerful agents for the alleviation of disadvantage, focused on the distribution of resources. Its area-based focus, however, meant that a major criticism could be levelled against the policy because greater numbers (although relatively lower concentrations) of disadvantaged students attended schools outside EPAs than within them.

My involvement in a major revision of the then Inner London Education Authority (ILEA)'s Education Priority Indices (EPI) in 1981 required consideration of many of the issues which became a focus of school effectiveness research during the next two decades. These include the nature of the links between particular background characteristics and student attainment, the impact of compositional or contextual effects related to the concentration of socio-economically disadvantaged students in some schools, and the need to consider social as well as academic outcomes and the links between the two.

The ILEA EPI research represented an advance on the earlier area-based EPA policy by its attempt to provide a better empirical and theoretical basis for the distribution of extra resources to schools. It focused on individual student level data, rather than aggregate information about schools' intakes and highlighted the necessity of using information about student intakes when examining the links between student background and attainment to avoid the potential danger of financially rewarding schools which under-performed. The major limitation, however, was the failure to examine explicitly the schools' contribution to student outcomes. The ILEA, while using the school as the means to distribute extra resources – in common with other LEAs at the time – did not investigate the way such extra resources were used and their impact, if any, on student achievement in general or for disadvantaged groups in particular. Only in this way could a proper evaluation of the impact of the educational priority policy be conducted.

A focus on developing methods to enable the use of intake information in the study of educational outcomes lies at the heart of developments in school effectiveness research during the 1980s and 1990s. The UK, of course, was not alone in the attention given to educational priority initiatives during the 1970s and early 1980s. For example, the Netherlands also has a strong tradition in this and, as in my own case, some educational researchers involved in these studies were also drawn towards the school effectiveness field in the 1980s (van der Werf, 1995).

The paper describing the ILEA's EPI research was co-authored with Peter Mortimore and Florisse Kysel. It is included in this volume in part because it was the gateway through which I entered the school effectiveness field. It stimulated my interest in the task of disentangling the powerful influences of student background from those of the school in shaping students' educational outcomes. This route led me from the geographer's interest in the spatial manifestations of educational and social phenomenon into the educationalist's concern with the ways in which individual and institutional factors combine to promote or inhibit students' learning, development and progress, and thus their educational outcomes.

The main reason for including the EPI paper, however, is not biographical but rather because it links in with the two driving forces which still underlie much school effectiveness research – the concerns with equity and excellence – in particular with promoting the achievement of the urban poor evident in early studies in the US and UK. Although the concept of educational priority fell into disuse during the 1980s in the UK and despite the fact that at one point it was considered politically unacceptable to highlight the links between social disadvantage and educational achievement, the situation changed radically after the election of a new Government in May 1997. The need to combat social exclusion and alienation has become a major concern of current Government policy, and in education the concept of Educational Action Zones is proposed to help raise standards in schools serving the most disadvantaged groups. Over the intervening 18 years, therefore, it is perhaps ironic that a return to area-based solutions is once more receiving political and policy attention. Despite the new interest in using education as perhaps the key focus of the current UK Government's strategy for combatting disadvantage via the Raising School Standards initiatives, the system for financing educational need, however, remains very confused, resulting in a 'geographical lottery'. At present there is no consistently applied system for allocating additional funds. In different areas students with similar characteristics attract different levels of resources to their schools in the absence of a national funding formula, a topic I reviewed in more detail in early 1990s when local management of schools (LMS) was instituted (Sammons, 1993).

The concept of educational priority and the funding of educational need, although ignored over much of the intervening period, remains, I believe, of considerable theoretical and practical relevance to those concerned with combatting educational disadvantage and raising standards. The ways in which school effectiveness research can contribute to the debate about the role of schools in promoting equity and can also enhance understanding about the limits to their capacity to effect social change will be returned to in the third section of this Volume which examines the uses of school effectiveness research in the context of education for the 21st century.

The origins of educational priority indices

In 1967 the Plowden Report recommended that priority should be given to schools where educational handicaps were reinforced by social handicaps. The report provided suggestions of objective criteria which could be used to identify both schools and areas which needed special help and, to determine how much assistance should be given to them. It provided a major stimulus for the development of policies of positive discrimination in the distribution of educational resources and, in particular, for the use of educational priority indices.

Halsey (1972) described the original index devised by the ILEA as the most sophisticated of those employed by any authority. This and subsequent indices have been used to distribute substantial sums of monies to schools over the last decade as a means of effecting positive discrimination in favour of schools with disadvantaged intakes. The design and methodology of the original ILEA index has been fully described by Little and Mabey (1972). These authors noted that, in constructing an educational priority index, they were limited by the criteria listed in the Plowden Report, time and the availability of suitable data. Conceptual problems were also encountered because the Plowden Report slid easily, in its discussions of educational priority, from areas to schools and to pupils as if those terms were interchangeable. Some data, used in the original ILEA index, were collected about pupils (for example, the proportion in receipt of free meals). Other data were obtained on an area basis (the percentage of males in semi- or unskilled manual work living in catchment areas ascribed to schools). The school, however, remained critical in the distribution of resources because the Plowden strategy had recommended using the experience of school as a means of compensating children for their disadvantages.

Criticisms of the ILEA index
The original ILEA index has been criticised on a number of grounds. The most widely publicised criticisms were made by Barnes and Lucas (1974) who demonstrated that the majority of disadvantaged children were not concentrated in a few areas and, therefore, that more disadvantaged children attended schools which were not designated for educational priority, than attended designated priority area schools. Analyses of census data by Holterman (1975) produced similar findings indicating that, at the small area level, although there was some concentration of the deprived, only a small proportion lived in areas covered by priority area policies.

On the basis of their findings, Barnes and Lucas (1974) claimed that the ILEA index failed as a means of ensuring positive discrimination because it did not direct extra resources to schools attended by the majority of disadvantaged children. However, the criticism that more disadvantaged pupils were outside educational priority schools than inside them, is not a valid criticism either of the method of constructing, nor the concept of, an educational priority index as such. Rather, it is a criticism of the way in which the first indices were used.

Originally the ILEA index was employed to distribute resources on an 'all or nothing' basis. Thus, some schools received priority area status because of their index score and were entitled to additional resources, while others did not receive such status. This method of using the index penalised schools which had above average numbers of disadvantaged pupils, but which just failed to reach the 'cut off' point used to ascribed priority status. Since 1974, however, the ILEA indices have been employed to distribute additional resources to all schools. The allocations are calculated on a formula combining pupil roll and the school's index score (see Shipman & Cole, 1975, for details). Additional resources are therefore allocated to schools according to the incidence of pupil disadvantage as measured by the indices.[1]

---

1 Substantial funds are currently allocated to a pool for distribution on this basis. In 1982–83 the additional resources allocated to the top ten and bottom ten primary schools according to index score were on average £84 and £25.5 per pupil respectively. For secondary schools the equivalent figures were £55 and £36 per pupil.

A more serious criticism is that levelled by both Barnes and Lucas (1974) and later by Shipman (1980) who claimed that the use of a school-based programme was a blunt instrument for implementing change. They argued that educational priority resources were wrongly directed at schools or areas, and should be given directly to the disadvantaged children or their parents. There is no evidence, however, that a policy of giving resources directly to individual disadvantaged children or to their parents, as advocated by Shipman (1980), would be more likely to raise these children's achievements – the aim of the Plowden initiative – than allocating resources to schools. In addition, such a method of distributing resources entails problems of the labelling of children, as well as of ensuring full take up, as Mortimore and Blackstone (1981) have recently noted. These difficulties beset all individually based means-related benefits.

An individual policy would also ignore the possibility of 'externality effects' (problems caused by concentrations of disadvantaged children in some schools which can influence all pupils' attainments).[2] At the area level Holterman (1977) claimed that externality effects (due to concentrations of the disadvantaged) may be of importance and Smith (1977) noted the existence of externality effects in his analysis of the performances of children in educational priority schools. Smith claimed 'There are fewer children than expected in high-scoring categories. The whole range of scores is depressed – not one particular category' (p. 276). Thus, even non-disadvantaged children in schools where there are high concentrations of disadvantaged pupils, do less well than their peers attending different types of schools.

At the school level, Little and Mabey (1973) also identified externality effects in an analysis of the reading attainments of primary pupils. These authors noted that pupils' attainments were influenced by the social composition of their schools. Similarly, at the neighbourhood level, Panton (1980) produced evidence of an externality effect in a study which indicated that the attainments of pupils of similar social backgrounds varied according to the social composition of their home area. These results support the thesis that the attainments of pupils in schools containing high proportions of disadvantaged pupils are likely to be lower than average. For this reason, allocating additional resources on a school basis, using educational priority indices, can take into account the possible negative externality effects on all children's attainments in schools where a high proportion of the pupils are disadvantaged.

The arguments for positive discrimination in the distribution of resources to schools have been reiterated by Tunley, Travers and Pratt (1979) in their analysis of the pattern of educational provision in Newham. This borough, although serving an area similar to the ILEA, had not implemented a policy of positive discrimination. Tunley, Travers and Pratt conclude that 'children from socially disadvantaged areas tend to attend schools which are less well provided than

---

2 An externality effect is defined as the influence of a concentration of disadvantaged children in a particular setting (e.g., a school or neighbourhood) upon the attainments of all children in that setting.

those attended by less disadvantaged children and that institutional mechanisms reinforce this relationship (p. 114). It thus appears that, where a policy of positive discrimination is not instituted, disadvantaged children may well attend schools which are less well provided for than schools attended by their more advantaged peers.

Revisions of the ILEA indices

Although the concept of educational priority indices has been accepted on grounds of social justice (see Glennerster & Hatch, 1974) and a school based policy can be justified, there is less agreement as to the criteria and data which should be used. The ILEA adopted a primary index in 1971 and a secondary index was agreed in 1973. During the last decade, biennial reviews of these indices have been used to adapt the criteria, alter measures and enable the collection of up-to-date information to reflect changes in schools' circumstances. The measures used in the original index are given in Table 1.1.

The Plowden report's recommendations and the availability and reliability of data were the major influences upon the criteria and data selected, although the requirement that efficiency should not be penalised was also taken into consideration.[3] The most fundamental change produced by these earlier revisions was the move away from an area to a pupil base for the collection of certain information. Census data were originally used to obtain information about the proportions of pupils from semi- or unskilled manual backgrounds and those living in poor housing conditions. However, these data did not provide a very reliable guide to pupils' occupational backgrounds or housing conditions because primary catchment areas are erratic in size and shape and it is, therefore, impossible to ascribe catchments to schools accurately without mapping all pupils' addresses. In addition, 1971 Census data rapidly became obsolete and, following the cancellation of the 1976 mid-term census, this source was no longer

Table 1.1   Measures Included in the Original Educational Priority Index.

- Social class composition
- Overcrowding
- Housing stress
- Family size
- Free meals
- Absenteeism
- Immigrant children
- Retarded/handicapped pupils
- Teacher turnover
- Pupil turnover

Note. Information on these criteria was originally obtained from census figures.

---

3   Thus, for example, a measure of teacher absence used in the original index was deleted because the inclusion of this factor could have the effect of penalising schools which maintained high staff morale despite difficult circumstances.

used. Data about occupational backgrounds were obtained directly from schools and, because information was not available, the housing measures were omitted from the indices.

The institution of biennial reviews resulted in some improvements in the measures and sources of data in constructing the ILEA's educational priority indices. However, the reviews were conducted on an *ad hoc* basis and no systematic examination of the value of the criteria, theoretical basis or method of constructing the indices was undertaken. In 1981, the Schools Sub-Committee, therefore, requested a major review of the educational priority indices used by the ILEA. A working party of heads' and teachers' representatives and officers of the authority was convened to examine the measures and methods employed, and to make recommendations for the construction of future indices.

Measures included in the primary and secondary indices in use in 1980 and those collected for the purpose of the revision exercise in 1981 are given in Table 1.2. The absence in the 1980 indices of any measure to take into account pupils who were not fully fluent in English (a criterion listed in the Plowden Report) and of ethnic family background – factors which have been found to be associated with educational outcomes (see Mabey, 1981) – penalised schools with high numbers of pupils with these characteristics. Data were therefore collected on these two characteristics in 1981.[4] Building factors were included in the sec-

Table 1.2 Measures Included in the 1980 EPA Indices and in the 1981 EPA Survey.

Primary index 1980	*Primary data collection 1981*
Eligibility for free meals	Eligibility for free meals
Large families	Large families
One-parent families	One-parent families
Parental occupation	Parental occupation
Behaviour	Behaviour
Verbal reasoning band	Pupil mobility
Pupil turnover	Fluency in English
	Ethnic family background

Secondary index 1980	*Secondary data collection 1981*
All measures included in the primary index and five building factors	All measures included in the 1981 primary data collection and verbal reasoning band
Split site	
Overcrowding of building	
Overcrowding of site	
Old buildings	
Tall buildings	

---

4 It should be made clear that we do not in any way regard 'ethnicity' as a disadvantage. The inclusion of membership of a minority ethnic group as a factor was decided on the criteria of underachievement by such pupils. Clearly under-achievement is related as much to the school system as to the performance of individuals.

ondary index in 1980, but had been removed from the primary index in 1978. These factors correlated very poorly with index scores at the school level and were of questionable relevance in an index designed to take into account social disadvantage. It was agreed therefore, that problems caused by poor school buildings should be dealt with by specific measures rather than be included in the educational priority indices.

Shipman (1980) had argued that the inclusion of measures of low attainment and disturbed behaviour penalised schools which managed to achieve high attainment or good behaviour in the face of difficult circumstances. However, this is only the case where attainment or behavioural data relate to school outcomes, rather than to their intakes. To overcome difficulties associated with the sources of data and measures used, information was collected in 1981 for the full intake years of infant, junior and secondary schools. This removed any confusion about some measures, such as behaviour, which could also represent school outputs. In addition, information was also collected from a sample of pupils in other years. This procedure provided a more representative information base. (In the past, indices were constructed from data collected mainly for fourth year junior pupils.)

Because of the strong research evidence of association with educational outcomes, the measures of pupils' social backgrounds included in previous indices (entitlement to free meals, large families, single parent families and parental occupation) were retained.[5] Fluency in English and ethnic family background were added. Behavioural data were collected for all pupils in the intake years and information about pupil mobility was obtained on a school basis.

In their description of the original ILEA index, Little and Mabey (1972) claimed the major problem was 'what weight should be given to each factor since almost certainly all are not equal' (p. 83). However, because they could find no empirical justification for a differential weighting scheme, Little and Mabey decided upon a system of equal weightings for each measure. Carley (1981), in a review of the use of social indicators, argued that such a decision also entailed problems because, 'If no differential weighting scheme is used, that is each indicator is equally weighted, the prior choice of indicators becomes all important and thus, in effect, simply transfers the value weighting to the choice of indicator' (p. 80).

Additional difficulties are found in Little and Mabey's attempt to ensure equal weighting of the factors. Their method was designed to enable each factor to contribute an equal amount to the final index. In practice, however, their method did not treat each factor as of equal importance at the school level.[6] Although the method of constructing the ILEA indices was later slightly revised, it remained a method based on standardisation. The percentage of pupils scoring on each measure was converted into a standardised Z-score using the formula:

---

5 See, for example, Douglas (1964), Davie, Butler and Goldstein (1972) or Rutter and Madge (1976).
6 Little and Mabey (1972) calculated a school's score on each factor as follows: →

$$Z = 50 + \left(\frac{X-\bar{X}}{S} \times 20\right)$$

where $\bar{X}$ = Mean score for all schools on factor X
X = Score of a school on factor X
S = Standard deviation for factor X

The school's final index score was the mean of these Z-scores. This method ensures that a high score on a measure with a low mean and small standard deviation will count more than a high score on a measure with a high mean and standard deviation. Thus, to have 50 per cent pupils from one-parent families would increase a school's score more than to have 50 per cent pupils eligible for free meals.[7] This effect is not desirable because it takes no account of the relationships between educational disadvantage and these measures. Results of

---

If $\bar{X} + 2S >$ X score = 0
If $\bar{X} + 2S <$ X score = 100

$$Y = \left(\frac{X-(\bar{X}-2S)}{4S}\right) \times 100$$

Thus, if for free meals $\bar{X}$ = 35.7
  S = 13.2
and for one-parent families $\bar{X}$ = 23.4
  S = 8

a school with 50 per cent pupils eligible for free meals would obtain a score of 77.1, while a school with 50 per cent from one-parent families a score of 100.

7 Using a standardised scores method a school with 50 per cent pupils eligible for free meals would obtain a lower score than one with 50 per cent from one parent families.

$$Z = 50 + \left(\frac{50-35.7}{13.2} \times 20\right) = 71.67$$

Free meals $\bar{X}$ = 35.7
  S = 13.2
One-parent families $\bar{X}$ = 23.4
  S = 8

$$Z = 50 + \left(\frac{50-23.4}{8} \times 20\right) = 116.50$$

analyses of pupil based data collected in 1981 indicated that free meals was more closely associated with low attainment than one-parent families. Research reported by Essen and Wedge (1982) has produced similar findings.

Another disadvantage of using the 'standardised scores' method of constructing educational priority indices is that it provides a relative rather than an absolute measure of disadvantage. Because a school's priority score is influenced by the average for all ILEA schools, its score could change from year to year if the ILEA average changed – even though the proportions of disadvantaged pupils in the school had not altered.[8] The use of standardised scores to construct the indices therefore obscures any increase in absolute levels of disadvantage amongst school pupils. In a time of economic recession, when the proportions of pupils with parents unemployed and those eligible for free meals are rising, this is particularly undesirable.

In addition, the calculation of educational priority index scores using the standardised method takes no account of the incidence of multiple disadvantage. Many studies have suggested that multiply disadvantaged pupils have a much higher risk of poor educational achievement than non-disadvantaged children. In the past, the 'standardised scores' method was employed because not all the measures were in the same form. Originally individual data were not available and, therefore, it was not possible to assess the extent of multiple disadvantage, nor to devise an acceptable method of weighting the different factors.

The 1980 indices were, of course, open to the criticisms that can be applied to the choice of any criteria. However, more fundamentally perhaps, the lack of any theoretical basis on which the choice of criteria could be made and the method of combining the criteria were seen as major weaknesses. In order, therefore, to provide a better foundation, all measures that could be used as criteria in the indices, were subjected to a series of analyses designed to test the relationship between the measure and educational outcome.

---

8 Thus, in 1980 the mean proportion of pupils eligible for free meals was 28.5 with a standard deviation of 12.6. A school with 50 per cent pupils eligible for free meals would receive a higher score than a school with 50 per cent pupils eligible for free meals in 1981 (because of a change in the mean and standard deviation for schools as a whole).

$$Z = 50 + \left(\frac{50 - 28.6}{12.6} \times 20 \right) = 83.97$$

$$Z = 50 + \left(\frac{50 - 35.7}{13.2} \times 20 \right) = 71.67$$

1980 $\bar{X} = 28.6$
$S = 12.6$
1981 $\bar{X} = 35.7$
$S = 13.2$

During the 1981 revision exercise the educational priority data were collected and analysed on an individual basis. It was thus possible to examine the strength of associations between the various measures and to investigate their relationship with educational outcomes and pupil behaviour at the individual level. The extent to which pupils experienced multiple disadvantages was also established and a method devised to take into account the effects of multiple disadvantage in constructing future educational priority indices.

Analyses of the educational priority data

Measures of the associations between the various educational priority measures at the pupil level indicated that, although many of the measures were significantly associated, the strength of the associations was not great. Thus, for example, although 69.8 per cent pupils from one-parent families were also eligible for free meals, this group represented only 11.1 per cent of the total group eligible for free meals. Because none of the measures was very strongly related with any other measure at the pupil level, it was not possible to exclude any one measure from the calculation of future educational priority indices on the basis of duplication.

In order to establish the extent of cumulative disadvantage, the percentages of pupils affected by different numbers of the educational priority criteria were examined for the full intake years of infant, junior and secondary schools. The results, presented in Table 1.3, indicated that the percentages of infant, junior and secondary pupils experiencing different numbers of factors were similar. Thus, only a minority of pupils (less than a quarter) were classified as not experiencing any of the educational priority measures, while nearly a third were affected by three or more of the criteria.

The percentages of secondary pupils affected by different numbers of factors scoring in the bottom quartile of a test of verbal reasoning (an indicator of educational performance) were calculated and figures presented in Table 1.4. These figures were compared to establish the relationship between the effect of multiple disadvantage and educational outcome. A very strong association was found. Only 11 per cent of pupils not affected by any factor were in the lowest verbal reasoning band, compared to 92 per cent of those affected by all measures. A similar relationship between multiple disadvantage and disturbed pupil behaviour was identified. In both cases the effect of the educational priority measures appeared to be cumulative.[9]

---

9 As part of the secondary transfer process all pupils take a verbal reasoning test in the last year at primary school. All pupils are assigned to one of three bands on the basis of test scores and teachers' judgements. It was considered justified to use VR band as a guide to the level of educational outcome because, although only a crude measure, it is strongly predictive of later public examination performance (see Byford & Mortimore, 1981). However, if more complete attainment data had been available, it would have been desirable to conduct more sophisticated analyses of the relationships between the priority criteria and educational performance.

Table 1.3 The Percentages of Infant, Junior and Secondary Pupils Affected by Different Numbers of Factors.

Number of factors*	Infant pupils	Junior pupils	Secondary pupils
0	22.7	22.1	20.0
1	22.4	22.7	25.0
2	19.7	18.9	21.9
3	17.8	17.0	16.4
4	11.3	11.7	11.0
5	5.0	5.9	4.6
6	1.1	1.7	1.1
7**	0.1	0.1	0.1
	$n = 17368$	$n = 17540$	$n = 17216$

Note. Pupils for whom any item of data was missing were excluded.
* VR band was excluded. This was available for secondary pupils only.
** Very few pupils experienced seven of the priority measures because few pupils not fully fluent in English came from one-parent families (a reflection of the traditions of certain ethnic groups).

Table 1.4 Percentages of Secondary Pupils Experiencing Different Numbers of Educational Priority Criteria by Measures of Educational Outcome and Behaviour.

Number of factors*	% Pupils VR Band 3	% Pupils with disturbed behaviour
0	10.8	5.6
1	16.7	9.4
2	25.5	14.7
3	32.2	20.4
4	38.6	25.1
5	49.1	28.6
6	61.5	32.7
7	91.7	42.3
$n = 22241$		

*VR Band was included in the number of factors when the relationship with disturbed behaviour was investigated. Disturbed behaviour was counted as a factor when the relationship with VR Band was examined.

The results are presented for secondary pupils because a measure of educational outcome was available for this group. Very similar associations were found, however, between the number of factors and disturbed behaviour for infant and junior pupils. It is probable, therefore, that the relationships between multiple disadvantage and educational outcome at the primary level would follow the same pattern as that for secondary pupils. Unfortunately, it was not possible to test this hypothesis because no measure of educational attainment was available for the infant or junior groups.

A new method of constructing educational priority indices

In previous years the methodology used to construct the indices did not allow the measures to be weighted according to the contribution made to educational outcome. No account was, therefore, taken of the numbers of factors, nor of the relationship between different combinations of the educational priority criteria and a pupil's educational performance. It seemed probable that certain combinations of the educational priority measures might be more closely associated with educational performance than others. The results of analyses indicated that, when experienced in isolation, certain of the factors were better predictors than others of poor educational performance. Thus, for example, of pupils scoring on the free meals criterion alone, 21.1 per cent were in VR Band 3. By contrast, 13.1 per cent of pupils scoring only on the one-parent family factor were in the lowest achievement band. The figure for the one-parent family group confirms the results of analyses by Essen and Wedge (1982) who noted that when adjusted for housing and income, 'children in one-parent families have similar average scores to those in small two-parent families' (p. 111). However, in combination with other factors, membership of a one-parent family did increase the probability of low achievement.

Although the effects of the various educational priority measures are cumulative, they are not strictly additive, as figures in Table 1.5 indicate. Thus, 21.1 per cent of pupils scoring only on the free meals factor were in VR Band 3 and 18.4 per cent of those scoring only on the measure of social class. However, 26.7 per cent of those scoring on both the large family and the social class categories were in the lowest achievement band.

Few projects have attempted to predict educational failure as a means of guiding a policy of intervention. Alberman and Goldstein (1970) looked at the performance of an 'At Risk' register of children who might be disproportionately susceptible to specific handicaps because of perinatal circumstances.

Table 1.5 Percentages of Secondary Pupils with Different Combinations of Characteristics in VR Band 3.

Combination of characteristics	% Pupils in VR Band 3
Free meals only	21.1
Large families only	13.0
Social class* only	18.4
Behaviour only	35.8
Social class and free meals	26.7
Social class and large families	29.1
Free meals, large families and social class	31.8
Free meals, large families, social class and behaviour	55.7

*Pupils with parents in semi- or unskilled manual employment and those with parents not employed were included in one category because the proportions of pupils in these groups in VR Band 3 were very similar.

Similarly, Hutchinson, Prosser and Wedge (1979) attempted to identify a group of 'at risk' children using analyses which predicted pupil attainment at age 16 on the basis of characteristics at ages 7 and 11. Although social class, large families and poor accommodation were factors associated with lower attainment, performance at age 11 was the strongest predictor of later attainment. However, as the authors noted, this was not a complete explanation 'since we are still left to ask what causes eleven year old attainment' (pp. 74–75).

It is possible to use the concept of predicting the probability or 'risk' of low attainment, on the basis of social characteristics, as part of a policy of positive discrimination. A new method of constructing the priority indices was developed based on the probability or 'risk' of a child with a specific combination of characteristics being in the lowest attainment band at age 11. For each combination of the various measures this method calculates the probability or 'risk' that a pupil with a specific combination of characteristics would be assigned to VR Band 3. Using the 1981 data each pupil in the intake and sample groups was given the appropriate weight on the basis of the combination of measures on which he or she scored. The sum of these individual pupil weights is expressed as a percentage of the number of pupils, giving an index score for the school.

$$\text{School's Index Score} = \left(\frac{\sum Wt_i + \sum Wt_s}{N_i + N_s}\right) \times 100$$

Where: ΣWt_i = Sum of pupil weights for the intake year (calculated as the probability of a pupil being educationally disadvantaged given his or her score on the educational priority measures)
ΣWt_s = Sum of pupil weights for the sample
N_i = Number in intake group
N_s = Number in sample group

Because individual mobility data had not been collected it was not possible to examine the relationship between mobility and the other educational priority measures at the individual level. Therefore, the mobility measure could not be treated in the same way as the other measures in constructing the new index. It is possible that mobility only becomes a disadvantage for pupils when it is experienced in combination with other disadvantages, but it was not possible to test this hypothesis given the data available. However, it was possible to establish the relationship between mobility and the other priority measures at the school level. This was because mobility data was collected for the sample year groups for whom data had been collected about pupils born on specific dates.

Correlation coefficients were calculated to establish the associations between the different measures at the school level, and the figures for secondary schools are presented in Table 1.6. The results for infant and junior schools were similar and are therefore not included.

Table 1.6 Correlations Between the Priority Measures for Secondary Schools.

	FM	LF	FS	PO	Not fluent	Mobility	VR 3
FM	1.000	0.508	0.570	0.679	0.427	0.383	0.622
LF		1.000	0.194	0.441	0.471	0.360	0.457
FS			1.000	0.402	0.042 n/s	0.214	0.350
PO				1.000	0.414	0.324	0.501
Not fluent					1.000	0.509	0.429
Mobility						1.000	0.335
VR 3							1.000

Where: FM = % pupils eligible for free meals
LF = % pupils from large families
FS = % pupils from one-parent families
PO = % pupils with parent(s) in semi/unskilled manual work or not employed
Not fluent = % pupils not fully fluent in English
Mobility = % pupils mobile over previous 12 months
VR 3 = % pupils from Verbal Reasoning Band 3.

*Behavioural data were not collected for pupils in the sample years.
$p < 0.05$ unless indicated n/s

Mobility was significantly positively associated with all the educational priority measures at the school level of analysis. The strongest association was with the proportion of pupils not fully fluent in English and it is possible that some schools where many pupils join over a twelve-month period may be admitting children from abroad. The new method of calculating index scores for schools based on the sum of pupil weights in the intake and sample years took no account of variations in mobility between schools. Regression analysis was used to establish the relationship between mobility and educational outcome at the school level, controlling for the impact of each of the other measures. It was necessary to examine the relationship between mobility and educational outcome when the effects of the other measures have been taken into account because the contribution of these measures has already been recognised in the calculation of pupil weights.

The results of a hierarchical regression analysis indicated that mobility accounted for less than one per cent of the variation in the proportion of pupils in VR Band 3 when the combinations of the other measures had been taken into account.

Although these results cannot be applied to individual children, Lacey and Blane (1979) have argued that, geographic mobility may have little direct effect on educational achievement, (p. 205) and Mathieson, Mico and Morton (1974) noted, in a study of pupil movement, 'If high mobility is not associated with social problems then it does not appear to be a great problem for the school' (p. 8).

However, because a few schools experienced very high mobility it was decided to use the appropriate regression coefficient to adjust the sum of the weights assigned to pupils in the sample years to take into account pupil mobility in these years.

$$\text{School Index Score} = \left(\frac{\sum Wt_i + \sum Wt_s + B(M - \overline{M})N_s}{N_i + N_s} \right) \times 100$$

Where: B = B co-efficient calculated from regression analysis
$\underline{M}$ = proportion of mobile pupils in school
$\overline{M}$ = mean proportion of mobile pupils in all ILEA schools.

Conclusions

Despite criticism of the concept of using educational priority indices as a means of implementing positive discrimination in the distribution of resources to schools, we fear that, in the absence of such a policy, schools attended by disadvantaged pupils may receive poorer resource provisions than schools attended by more advantaged groups. The opportunity of undertaking a major review of the measures and methods used in constructing the educational priority indices enabled thorough analyses of the relationships between the criteria used and educational outcomes, and suggested improvements in the measures and method of combining data.

Although the concept of priority indices has been criticised on a number of grounds, little criticism has been made of the way in which the various priority criteria have been combined in the construction of the indices. Serious theoretical and empirical disadvantages in employing a method based on standardised scores were identified during the revision exercise. Such a procedure took no account of the relationships between different measures of disadvantage and educational performance, and ignored the effects of multiple disadvantage. In the past it was not possible to assess these relationships because data were not available for individual pupils. In 1981, however, individual data were collected and a new method of constructing educational priority indices was devised. This method has the advantage that it weights each measure according to its contribution to educational outcome, and takes into account the cumulative effects of social factors. The new indices therefore, conform to the Plowden proposal that educational priority should be given to schools where social handicaps reinforced educational handicaps. They also provide a theoretical and empirical justification for the inclusion and weighting of the different criteria used to determine educational priority.

2

Contextualising School Performance

Introduction

The Educational Priority Indices research summarised in the last chapter drew attention to the relationship between students' background characteristics and their educational outcomes. It is an example of applied research conducted to inform policy (in this case by the Local Education Authority) and provide a rationale for additional resource distribution to schools.

As well as an interest in Educational Priority policies, the value of evaluating school performance in the light of information about their student intakes was also recognised by the Inner London Education Authority (ILEA) during the 1980s. The ILEA was at the forefront of research which explicitly linked intake and outcome data (at an aggregate level) in the analysis and publication of secondary schools' public examination results (see Nuttall et al., 1989, Nuttall, 1990). Despite this innovation at the Local Education Authority level, however, such approaches were rare and no national system for the exploration of school performance which recognised the impact of intake was developed in the UK during this time. Indeed, the political climate under the Conservative administration of the Thatcher Years specifically sought to downplay and discredit evidence of the impact of social disadvantage on students' achievement.

The 1988 Education Reform Act represented a major change in the UK's national education policy with the introduction of a National Curriculum and associated National Assessment at Key Stages (KS) in students' school careers.

Financial delegation under the Local Management of Schools (LMS) policy was instituted and a greater emphasis placed on parental choice via open enrolment. National publication of performance tables for schools was also introduced in 1992 for secondary schools in England and Wales. These, popularly christened 'league tables', showed schools' raw (i.e., unadjusted for any intake differences) public examination results at General Certificate of Secondary Education (GCSE) and Advanced (A) level. From 1996 students' raw KS2 results were likewise published for primary schools showing core curriculum (English, mathematics and science) results for students in their last year of primary education at age eleven.

A system of regular national inspection of all schools (initially on a four year cycle), was also introduced from 1993 with the creation of the Office for Standards in Education (OFSTED) and a national framework for inspection was developed and published.

In combination, these changes were intended to encourage the application of market forces to drive up educational standards. In Choice and Diversity (1992) the Government made clear that the publication of performance tables was intended to inform parental choices and thus lead to the 'withering away' of so called 'poor' schools (those with low raw results) if they failed to improve.

One of the key tenants of the Conservative education policy was a belief that raw results 'speak for themselves' and did not require contextualisation. The publication of raw league tables proved highly controversial and was heavily criticised by many educationalists. School effectiveness researchers provided some of the most powerful critiques of this decontextualised approach to judging school performance (McPerson, 1992).

The Assessing School Effectiveness research, summarised in this chapter made an important contribution to this debate. It was commissioned by the then Chief Inspector of Schools in England (Sir Stewart Sutherland) as the first piece of research funded by OFSTED for the explicit purpose of providing a better evidence basis to assist inspectors in their task of evaluating schools and was conducted over a tightly constrained six months time frame from 1993 to 1994. At the time it showed a commendable foresight by the new national inspection authority in seeking to advance the study of school performance.

In the foreword to the final report of the research the HMCI commented:

> *Assessing school performance and standards lie at the centre of OFSTED's responsibilities and its programme of school inspection. OFSTED also needs to draw on the results of educational research. Recent school effectiveness research has underlined the importance of a "value added" approach to measuring school performance, using details about pupils' performance at entry to the school as the baseline from which to measure progress. Unfortunately, such data does not yet exist at national level. In the interim, it is important for OFSTED to have some way of comparing the achievement of schools with broadly similar characteristics – not just for schools with below average performance, but to help in identifying schools*

> *where the achievement falls well above or well below that achieved by similar schools.*
>
> *For this reason OFSTED commissioned researchers from the Institute of Education London to see how far it was possible to use existing data to help us take this interim step, specifically for our own use and for inspection purposes.*
>
> *OFSTED is exploring how to incorporate the findings of the project into the statistical data already provided to inspectors. The techniques developed may well have wider application for schools looking for ways of "bench-marking" their own performance against broadly similar schools (extract from Professor Stewart Sutherland's Foreword to* Assessing School Effectiveness, *Sammons et al., 1994b).*

Given the political climate of the time, the research report proved highly controversial. Inevitably there was a conflict between the Government's adherence to raw league table results and an approach which explicitly recognised the impact of socio-economic disadvantage.

A further year's work testing the model proposed in Assessing School Effectiveness using data for several years and all non-selective schools was conducted by officers within OFSTED and a working party of academics and practitioners. At the moment the research approach was due to be incorporated officially into its Pre-Inspection Context Indicator reports, however, it was blocked following a provocative leader in a national paper, the Times Educational Supplement.

Under its new Chief Inspector (Chris Woodhead), OFSTED's role had become far more politically controversial during 1995. In the run up to a general election the contextualisation of results was seen to weaken the Conservative Government's crusade to raise standards. Chris Woodhead's high profile approach identified poor teaching as the major factor responsible for the under-performance of certain groups. Low expectations rather than disadvantage was highlighted. The publication of the names of failing and at risk schools identified following OFSTED inspection proved highly controversial. The front page of the Times Educational Supplement (23.2.96.) was headed 'Inspectors to take account of deprivation' and commented on 'OFSTED's attempt to make comparisons less odious'. The leader printed under a heading Like with Like (and reproduced below) suggested the research was an important challenge to Government orthodoxy and the resulting attention led to the publication of a denial by the Chief Inspector that OFSTED planned to use the Assessing School Effectiveness methodology.

> *Has Chris Woodhead been caught secretly acting on the side of the angels? The development by the Office for Standards in Education of fairer comparisons between schools will be welcomed by those struggling valiantly against the odds. And any in more favourable circumstances who have hitherto been complacent about their results should be put on their mettle at last as these background measures*

of intake help to put school results into some sort of social and economic context.

The University of London Institute of Education helped OFSTED devise school indicators that enable like to be compared with like. These are not "value-added" measures, as they do not include attainment on intake, but they go some way towards levelling the playing field.

The approach is not perfect by any means. There is a danger, recognised in the OFSTED's own analysis of the Institute's work, that pupil background might be used to excuse lower achievement in schools with a high proportion of pupils on free meals: the objection to social adjustments of exam league tables voiced by Education Secretaries since Kenneth Clarke.

But as a tool for helping to judge and improve individual schools, the "like" schools approach is a distinct advance on crude comparisons with national averages. Any danger that they will legitimise failure will be far outweighed by the likelihood that all underperforming schools will be highlighted, that inspectors will be able to say with greater confidence when a school is failing, and that teachers in lower-achieving schools will see any improvement properly recognised.

Well done OFSTED, then, particularly if this is now to replace the expectation that inspectors judge achievement against their own apparently miraculous powers of recognition of underlying pupil ability. Applied on a school-by-school basis, with inspectors and heads checking the validity of the social indices used, this is a real breakthrough.

It is such good news it seems churlish to carp. But why does OFSTED persist in hiding its enlightenment? Why is it acting so mysteriously over this approach, apparently used by Chris Woodhead in selecting outstanding schools for his recent annual report, and broadly supported by the gaggle of statistical experts assembled before Christmas to give it a seal of technical approval?

The new indicators are going to slip quietly into the pre-inspection context and school indicator (PICSI) reports from April 1 when the revised approach to inspections begin.

Of course, using social indicators contradicts previous government pronouncements; that may explain the sensitivity. But this is not an occasion for holding back or doing good by stealth.

These indicators should open to public debate. And all education authorities and schools – not just those being inspected – need to know the scores of other like schools for comparison if they are to act as spurs to improvement and the means to target extra support.

The controversy surrounding the contextualisation of schools' results remains in evidence although after the election of a new Government in May 1997 OFST-

ED in fact adaptated – the Assessing School Effectiveness method to assist it in reporting on the performance of LEAs. The new Labour Government adopted a policy which combined the use of schools' raw results with that of bench marking and target setting attempting to use data about student intakes to put schools' results in context. Information about students' backgrounds, (social disadvantage, gender) is now officially accepted as relevant in evaluating school performance. Nonetheless, the need to avoid lowering expectations for specific groups of students, while recognising the strong evidence of the impact of background factors on educational outcomes remains an important topic of debate, particularly in relation to the current Government approach to target setting. The Assessing School Effectiveness research took as its starting point the premise of school effectiveness studies clearly summarised by the late Desmond Nuttall

> Natural justice demands that schools are held accountable only for those things that they can influence (for good or ill) and not for all the pre-existing differences between their intakes (Nuttall, 1990, p. 25).

The debate about the role of social disadvantage in educational outcomes and the challenges facing schools of high poverty in inner city areas in particular remains contentious in the UK (Mortimore & Whitty, 1997) and is likely to continue to feature high on the research and policy agenda in education into the 21st century. It is a topic which will be returned to in the third section of this volume. The Assessing School Effectiveness research thus provides an introduction to that debate in the context of the UK in the mid 1990s.

In October 1993 the recently created Office for Standards in Education (OFSTED) responsible for a new system of regular four year inspections of all schools in England and Wales commissioned a six month research project to examine ways of 'putting schools' performance more securely in context'. The project's remit was to establish the utility of using nationally available data sources to obtain selected indicators of schools' probable intakes for the purpose of grouping secondary schools with similar kinds of intakes. The intention was to enable comparisons of GCSE performance to be placed in better context so that schools would be compared in 'like with like' terms.

The project specification explicitly recognised that, ideally, a 'value added' approach which employed baseline measures of students' prior attainment would provide the most appropriate basis for evaluating school performance. However, in the absence of nationally available prior attainment data, the project was set up to investigate the usefulness of developing other less sophisticated ways of contextualising performance in the interim. It was intended that any grouping method developed would be of use in the short to mid-term and would be superseded by value-added methodology when prior attainment data-bases were instituted.

Over the last twenty years or so increasing academic interest has been devoted to the related research fields of school effectiveness and improvement. Considerable evidence has accumulated at both the primary and secondary levels of the existence of significant differences in schools' effects on students' educational outcomes (e.g., see reviews by McPherson, 1992; Reynolds, 1992; Scheerens, 1992; Mortimore, 1993). Sophisticated methods (multilevel modelling) have been developed for the appropriate analysis of data at different levels and these are now recognised as necessary for the proper study of school effectiveness (e.g., Goldstein, 1987; Paterson & Goldstein, 1991) and broad academic agreement has been reached concerning the appropriate design and methodology for such studies.

School effectiveness research clearly demonstrates the need for comparisons of schools to be made on a 'like with like' basis to ensure that they are fair and valid (Mortimore et al., 1988b; Sammons, 1989; Nuttall et al., 1989; Mortimore, 1992; Scheerens, 1992). Prior attainment data collected in a standard form at entry to the school (e.g., at secondary transfer for studies of secondary schools) is vital for value-added analyses of students' relative progress over time. In addition, however, research has provided evidence of strong links between students' individual background characteristics and school performance, even when prior attainment data are available (Nuttall et al., 1989; Willms, 1992; Sammons, Nuttall & Cuttance, 1993; Thomas, Nuttall & Goldstein, 1993; Thomas, Sammons & Mortimore, 1994). In particular, measures of sex, social class/socioeconomic status, low family income and ethnicity have been shown to be important predictors of attainment at both primary and secondary levels. Such measures are also related to prior attainment and, in the absence of measures of prior attainment, assume much greater importance as control measures in the analysis of schools' educational outcomes (Sammons, Nuttall & Cuttance, 1993).

The need for further value-added research has been officially recognised and the Government has accepted Sir Ron Dearing's recommendations on this topic in his interim report on 'The National Curriculum and its Assessment' (Dearing, 1993). However, the data requirements for value-added analyses of school effectiveness are considerable. Such research needs to be longitudinal, can be time consuming and is relatively expensive. At present appropriate data for value-added analyses for statutory aged students[1] are not available on a national basis in the UK, and are unlikely to be available within the near future. Given this situation, what can be done in order to improve the quality of Pre-Inspection Context and School Indicator (PICSI) information and to place school performance 'more securely in context'?[2]

---

1 Clearly value-added analyses is possible for the post 16 phase (i.e., GCSE to GCE A-levels).
2 PICSI reports are designed to provide a wealth of information about different aspects of a school as background context for inspectors prior to conducting an inspection. Reports cover a variety of aspects including the social context of the school's immediate locality, percentage of pupils with statements of special educational need (SEN), percentage receiving free school meals and detailed information concerning standards of achievement, as well as information pertaining to curriculum, staffing and organisation. Information about the individual school is provided in the context of LEA averages, and averages for relevant groupings of LEAs (e.g., Shire counties, England).

The most sophisticated model – identified by the specially designed research studies cited earlier - takes account of both prior attainment to indicate the relative *progress* of a pupil and of his or her background at intake to ensure that the effects of extraneous factors have been taken into account before multilevel modelling is used to identify an estimated school effect. Both intake and outcome measures are collected on an individual basis and aggregations at the school-level are statistically corrected for bias to do with clustering.

In contrast to raw results, where rough comparisons between schools are likely to be made according to factors such as school type (e.g., selective, non-selective), multilevel analysis can provide an estimate of each school's expected performance, given the intake. These results can be used to identify those that are under-performing, by reference to the associated confidence limits calculated for each school. A less sophisticated version of this model utilises either individual or aggregated outcome measures but makes do with aggregated school level background information culled from school records or from the latest census data. The most simple version based on this conceptual approach operates in the absence of any systematically collected, standardised intake information and does not use statistical techniques to relate outcome and intake data directly. In such a version, schools are allocated to groups or clusters based on a set of social indicators about their intake derived from census data (at ward or enumeration district [ED] level related to schools' probable catchment areas).

Any nationally available school-level intake data (e.g., concerning free school meals or ethnicity) could also be included in the analysis. Schools are compared in terms of outcome (e.g., KS3 NCA or GCSE examinations) measures only *within* each group. The group model was originally thought to be the only method available that would fit the criteria detailed in the OFSTED specification and which took account of the limitations related to time available for the research, the requirement to use nationally available data and produce a methodology which could be readily applied in-house by OFSTED for use with schools in autumn 1994.

The disadvantage of adopting the grouping model (described above) is that the relationships between the grouping factors and the outcome measures is theoretical, rather than statistically-based. After commencing the OFSTED project it was established that individual level GCSE outcome data could be made available for a national sample of schools. Given this, it proved possible to utilise both multilevel and less sophisticated (Ordinary Least Squares) regression techniques to relate intake and outcome information and explore, in detail, the statistical relationships between the two.

It is important to recognise that *any* model – including the most sophisticated version described above – is likely to be criticised. On the one hand it has been claimed to be too subjective (Lawlor, 1993), whilst on the other, even our colleagues have argued that it can only lead to oversimplistic league tables (Goldstein, 1993).

Our view of the most sophisticated (the value-added) model is that both sets of criticisms are unfair and need to be qualified. The model is *not* subjective since its basis is an established relationship between intake and outcome: factors (prior attainment or social background) are only taken into account if they have been *shown* to have a *systematic* association with outcomes. Likewise, whilst we share Goldstein's view that a slight change of factors can alter the ranking of average school outcome, we do not consider that this, necessarily, invalidates the model. Rather we think it illustrates its limitations. The model has value as a screening instrument indicating the most extreme points with a fair degree of confidence rather than as the basis of detailed league tables. Both the less sophisticated and the most simple version of the model we have described, are likely to be more open to criticism than this. The empirical testing of the validity of the grouping model, therefore, will be extremely important as an indication of its validity as an interim measure for screening schools' examination or NCA (KS3) performance.

The model of grouping 'like' schools on the basis of selected background and other relevant indicators only provides an *indication* of possible variations in schools' effectiveness and it is vital that within-group variations in schools' examination and NCA assessments be looked at over several years to establish the extent of stability in patterns of performance and trends over time. The importance of inspectors' professional judgement concerning the possible reasons for apparent differences in effectiveness will remain vital. In testing the applicability of the grouping model it is essential to investigate the validity of allocating catchment areas to schools (e.g., in terms of ward or ED in which a school is located and adjacent wards or EDs).

Aims and objectives
The principal aim of the proposed study outlined in the specification for tender, was:

> 'to develop measures that can be reliably used to group schools into broadly similar categories for the purpose of assessing school performance, so that in any comparison "like is compared with like".'

The two complementary objectives were also noted:
(1) to identify possible measures of school background and intake factors;
(2) to test the validity of any such measures or indices against existing measures of school performance.

Research design and methodology

A number of important limitations affected the research design and therefore the methodology adopted for the project. They concerned the availability of data and time-scale of the project.

Data limitations
A necessary limitation for the purpose to which OFSTED wishes to employ the research is the requirement to use only data concerning the characteristics of

school intakes which is *nationally* available. Although schools in many LEAs may hold some background data and information of various kinds concerning students' prior attainment at entry to secondary school, the only nationally available data concerning intakes and outcomes at the school-level relevant to this project concerns GCSE examination results, the percentage of students taking free school meals (a crude indicator of low family income) and the percentage with statements of Special Educational Needs (SEN). For the purposes of the research individual student-level data concerning GCSE results was available for a ten per cent sample of schools, and data concerning sex and age of examination candidates was also included.

OFSTED was able to provide the research team with two examination data files. It was decided to utilise GCSE data for 1992 since this was considered to be of a higher quality than 1991 and this was only one year after the collection of the 1991 Census (the largest national data set concerning socio-economic, ethnic and other characteristics of the UK population). OFSTED already utilises three Census-based measures as indicators of specific socio-economic aspects of schools' likely catchment area in producing its PICSI reports. The research project was intended to examine the possibilities for further development of school background indicators and provide a methodology for using indicators to group schools for the purpose of assisting in the process of evaluating schools' GCSE examination performance.

Selection of indicators
A review of relevant literature was used to guide the selection of indicators.

Educational disadvantage/educational priority
In 1967 the Plowden Report recommended that priority should be given to schools where educational handicaps were reinforced by social handicaps. The report provided suggestions of criteria which could be used to identify both schools and areas which needed special help. It provided a major stimulus for the development of policies of positive discrimination in the distribution of educational resources and, in particular, for the use of educational priority indices (Halsey, 1972; Little & Mabey, 1972; Sammons, Kysel & Mortimore, 1983). Early educational priority indices commonly utilised area-based measures derived from census data concerning schools' estimated catchment areas. For example, measures of low social class composition, overcrowding, housing stress and family size were extracted for the original ILEA index (Little & Mabey, 1972)

A variety of criticisms of the early educational priority work have been made (Barnes & Lucas, 1974; Shipman, 1980) particularly concerning the failure to acknowledge the differences between examining pupil, school and area level data (Sammons, Kysel & Mortimore, 1983). The major significance of the Plowden Report and associated work concerning educational priority indices was perhaps its application of research findings (from a well established tradition of sociological/educational enquiry) concerning links between a variety of personal, ethnic, family and socio-economic factors and educational achievement to the formulation of policy, particularly to the resourcing of schools.

Major longitudinal studies have consistently demonstrated the existence of significant differences in attainment at all levels, but particularly in terms of public examination results and entry into higher education, for those of different social class backgrounds. In addition, measures of low income (e.g., unemployment, eligibility for free school meals, receipt of clothing grants etc.), large family size and (to a lesser extent) one parent family status, and poor housing conditions have also been found to be powerful predictors of academic attainment (e.g., Douglas, 1964; Davie, Butler & Goldstein, 1972; Rutter & Madge, 1976; Essen & Wedge, 1982; Mortimore & Blackstone, 1982).

The British sociological/educational research tradition focused very much on identifying the factors associated with poor educational attainment, the prediction of educational failure and those at risk of under-achievement (Hutchison, Prosser & Wedge, 1979, Sammons, Kysel & Mortimore, 1983) and laid an emphasis on the concept of educational deprivation/disadvantage. Rather less attention was paid to factors associated with high achievement although it is important to take account of such factors to produce a full model of attainment. High social class, gender, greater level of parental educational experience and qualifications, cultural capital in the home, and membership of certain ethnic groups have all been found to be associated with higher educational attainment and, except for gender, participation in higher education (see Sammons, Mortimore & Varlaam, 1985; Mortimore et al., 1988a; Nuttall et al., 1989).

The strength and persistence of the impact of individual, family and socio-economic factors on educational attainment at all stages in students' school careers has been widely recognised. A recent useful research review of these issues has been provided by Brown and Riddell (1992) covering the topics of social class (Paterson, 1992), gender (Riddell, 1992) and ethnicity (Gillborn, 1992). Whilst the evidence of distinctions between the social classes is clear cut, in connection with gender and ethnicity the situation is more complex. Girls tend to outperform boys at primary level (APU, 1980, 1982; Mortimore et al., 1988a; Tizard et al., 1988, Thomas, Nuttal & Goldstein, 1992) and in terms of overall examination performance at secondary school at GCSE (Daly, 1991; Thomas, Nuttall & Goldstein, 1992, 1993, 1994). However, sex differences in subject entry and performance remain significant at GCSE and A-level, and girls until recently remain less likely to enter higher education (although the DFE's latest statistical bulletin [DFE 1993] shows an increase in participation by females and higher levels of attainment at school).

For ethnicity, past research has provided evidence of black educational underachievement for Afro-Caribbean students (Rampton, 1981; Swann, 1985; Mortimore et al., 1988a; Drew & Gray, 1991) even after controlling for social class. For the 'Asian' category there is evidence of significant differences between students of Indian, Bangladeshi and Pakistani background (Nuttall, 1990b). In considering ethnic differences in attainment it should be recognised that underachievement of specific groups may be related as much to the school system, as to cultural/individual differences. Moreover, recent research suggests that ethnic minority groups' performance may now exceed that of majority groups. For three consecutive years value-added analyses of GCSE results conducted for the AMA has shown that no ethnic group performs worse that the 'white' classification (Thomas, Nuttall & Goldstein, 1992, 1993, 1994).

In British studies the impact of low income (a euphemism for poverty) has been found to have an additional impact to that of social class and unemployment (e.g., Essen & Wedge, 1982; Sammons, Kysel & Mortimore, 1983, Sammons, Mortimore & Thomas, 1993b). A useful discussion of the long term impact of poverty on educational attainment in the US context has been provided by Croninger (1993). This has shown that effects on reading and mathematics were significant and that the impact was common across ethnic and gender groups.

School effectiveness
Because of the strong evidence of links between background factors and educational attainment (whether at the individual or the cruder group level of analysis) studies of school effectiveness have attempted to take account of such factors in investigating variations in school performance.

Major British school effectiveness studies covering England, Wales, Scotland and Northern Ireland have consistently demonstrated the need to take account of student background factors such as age; sex; prior attainment; low income; social class/SES; family size/structure; parental unemployment; fluency in English; ethnicity (e.g., see Reynolds, 1976; Rutter et al., 1979; Marks, Cox & Pomian-Srzednicki, 1983, 1986; Marks & Pomian-Srzednicki, 1985; Mortimore et al., 1988a; Tizard et al., 1988; Nuttall et al., 1989; Smith & Tomlinson, 1989; Willms & Raudenbush, 1989; Daly, 1991; Marks, 1991).

For the purpose of comparing school effectiveness, the availability of baseline prior attainment data is crucial (see Cuttance, 1986; Mortimore et al., 1988b; Fitz-Gibbon, 1991; Jesson & Gray, 1991, McPherson, 1992; Goldstein et al., 1992, for a further discussion of the issue of valid comparisons of schools). Nonetheless, even after control for prior attainment student background characteristics have been shown to have an additional impact (Daly, 1991; Thomas, Nuttall & Goldstein, 1992, 1993, 1994; Willms, 1992; Sammons, Nuttall & Cuttance, 1993). Sammons, Mortimore and Thomas (1993) note

> Where no or only crude measures of prior attainment are available, the estimates of the impact of background factors upon pupils' later attainment are likely to remain large (p. 389).

At present, no relevant nationally available prior attainment data at entry to secondary school is available to allow value-added estimates of schools' GCSE performance to be calculated. For this reason it is particularly important to investigate the extent to which various indicators of intake characteristics can be used to help put school performance in context for the OFSTED project. The need for inspectorate judgements of schools' performance to be placed in context was clearly demonstrated by Gray and Hannan's (1986) analysis of the presentation and interpretation of schools' examination results in HMI reports.

Ideally, individual-level data related to students' personal, family and socio-economic, ethnic and language characteristics', prior attainment data and GCSE outcome data is required to establish the net impact of specific factors and to take account of their combined effect on GCSE attainment before calculating estimates of schools' effects using multilevel modelling techniques (Goldstein & Paterson, 1991). Examples of studies which have adopted this approach include Smith & Tomlinson, 1989, Nuttall et al., 1989 and, at A-level, Thomas, Nuttall & Goldstein, 1992, 1993; Fitz-Gibbon, 1991.

Where individual-level data are not available the use of aggregate data entails some inaccuracies or mis-specification of models of schools' effects (Goldstein, 1987). The use of aggregate-level data also entails the possibility of the ecological fallacy (see Robinson, 1950). In other words, patterns of association (correlations) at the area level between two factors do not necessarily apply at the individual level. Nonetheless, in the absence of individual level information, aggregate data still have some explanatory power in analyses of variations in examination performance. For example, work by Gray and Jesson (1987) used aggregate data in a study of LEA differences in examination performance. Four factors (high social class; low social class; children from a non-White ethnic group; lone parent) were found to be important and accounted for 75 per cent of the variation in examination success at the LEA level.

A study by McCullum (1993) also found strong links between LEA performance in SAT and GCSE performance and various contextual variables derived from the 1991 and 1981 Census (see Table 2.1). The high social class factor was found to be particularly important but it must be recognised that there is a tendency for aggregate data to produce higher levels of association than are identified using individual-level data.

Additional educational needs component of standard spending assessments
The DoE uses an Additional Educational Needs (AEN) component in calculating LEAs' Standard Spending Assessments (SSAs) based on census measures and the proportion of children of parents on income support. Proposals in the November 1993 budget, implemented in April 1994, reduce the overall AEN weighting in calculating SSAs and also reduce the weighting given to the ethnicity factor. At present three factors are covered by the AEN component: income support, lone parent and ethnicity.

Table 2.1 Contextual Variables (McCullum, 1993) Related to SAT and GCSE Performance.

High social class	Unemployed
Home owners	Overcrowded housing conditions
Higher educational qualifications	Without a car
White	New Commonwealth + Pakistan

The AEN index, in contrast to educational priority indices used by some LEAs in response to the Plowden report, does not relate the educational needs factors to measures of educational attainment or the 'risk' of under-attainment directly. In a critique of proposals to reduce the AEN component of SSAs, West et al. (1993) note an R^2 figure of 0.47 between the proportion of children with no graded examination result and the components of the current AEN index (p. 16). They argue that there is a case for considering measures based on returns from LEAs using school-based data.

Problems in the DoE AEN formulae identified by West, West and Pennell, 1993, concerning the use of census data include the following:

- Likely to be affected by under-recording especially in inner city areas such as inner London which are most likely to be amongst those in the community who are most disadvantaged.
- The ten year time lag between each census means the data is not responsive to demographic and other changes.
- Census data may not show the level of need in LEAs where significant proportions of the population choose to educate their children privately (state school populations will therefore be more in need than the population as a whole in some LEAs).
- The absence of data concerning refugee status, homelessness and travellers.

Local Management of Schools (LMS) formulae and additional educational needs
Little association exists between the measures used in the AEN calculations and the ways LEAs define special educational needs for the purpose of allocating additional resources. For example, Lee (1991, 1992) notes from his survey of LEAs' LMS formulae that many adopt a very simple approach using data concerning uptake (or eligibility) of free school meals as the sole factor.

In a detailed analysis of five case studies of inner London LEAs, Sammons (1993) found that more sophisticated approaches were adopted by some LEAs, possibly a reflection of the ILEA tradition of using an educational priority index, for the distribution of additional resources via their LMS formulae. For example, Hackney (one of the most disadvantaged LEAs in England) utilised school-based data concerning individual students.

Table 2.2   Hackney's EP1 Factors (Alston & De Vaney, 1991).

- Eligibility for free school meals/or receipt of Family Credit
- Parental occupation (semi- or unskilled manual or unemployed)
- English as a second language (beginner in English defined as less than two years in school with English as the medium of instruction)
- Special circumstances (traveller, refugee, in care, living in temporary accommodation)
- Cumulative disadvantage
- Pupil mobility
- Low Reading Test results at secondary transfer (secondary schools only)

The Hackney index provides another example of the application of research findings and a theoretical framework to justify the allocation of additional resources – namely the known statistical link with poor attainment and takes some account of the concept of cumulative disadvantage. It also includes the factor 'special circumstances' (related to homeless families, children of travellers and refugees).

Urban deprivation
Many studies have used census data to monitor the extent and location of social deprivation, often to assist funding policy (e.g., for health care or Local Authority grants). Davies, Joslin and Clarke (1993) rightly note the content and derivation of measures of deprivation has been the subject of much discussion with a tendency for analysts to trawl the census variables for information on deprivation rather than selecting variables associated with a defined concept of deprivation. They argue that indices (e.g., Morris & Carstairs, 1991; Forrest & Gordon, 1993) can be thought of as attempting to measure poverty using proxy variables for income. Forrest and Gordon constructed two deprivation indices: one of material deprivation – a combined score on four indicators (three relating to housing conditions, overcrowding, lacking basic amenities, no central heating and households without a car). The second – of social deprivation – combined six indicators (unemployment, lone parenthood, youth unemployment, single pensioners, long-term limiting illness and dependants in household). Both indices are calculated as the sum of indicator variables.

Davies, Joslin and Clarke's (1993) analysis attempts to examine the links between income levels and measures commonly combined into indices of deprivation. Using ordinary least squares regression they report that seven socio-economic variables (as well as controls for age and region) can explain about '35–45 per cent' of the variation in household income. They note that their 'regression parameters provide one solution to the problem of finding weights for a deprivation index' and argue that their data 'offer no support for the practice of assigning equal weights to the indicators'. The factors found to be important cover:

- car access (no car)
- housing tenure (rented)
- unemployment
- lone parenthood.
- number of dependent children
- single person household
- economic activity

The Department of the Environment (DoE) commissioned research using 1991 Census data (and exploring other non-census measures) to create a combined index of deprivation at the enumeration district (ED) level (DoE, 1993 draft). The patterns of some 50 variables at different spatial scales ED, Ward and Districts were examined. The list was subsequently reduced to fewer than 20 variables 'on the grounds of statistical or policy considerations, e.g., robustness, coverage of issues'. In selecting the variables the researchers noted that they

> 'avoided measures of vulnerable groups *per se* (such as the elderly, lone parents, ethnic groups) whose members may or may not be deprived'.

The DoE researchers rightly argue that

> 'deprivation is a multidimensional concept. There is no independent or objective way to check whether we are measuring deprivation "correctly"' (para 2.41).

The DoE index utilised correlational techniques and principal components analysis (PCA) for the multivariate grouping of variables. In the ED analysis six variables were included and two principal components were identified; the first largely an economic-related component (E), the second a housing-related component (H).

The DoE researchers concluded that variables should be used in their unweighed form 'on the grounds of transparency' but that there is ultimately no 'correct' method of weighting (para 2.5.4). However, in fact the use of transformations in this research (either Z scores or logarithmic transformations of chi square) in effect does weight factors differentially because it is intended to reduce the impact of very high values from highly skewed distributions. A disadvantage of using standardised scores has been noted in connection with the calculation of EPI indices (see Sammons, Kysel & Mortimore, 1983) standardisation

> 'ensures that a high score on a measure with a low mean and small standard deviation will count more than a high score on a measure with a high mean and standard deviation' (p. 32).

This effect is undesirable *unless* it can be demonstrated that the variables in question are of lesser or greater importance in determining the concept in question (whether this is educational attainment or disadvantage/deprivation).

Another possible disadvantage of using the standardised scores method of constructing indices is that it provides a *relative* rather than an absolute measure of disadvantage because index scores will be influenced by the average for all units (EDs, Wards or Districts) considered. This may be problematic in a time of changing incidence of different measures. For example, if average unemployment increases over time but in a given ED the rate remains constant, it would be given a lower index score that at a previous time point, yet in absolute terms the level of disadvantage (i.e., unemployment) has not changed.

In addition, in considering the use of indices of deprivation, Carley (1981) rightly argues that

> 'If no differential weighting scheme is used, that is each indicator is equally weighted, the prior choice of indicators becomes all import-

Table 2.3 Indicators Proposed for the DoE (1993 draft) Urban Deprivation Index 1991.

Issue	Indicator	Spatical scale		
		ED	Ward	LA
Environment/Housing:				
Health	–Standard Mortality Rates		*	*
Shelter	–Lacking amenities	*	*	*
	–People in mismatched accommodation	*	*	*
	–House conditions			*
Security	–Car insurance premiums			*
People:				
Education	–17-year-olds no longer in full-time education		*	*
	–Births outside marriage (not jointly registered)			*
Family	–Persons per room	*	*	*
Economic:				
Income	–Households without a car	*	*	*
	–Children in non-earning households	*	*	*
	–Adults on income support			*
Jobs				
	–Total unemployment (%)	*	*	*
	–Ratio of long-term to total unemployed			*

ant and thus, in effect, simply transfers the value weighting to the choice of indicator' (p. 80).

For the purposes of the OFSTED research it is important to examine the issues of data trans-formation and weighting of factors to ensure that any decisions made have a theoretical and statistical basis. It is argued that, predicting educational attainment (at the individual school or LEA level) should guide the selection of factors given the prime purpose of the research to assist in putting schools' examination performance in context. For this reason variables such as lone parent and ethnicity (which have been found to be relevant in many studies) are examined. In addition, factors relating to educational advantage are considered as well as those related to educational disadvantage.

Social deprivation and school examination performance
Reid (1991) undertook an analysis of pupil-based post code data related to the 1981 Census on behalf of the City of Bradford using a post code classification system which uses a variety of social measures including those from the Census. The 54 variables cover six categories: Demographic; Household and Age (census); Socio-economic (census); Financial; and Housing. Every post code in the country is allocated a MOSAIC type designation. It is claimed that because

'the average number of households in a post code is 15–17 and given the level of residential segregation and differentiation which characterises our society this system provides for a relatively high and accurate level of identification of lifestyle'.

Reid concludes

> 'Obviously, this procedure is ideal for identifying and comparing school populations, especially where these are not derived from distinct school catchment areas' (p. 8).

For the Bradford analysis the national system was recalculated and reduced using data for all school pupils for whom complete post code data was supplied (77320). Key factors were identified from the general research base as being related to both educational performance and social disadvantage.

(1) Percentage in social class 1.
(2) Percentage of non-manual classes.
(3) Percentage of unskilled manual class.
(4) Percentage of households who moved on during the previous year.
(5) Percentage unemployed.
(6) Percentage of households owning a car(s).
(7) Percentage of households sharing household amenities.
(8) Percentage of households with overcrowded accommodation.
(9) Percentage of households with four or more children.
(10) Percentage born in India.

The major purpose of Reid's Bradford analysis was to establish a base for the comparison of the educational performance and needs of schools. The 35 Bradford Lifestyle Types (BLTs) were ranked from lowest (least socially deprived/most advantaged) to highest (most socially deprived). Reid did not weight the different variables used arguing that little is known about the exact levels of specific aspects of social advantage or disadvantage which relate to specific increments of educational achievement or of cumulative deprivation. In effect this transferred the weights to the combination of variables used.

> 'The central role of social class or socio-economic position... is incorporated since three of the ten variables used are direct measures of this' (p. 29).

The ten separate variable rankings for the BLTs were simply conflated into overall rankings. From the overall ranking of BLTs a social deprivation score was computed for each school.

Reid related schools' social deprivation scores to the schools' rank in terms of mean GCSE points score for the Year 11 age group in 1990. He found a fairly close relationship ($r = -0.81$) which (for this sample) was considerably higher than for the other traditional measure of social deprivation – eligibility for free school meals.

In 1993 Reid undertook a 'socio-environmental analysis' of 1991 and 1992 GCSE results of 23 Bradford Upper Schools, which he argued provided a basis for 'fairer analysis and comparison than can be made with the crude, raw scores only.' (p. 5). This used actual average GCSE scores for school comparisons. Predicted average GCSE scores were computed on the basis of the proportion of pupils in each school from each of the 35 BLTs and their average actual GCSE scores. Whilst reporting the difference over two years between actual and predicted scores, Reid claimed that the significance of a difference between a school's predicted and actual score is not easy to determine (a reflection of the statistical techniques adopted and absence of individual-level data).

Despite obtaining detailed pupil-level post code data the analysis suffers from a number of defects. The first relates to the age of the data (1981 rather than 1991). The problems associated with relying on data collected on a ten yearly cycle are likely to be particularly problematic at the small-area level and in relation to the analysis of the characteristics of schools' intakes. However, these difficulties will be faced to a greater or lesser extent in any analysis using Census data (although changes at the ward level are likely to be relatively less marked over time).

More seriously, the analysis does not examine the relative importance of different factors used to construct social deprivation scores, despite the strong body of evidence noted earlier in this literature review concerning the likely importance of different factors as predictors of attainment. In addition, no individual level analyses are reported. Due to the limitations of the available data, the statistical analyses are conducted at a crude level (based on rank orders and average rank orders to create social deprivation scores). In addition, given the acknowledged importance of student characteristics such as gender and ethnicity, it would have been important to incorporate additional student data into the analysis.

Thus, whilst the census indicators utilised in Reid's work are undoubtedly relevant to the OFSTED project's remit, and the potential of utilising post code data to obtain information about schools' intakes is worthy of further consideration, the methodology employed in his analysis is considered to be inadequate for the purpose of providing a more secure base for putting school's performance in context.

Choice of indicators

The review of relevant literature, concerning educational priority/disadvantage, school effectiveness and urban deprivation, produced evidence of strong links between student attainment and a relatively small number of background factors (including both measures of advantage and of disadvantage). It is theoretically desirable to utilise only measures for which there is evidence of strong links with attainment to ensure that only relevant factors are taken into account in the analysis of school context. The background factors identified can be classified into six broad groups (see Appendix 2.1). Individual pupil-level data on personal characteristics covered by Group 1 (age, sex) is available for the NCER 10 per cent sample of schools made available by OFSTED. It has proved possible to identify 1991 Census surrogates for most of the relevant variables covered under Groups 2–6, although school-level data are preferred if available (e.g., free school meals).

Unfortunately, measures of the best predictor of public examination performance (prior achievement) are not, as yet, nationally available. It is not possible, therefore, to adopt a 'value-added' approach in investigating ways of putting schools' performance in context. The project team recognises the limitations imposed by the absence of such data, and hope that the project results will be viewed as *interim* measures of utility only in the absence of proper value-added analyses.

In addition to pupil-school level data available on the OFSTED files selected data were matched at the Ward level from the 1991 Census for a total of 66 variables covering 13 factors.

- large family
- no earner
- car ownership
- tenure
- housing conditions —overcrowded
  —lack of amenities
- ethnicity
- language[3]
- population mobility
- population density
- lone parent families
- unemployment —total
  —males
  —females
- Social class (Registrar General's Classification)
- higher education

Where possible variables were calculated separately on the basis of total households, total persons and total households with dependent children. It was considered possible that measures calculated on the basis of households with dependent children might be more representative of the possible intakes to schools, whereas those based on total households might be more typical of the neighbourhood context (which has been argued may affect students' school attainments (e.g., Panton, 1980; Sammons, 1985). Measures based on total households are more commonly used in indices of urban deprivation and for the calculation of current PICSI indicators.

Census data were matched to the identifier for the ward in which each school was located for both the full examination and the ten per cent sample files. In addition, data were matched to the identifiers for three neighbouring wards for

---

3 Only the indicator persons aged 0–17 in households with a head born in the New Commonwealth is available in the Census – a very inadequate indicator of potential fluency in English.

schools in the ten per cent sample file (no identifiers were available for neighbouring wards for schools in the full examination files). This was done so that any differences in the patterns of association identified between the selected census variables and measures of school examination performance using both a narrow and a broader measure of the schools' possible catchment areas could be established.

It is recognised that the use of the ward in which a school is located may provide a very inadequate indication of a school's actual catchment areas. A school may be located near a ward boundary and attract more pupils from another ward than that within which it is located. Within an individual ward, conditions may vary enormously and some schools may attract pupils disproportionately from atypical pockets of advantage or disadvantage. Ward level data can therefore only provide a very rough guide to the characteristics of schools' pupil intakes.

Analysis of indicator and examination data

Screening the indicators (correlation with examination results)

Although there are arguments in favour of the transformation of variables which are highly skewed[4] for the purpose of parametric analysis, there are also arguments against transformations. Variables will be differentially affected by transformations and the results of any statistical analysis may be less readily interpretable. Whilst data subject to parametric analysis should strictly adhere to the requirements of normality, there is evidence that correlation and regression techniques are fairly robust with respect to violations of normality. A disadvantage of the use of transformed data is that the results will be on the transformed and not the original data. Thus interpretations and conclusions drawn from the results can refer only to the transformed data and back-transformation is not always straightforward. Also, as noted earlier, the use of transformed data concerning census indicators means that such variables become relative rather than absolute measures of disadvantage/advantage. This can be problematic if the incidence of some factors is increasing or decreasing in absolute terms over time (e.g., single parenthood, unemployment, council house tenure). Given these considerations, data were retained in the form of raw percentages.

Correlations were calculated at the school-level between the percentage of the relevant 15 plus age group obtaining five or more GCSE Grade A–C passes and each of the census variables for all schools in England ($N = 3000$)[5]. In addition, for the NCER ten per cent sample of schools correlations ($N = 416$) were calculated for the five grade A–C measure and for the average total GCSE performance score for the 15 plus age group (based on pupils who entered any GCSE

---

4 Do not exhibit normal distributions.
5 Sixth form colleges were excluded from the analysis.

examination). The calculation of the GCSE performance score is described in Greenhill and Chumun, 1990. The results provide evidence of statistically significant associations between examination performance and many of the census measures (see Appendix 2.2). There are also marked variations in the strength of the relationship between individual measures and schools' examination results.

The strongest measures of association with school examination performance were with measures related to low income, social class and tenure. The variable percentage of no earner households with dependent children was negatively correlated ($r = -0.48$), as were variables related to unemployment ($r = -0.46$ percentage of total person unemployed). Conversely, associations between schools' examination results and the two high social class measures were strongly positive (e.g., $r = 0.45$ percentage households with head in Registrar General's category II). A measure of material conditions, car ownership (one or more), was strongly positively correlated ($r = 0.44$), as was the variable owner occupation ($r = 0.39$). Overcrowded housing conditions, by contrast, had a negative association ($r = -0.41$). The variable related to lone parenthood (percentage of total households) also had a significant negative correlation ($r = -0.38$). That related to percentage of population with higher educational qualifications was positive ($r = 0.38$). Correlations between the two examination performance measures and the various ethnic categories were weaker than those related to socio-economic and family factors. Three aggregated variables 'Black', 'Asian' and 'Other' ethnic minority were created.

An examination of scatterplots of the relationship between the different indicators and schools' examination performance for all schools revealed the existence of a cluster of schools for which the overall relationships did not hold. The vast majority proved to be grammar schools. Because such schools are selective it is inevitable that they will not attract a representative pupil intake in terms of the characteristics of the ward in which the school is located. The use of census-based data as indicators concerning intakes for such schools is, therefore, inappropriate.

As would be expected, given the nature of the census variables selected, many are themselves closely intercorrelated, demonstrating the composite/multifaceted nature of social disadvantage (and advantage). For example, the ownership of one or more cars (a crude income indicator) was negatively correlated –0.88 with the measure of no earners with dependent children, and positively with the measure of owner occupation 0.78. In order to establish the relative importance of the different measures a method of analysing the data which takes account of these patterns of association and determines the relative importance of each factor. The techniques of multilevel analysis were used.

Weighting the indicators
Multilevel methodology
Multilevel analyses (Goldstein, 1987) utilise techniques which explicitly take account of the hierarchical structure of data (the fact that pupils are grouped into specific classes, and classes into schools). These methods allow the calculation of estimates of schools' effects upon pupils' educational outcomes *after* con-

trolling for the impact of relevant pupil background characteristics (e.g., sex, age, social class, low income) and of prior attainment.

Multilevel analysis was used to establish the relative strength of the relationships between the various selected census indicators and students' total GCSE performance scores. The NCER ten per cent file was extended to include the selected census indicators. This file comprised 58628 cases covering 418 secondary schools. In addition to census ward-level variables, individual student-level data on total GCSE performance score, age (in months) and sex and six school-level variables were available. The six school-level variables were:

- total number of students on roll;
- percentage of total roll that were female;
- percentage of pupils receiving free school meals (an indicator of low income);
- percentage of pupils with special educational needs;
- percentage of students age 15 plus gaining five or more GCSE passes grades A-C;
- school type (selective/non-selective).

Two further variables related to the 15 plus were calculated – mean age and percentage female. The outcome variable total examination performance score ranged from 0 to 92, with a mean of 32.0.

Almost exactly a quarter of the total variation in students' total GCSE performance scores was found to lie between schools (was attributable to differences between schools rather than to differences between individuals). The potential explanatory variables were divided into seven blocks or groups: G1 ethnic; G2 socio-economic; G3 family; G4 housing; G5 mobility; G6 pupil; G7 education (of the age 18 plus population). Each block of variables was tested separately in the model. The socio-economic block accounted for a much higher percentage of the variation than other blocks, particularly at the school level. Individual variables found to be significant in the fixed effects analysis for each block were retained for testing in a combined model. Due to the inter-correlated nature of many of the census indicators, only a small number were identified as statistically significant in the analysis.

Selective schools
Given the likely relationship between selective school status and students' GCSE performance the multilevel model was expanded to incorporate school type (selective versus non-selective). This variable can be viewed as a very crude surrogate for students' prior attainment, given that students usually need to pass some form of attainment test at secondary transfer to enter such schools. The addition of this variable had a marked impact upon the percentage of school-level variation in GCSE scores accounted for (75.6 per cent in total). Overall the model accounted for 18.9 per cent of the total variation in students' GCSE performance. The inclusion of school type also had an impact upon the fixed estimates for the census indicators (see Table 2.4).

Table 2.4 Fixed Effects Estimates for the Impact of Pupil and School Intake Characteristics and School Type (Final Model – All Schools).

Variable	Estimate	Standard Error
FSMPCT*	−0.389	0.029
HHRGV*	−0.203	0.136
GIRLS*	4.215	0.0952
AGE-MTH*	0.249	0.0185
HIGHED	0.14	0.0829
TYPE2*	16.94	0.933
OTHERETH*	−0.126	0.0585
NONMAN*	0.0876	0.0399

Where:
FSMPCT = % pupils taking free school meals [school-level]
HHRGV = % total households with head in Registrar General's Group V (unskilled manual) [ward-level]
GIRLS = Student sex (female versus male) [pupil-level]
AGE-MTH = Student age in months (compared with average age for 15 plus cohort) [pupil-level]
HIGHED = % persons age 18 plus with higher educational qualifications [ward-level]
TYPE2 = School type selective [school-level]
OTHERETH = % total persons of other ethnic origins (i.e., not Asian, not White) [ward-level]
NONMAN = % total persons in households with head in Registrar General's Group I or II (non-manual) [ward-level]

Note. n of schools = 418; n of students = 58628.
* $p < 0.05$.

Because selective schools, by their very nature, will not attract a student intake representative of the ward (or adjacent wards) in which they are located, any census intake indicators for such schools are liable to be misleading. All students attending selective schools were, therefore, excluded from the NCER ten per cent sample file. This left a total of 56151 cases and 388 schools. Overall, only 16.9 per cent of the total variation in students' GCSE performance scores was found to lie between schools (a considerably lower figure than the 25 per cent found for the complete file) – a reflection of the removal of selective schools. Again, the results confirm the overriding importance of the socio-economic block of variables (Table 2.5).

All variables identified as significant for each block were tested simultaneously. A reduced set of significant variables was then identified and each remaining census indicator was tested with this set to obtain a final model which provided the best fit to the data. The final model accounted for 61.8 per cent of the school-level variance in students' GCSE performance (around 12 per cent of the total variance in GCSE performance scores).

These results again demonstrate the impact of the two pupil-level variables (sex and age in months). However, mean age of the 15 plus age group was not found

Table 2.5 Fixed Effects Estimates of the Impact of Pupil and School Background Characteristics on GCSE Performance Scores (Non-Selective Schools).

Variable	Estimate	Standard Error
HIGHED	0.163	0.0868
GIRLS*	4.246	0.14
SENPCT*	−0.336	0.147
AGE-MTH*	0.255	0.0191
FSMPCT*	−0.366	0.0293
NONMAN	0.106	0.0395
OTHERETH*	−0.163	0.0593
TNOR*	0.0031	0.000825

Note. n of schools = 388; n of students = 56151.
* $p < 0.05$.

to have an additional impact. Other school level variables found to be important in this analysis were the percentage of pupils classified as having special educational needs (a higher percentage having a negative effect), the percentage of all pupils receiving free school meals (also negative) and an indicator of school size (total number of pupils on roll). School size had a small, though positive, effect.

Of the census indicators tested only the higher education, non-manual and other ethnic minority variables were found to be important (all were significant or very nearly significant). Two measures of social advantage – the percentage of heads of household in occupations in the Registrar General's Group I and II (i.e., non-manual) and the percentage of persons age 18 to 65 with higher educational qualifications of the percentage-based variables showed a positive relationship. The estimate for the percentage of persons from other ethnic minority groups (not Asian or White) was negative.

The free school meals and SEN indicators (school-based measures) had higher predictive weights than the census-based measures

Variation in schools' GCSE performance
Schools' expected average GCSE performance scores
On the basis of data about the statistical relationships between the various variables included in the final model (Table 2.6) and students' total GCSE performance scores, an expected (predicted) score was calculated for each school. These expected scores can be compared with the schools' average total GCSE performance score (i.e., actual scores) and the difference between the two provides some indication of the extent to which the schools' performance conforms to that expected given overall relationships.[6] The correlation between the

---

6 Note that the difference between schools' actual and expected scores will not be exactly equal to the residual because elements of 'shrinkage' in the calculation of the residuals which takes account of the number of students in the analysis at the school level.

Table 2.6 OLS Estimates of the Impact of School Background Characteristics on GCSE Performance Scores (Non-Selective) Schools.

Variable	Estimate	Standard Error
FSMPCT*	−0.363	0.0298
HIGHED	0.158	0.0889
PGIRLS	0.354	0.12
(15+)*	−0.354	0.15
SENPCT*	0.109	0.0404
NONMAN*	−0.165	0.0603
OTHERETH*	0.00324	0.000837
TNOR*		

Where:
FSMPCT = % pupils taking free school meals [school-level]
HIGHED = % persons age 18 plus with higher educational qualifications [ward-level]
PGIRLS (15+) = % student cohort aged 15 plus who are girls
SENPCT = % pupils with statements of Special Educational Needs [school-level]
NONMAN = % total persons in households with head in Registrar General's Group I or II (non manual) [ward-level]
OTHERETH = % total persons of other ethnic origins (i.e., not Asian, not White) [ward-level]
TNOR = Total number of pupils on roll [school-level]

Note. n of schools = 388.
*p < 0.05.
Estimates for all indicators (except TNOR) represent the average difference in total GCSE points per percentage point. For example, schools showing a 10 point difference in the low income indicator (FSMPCT) are estimated to achieve, on average, a difference of 1/−3.63 total average GCSE performance score.

expected and the actual average GCSE performance scores at the school level was 0.76 indicating a significant positive but by no means perfect relationship. The range in schools' expected (predicted) GCSE performance scores was 12.4 to 43.0 compared with figures of 13.6 to 52.8 for schools actual average GCSE performance scores. Significant variation between schools in terms of GCSE performance which is *not* accounted for by the intake variables included in the model remains (this unexplained variation forms the schools' residuals).

Residuals give an estimate of the *difference* between expected and GCSE results. They can be interpreted as providing estimates of greater (positive residuals) or lesser (negative residuals) school effectiveness in terms of GCSE outcomes. The range in residual scores was substantial, −10.85 to 19.79. It is important to examine the confidence limits for each school's residual. Only those significantly different from zero (plus or minus two standard errors) are unlikely to reflect the result of chance variations. In all around half (196) of schools' actual scores were statistically significantly different from their expected (predicted) GCSE performance scores ($p < 0.05$). For 77 schools, nearly one in five of the sample, the residual was five or more GCSE points (e.g., one Grade C) higher or lower than that expected given the intake, a substantial difference in terms of average GCSE total performance scores. For a small number of these (9 or 2.3 per cent) the difference was ten or more GCSE points (e.g., two Grade C's).

The correlation between schools' residuals (estimated using multilevel techniques) and their average actual performance scores was 0.64 indicating that the measures of school effectiveness (i.e., the residuals) are less closely related to actual ('raw') GCSE scores than are the expected scores.

Aggregated analysis (ordinary least squares [OLS] regression)
One of the major pre-requisites of the OFSTED project's specification was to develop a system for putting schools' performance in context that utilises *nationally* available data, so that the results could be applied to all schools. Unfortunately, the NCER data base is a ten per cent sample. Individual student-level data concerning overall GCSE performance, age and sex were not available for all schools. The DFE school file provided by OFSTED only contains one outcome measure (per cent 15 plus age group obtaining 5 plus GCSE Grades A-C). However, it was considered feasible that aggregate data concerning percentage female, mean age and average total GCSE performance score for the 15 plus age cohort could be collected for individual secondary schools prior to OFSTED inspections. Given this, it would be possible to use standard (OLS) regression techniques to examine schools' GCSE performance using the same variable set as that employed in the multilevel model described earlier but on aggregate school-level data only (i.e., without the advantages of using individual student-level data about GCSE performance, age or sex).

For the NCER 10 per cent sample, the results of using standard regression techniques can be compared to the multilevel findings. It is of particular interest to establish the extent of any change in the weightings ascribed to the explanatory variable set and the extent of similarity/dissimilarly in the expected (predicted) scores obtained using the much less satisfactory OLS techniques. Unfortunately, it is not possible to calculate confidence limits for individual schools' residuals using the OLS regression method, because only aggregate school-level data were available for analysis.

The results (shown in Table 2.6) indicate that all the variables (with the exception of mean age of the year 11 age cohort) were significant in the OLS analysis. The weights assigned to each factor were very similar to those identified in the earlier multilevel analysis (for comparison see Table 2.5).
The range in expected (predicted) scores was very similar to that found for the multilevel analysis (12.40 to 43.01), and the expected scores obtained for the OLS and multilevel analyses were very highly correlated ($r = 0.99$). Although the range in the residuals (estimates of schools' relative effectiveness) was larger than that found in the multilevel analysis (−13.37 to 21.78), the residuals from both analyses were highly correlated ($r = 0.99$). The closeness in results for the two forms of analysis, in part, are likely to be a reflection of the limited quantity of individual student-

level background data (only sex and age) and notable absence of prior achievement data available for inclusion in the multilevel analysis. Given the very limited data available, therefore, it can be concluded that the use of the theoretically less satisfactory OLS technique makes little difference in practice to the expected GCSE performance scores calculated for individual schools. The major shortcoming is the absence of confidence limits attached to each schools' residuals which would allow the identification of schools with actual (raw) results that are significantly different from those expected taking account of their likely intake characteristics. Nonetheless, given the similarity between the OLS and multilevel predicted scores, it is considered justified to utilise the former for the purpose of forming groups of schools which have similar intakes and then investigate the extent of variations in schools' actual examination performance as an interim measure.

Grouping schools

Constructing an index on intake characteristics
The principal aim of the OFSTED project was

> 'To develop measures that can be reliably used to group schools into broadly similar categories for the purpose of assessing school performance, so that in any comparison "like is compared with like"'

During the project the possibilities of using cluster analysis to group schools into a variety of distinctive groups upon the basis of background indicators were considered. However, given that the better procedure is to examine directly the statistical relationships between intake and examination outcome measures, it was decided to use an index of expected (predicted) outcomes based on such relationships. The utility of the school groups was tested by investigating the extent to which such groups also differ in terms of overall patterns of schools' examination results. It was anticipated that significant within-group variation in examination results would be identified. Such variation would provide some indication of *possible* differences in effectiveness of relevance to inspection teams and pointers to possible areas for further investigation.

Schools' expected (predicted) average total GCSE performance scores were calculated on the basis of the overall statistical relationships between the selected intake indicators and schools' examination results. Thus the expected scores show what average GCSE results a school would be expected to produce given the characteristics of its student intake. Because the expected scores are *explicitly* calculated *taking account of statistical relationships with relevant intake factors*, they are considered to form the most appropriate basis for any method of grouping schools for the analysis of actual examination performance.

Grouping schools on the basis of expected GCSE scores
The range in schools' expected scores was found to be quite considerable (12.4 to 43.0). In order to enable the grouping of schools with similar kinds of intakes

any grouping method should cover a very small range in predicted scores. Therefore it is not necessarily appropriate to form groups of schools of equal numbers. Rather the grouping method adopted should form groups with little variation in expected results. This is particularly important at the extremes (top and bottom) of the distribution.

A solution which divided the 388 NCER sample schools into 23 groups on the basis of their predicted scores was explored. The number of schools in each group varied from a minimum of three to a maximum of 31 schools. The divisions between groups were intended to minimise variation in expected scores and, except at the extremes, 'cut offs' between each predicted score were chosen. The number of schools in each group is shown in Appendix 2.3. In addition, this table shows the range in actual GCSE performance in each group. These figures provide a better context within which schools' actual performance can be evaluated.

Although a more differentiated grouping is considered desirable, there is no 'right' solution. The divisions between groups in any categorisation are to some extent arbitrary. For some purposes, such as inspection, more broadly based groups may prove useful for distinguishing schools with different types of intake (as measured by expected GCSE scores). However, it needs to be remembered that the broader the grouping, the greater the within group variation in expected scores, and the greater the likelihood that schools with scores at the boundaries between groups may have an expected score closer to schools near the boundary of an adjacent group. Table 2.7 illustrates a 10 group solution.

Utility of grouping method
Some indication of the utility of the grouping method based on OLS expected results (described above) can be gained by comparing the results of the OFSTED analyses with those undertaken for a smaller sample of schools for which value-added estimates of schools' GCSE performance were available.

Table 2.7 A More Broadly Based School Grouping Based on OLS Expected Average Total GCSE Performance Scores 1992 (NCER Sample Non-Selective Schools).

Group	N of schools (N = 388)	Range in predicted GCSE scores	Range in actual GCSE scores	Average actual GCSE score for group
G1	7	12.40–18.69	16.86–20.96	18.51
G2	24	19.28–21.96	15.76–32.88	23.30
G3	36	22.07–24.97	13.56–44.82	22.43
G4	62	25.00–27.92	17.09–35.94	25.99
G5	69	28.02–30.96	18.45–39.88	29.66
G6	77	31.01–33.88	22.62–52.79	31.32
G7	53	34.08–35.97	21.44–52.56	35.39
G8	30	36.00–37.96	29.08–41.94	37.92
G9	23	38.10–39.86	35.36–51.02	40.23
G10	7	40.05–43.01	36.26–47.40	41.35

Individual GCSE total performance scores (1992) were available for 87 secondary schools in eight LEAs as part of value-added AMA study of examination performance (see Thomas, Nuttall & Goldstein, 1994). The AMA data-base includes both examination candidates and non-examination candidates for the 15 plus age group (the non-examination group forming approximately ten per cent of all 15 plus students). By matching relevant census measures identified by the analyses described in Section 3, it has proved possible to make comparisons of schools' expected (predicted) GCSE scores using more sophisticated multilevel analyses of the value-added (because individual prior attainment data (VR band) at secondary transfer was available for the AMA file) by schools compared with predictions based on aggregate intake data at the school or ward (using OLS techniques). Three separate models were tested using the AMA data. The results of these can be compared with those from the two models tested on the NCER data.

The AMA database was enriched by the addition of ward-level census measures and school-based information (it proved possible to match data for 81 schools). The AMA database contained individual-level background data concerning prior attainment (VR band a 3 category variable) ethnic group, eligibility for free school meals, sex and age. The six school-ward level variables identified by the OFSTED analyses were added to the AMA model but only the variable percentage taking free school meals (PFSM) was found to be significant (and negative in its impact) when individual-level background data are available. The factor PFSM, however, had a marked effect on the level of explanation achieved by the model

Overall, 19.1 per cent of the total variation in students' GCSE performance scores was found to lie between schools (a similar figure to the 16 per cent found for the NCER sample of non-selective schools). The AMA individual level prior attainment and background data accounted for 64.5 per cent of the school-level variation, the inclusion of per cent taking free school meals accounts for a further 14.6 per cent of the school-level variation. This model (model 4) accounted for 79.1 per cent of school level variation and 44.5 per cent of total variation in students' total GCSE performance scores. Details of the fixed effects results for the value-added multilevel model are shown in Table 2.8. These results provide one example of the type of value-added analysis possible using multilevel modelling techniques.

The fixed effects results (Table 2.8) demonstrate the very marked positive effects of the two measures of prior attainment (VR band 1, high ability; VR band 2 average ability) contrasted with VR band 3 (low ability) as predictors of later GCSE performance. In other words these two VR groups perform better than those allocated to band 3 at transfer. A much smaller but significant positive effect is recorded for GIRLS (rather over half a GCSE pass), and a slightly stronger negative effect for the individual measure of low family income (equivalent to one fewer GCSE pass Grade C). The effects for different ethnic groups are also striking. In comparison with White candidates those of all other ethnic groups except Black Caribbean were positive and, (apart from the Black Other

Table 2.8 Fixed Effects Estimates of the Impact of Student and School Intake Characteristics (AMA Sample).

Variable	Estimate	Standard Error
VR1*	27.92	0.444
VR2*	13.22	0.376
GIRLS*	3.126	0.3393
FSM*	−4.986	0.3395
BAFRIC*	4.422	0.9647
BCARIB	−1.021	0.5638
BOTHER	1.525	0.9364
INDIAN*	8.683	0.734
PAKIST*	7.242	1.021
BANGLA*	7.233	1.047
CHINESE*	11.78	1.474
OTHER*	5.705	0.6701
PFSM*	−0.343	0.0487

Where:
VR1	= Verbal Reasoning band 1 versus band 3 (pupil-level)
VR2	= Verbal Reasoning band 2 versus band 3 (pupil-level)
GIRLS	= Student sex (female versus male) (pupil-level)
FSM	= Student eligible for free school meals versus not eligible (pupil-level)
BAFRIC	= Student Black African ethnic group versus White (pupil-level)
BCARIB	= Student Black Caribbean ethnic group versus White (pupil-level)
BOTHER	= Student Black Other ethnic group versus White (pupil-level)
INDIAN	= Student Indian ethnic group versus White (pupil-level)
PAKIST	= Student Pakistani ethnic group versus White (pupil-level)
BANGLA	= Student Bangladeshi ethnic group versus White (pupil-level)
CHINESE	= Student Chinese ethnic group versus White (pupil-level)
OTHER	= Student other ethnic group versus White (pupil-level)
PFSM	= Percentage of pupils taking free school meals (school-level)

* $p < 0.05$.

group) significant. Chinese, Indian, Pakistani and Bangladeshi groups all recorded strong positive effects (equivalent to around a Grade A GCSE pass).

Contextual effects
The only school-level variable found to be important in the NCER analyses (per cent taking free school meals) had a significant negative effect over *and above* the impact of the individual student's eligibility for free schools meals.

The percentage of students taking free school meals is a crude, aggregate indicator of socio-economic characteristics of the student intake as a whole and entitlement to free school meals is considered to be a more accurate measure. These findings demonstrate that this aggregate indicator has a strong relationship with the GCSE performance of individual students even when individual information about entitlement is available. It takes account of a significant proportion of the school-level variation in students' results. The interpretation of this finding is not straightforward. Whilst the analysis clearly demonstrates that the overall 'poverty' level of the student intake has a powerful effect upon individuals' attainment at GCSE, the mechanisms underlying this relationship, which can be regarded as a major *contextual* or compositional effect, are not fully understood. Individual students' education and occupational aspirations, motivation and expectations are likely to be affected by the peer culture in their schools. Such cultures may be less academically-oriented in schools with a high proportion of disadvantaged students than in schools with a low percentage of such students. It is also possible that parental expectations, and ability to provide an educationally supportive home environment is diminished in schools with highly disadvantaged intakes, and this may have an impact upon students' overall achievement. It is possible that teachers' general expectations of students in schools with highly disadvantaged intakes may be lower than those of teachers in other schools, and that low expectations themselves have a negative effect on overall achievement levels, as argued by the recent report *Access and Achievement in Urban Education* (OFSTED, 1993). It is also possible that schools in areas serving disadvantaged intakes may have greater difficulties in attracting/retaining experienced staff and this may affect school organisation, teaching and learning and the quality of students' educational experiences and, ultimately, attainment levels.

Given the strength of the contextual effect related to the percentage of students' taking free school meals it is important to consider this factor in analyses predicting schools' examination performance. This is particularly relevant if adequate individual-level student background data are unavailable, as is the case for schools nationally. However, OFSTED inspectors should be aware of the complexities surrounding the interpretation of the free school meals effect (and the consequential reduction in expected GCSE scores for schools with high proportions of pupils taking free school meals). Errors in aggregated information (such as the percentage of students taking free meals) will have a substantial impact upon schools' expected scores. Discussion about school-intake information with headteachers at inspection may help to identify schools where such data may be inaccurately recorded. The need for inspectors to use their professional judgement to examine matters such as the quality of students' educational experiences in socio-economically disadvantaged schools, teachers' expectations, as well as peer group culture and levels of parental support which may be responsible for lower GCSE performance will remain crucial.

Schools' predicted average total GCSE performance scores
The correlation between expected and average actual total GCSE performance scores was higher ($r = 0.89$) than that identified for the NCER analysis (which was unable to control for prior attainment). This may in part be due to the fact that the AMA data-base is not a national sample and that, in terms of the within school balance of students of different levels of prior attainment, schools in inner London LEAs are likely to be more similar than those from more diverse areas. In particular, the former Inner London Education Authority's (ILEA) pol-

icy of using VR banding in an attempt to ensure balanced intakes to schools, may have affected these results. Also, the prior attainment measure available for the 1992 AMA group of schools (VR band) is very crude.

The correlation between school-level residuals and average total GCSE performance scores was 0.68, indicating a positive, but by no means perfect relationship. Overall, examining the schools' confidence limits showed that 33 of the 81 residuals were statistically significantly (2 plus standard errors) different from expected scores, representing 40.7 per cent of the total.

Aggregated analysis (OLS Regression)
In order to provide an example concerning the extent to which schools' expected average performance scores and associated estimates of their effectiveness (residuals) differed using a less sophisticated OLS model rather than a value-added multilevel model, additional analyses were conducted. An OLS model was applied to the AMA sample, using only aggregated data for average total GCSE performance scores and the six measures used in the multilevel analysis. This used the same variable set as that described in Section 4 for the OLS analysis of the NCER ten per cent sample. The results are shown in Table 2.9.

It is important to note that only two of the seven measures included in the OLS model were found to be statistically significant. These were per cent taking free school meals and per cent SEN (both measures obtained from school-based data). The small sample size ($N = 81$ schools) available for this OLS analysis affects the statistical significance of the estimates of these factors. It is also possible that the attribution of census-based measures of schools' intakes is less reliable in inner London than is the case nationally due to greater population mobility and under-recording in such areas. (In the larger [$N = 388$] and nationally representative NCER school sample all measures were found to be statistically significant or very nearly significant.) Given this, the residuals calculated from this model should be treated with caution. It is clear that for this relatively small group of schools ($N = 81$) the variables included in the national model do not provide an adequate fit to the AMA GCSE outcome data.

Table 2.9 OLS Estimates of the Impact of School Background Characteristics on GCSE Performance Scores (AMA Sample).

Variable	Estimate	Standard Error
TNOR	−0.003228	0.002158
FSMPCT*	−0.5467	0.07703
SENPCT*	−1.716	0.8178
HIGHED	−0.1423	0.1982
NONMAN	0.1731	0.1282
OTHERETH	−0.03983	0.09227
PGIRLS	3.709	2.177

* $p < 0.05$.

Nevertheless, for the purposes of this example, it is of interest to compare the results obtained using an OLS approach with those obtained from the technically more correct value-added multilevel approach.

The range of expected (predicted) scores obtained from the OLS analysis was less close to the expected range obtained by means of the value-added multilevel analysis (12.81–41.93 OLS compared with 15.46–49.41 multilevel value-added). The correlations between the OLS predicted results and actual average GCSE performance scores was 0.75, lower than that identified by the multilevel value-added analysis. The overall correlation between schools' expected results obtained from the OLS and value-added multilevel analysis was 0.89, indicating a fairly strong association. The correlations between the estimates of schools' relative effectiveness in promoting GCSE performance (residuals) obtained by the two analyses was slightly lower at 0.86. High overall levels of association, however, mask some important differences in the results obtained for some schools by the two methods of analysis as inspection of individual schools' predicted scores and residuals and (for the value-added multilevel results) confidence limits demonstrates. Whilst for nearly all schools the OLS residuals are similar for a fairly substantial minority (17 schools or 21 per cent) the two analyses given different indicators of effectiveness. To give an indication of the size of these differences the results for the ten schools for which differences are most notable are shown in Table 2.10.[7]

Table 2.10 Example of Comparison of OLS and Multilevel Value-Added Estimates of Schools' Relative Effectiveness (Residuals) (AMA Sample) for Ten Schools.

School[+]	Residual (value-added)	Confidence limit (value-added) 2-SE	Residual OLS	Average actual GCSE exam score
A	−2.718*	1.950	−0.496	27.53
B	−5.470*	2.368	−0.999	21.89
C	5.484*	3.219	0.188	39.90
D	−0.295	1.737	4.025	26.94
E	2.336	2.621	5.164	31.94
F	−1.046	2.594	3.943	24.89
G	−0.987	2.705	−6.006	26.25
H	−2.78	3.037	−8.213	27.98
I	2.116*	2.019	−2.000	29.16
J	0.846	2.044	5.008	32.96

* $p < 0.05$.

---

7 It should be noted that the AMA pupil sample includes non-examination students and excludes those for whom VR band information was missing. The student group is therefore not directly comparable to that used in the NCER analyses described in Section 3 which was based only on students entered for GCSE examinations and for which no exclusions on the basis of missing prior attainment data were made.

In addition, it is possible that with a better (more finely differentiated) measure of prior attainment, the difference in results between the two models (model 4 – value-added multilevel versus model 3 – OLS using aggregate data) would be greater.

Validity of using OLS techniques to group schools
As would be anticipated differences in the estimates of schools' GCSE performance obtained when more appropriate value-added multilevel techniques are applied (which take account of students' prior attainment at entry to secondary school) in comparison with OLS analyses using aggregate school-level data and no prior attainment information, were greater for the AMA than for the larger NCER sample. The availability of individual-level information, including prior attainment, allowing the calculation of confidence limits for each schools' results enables a much better picture of schools' relative performance (i.e., in the context of other schools serving similar intakes) to be obtained.

Nonetheless, OLS results and multilevel results based on some pupil-level data (e.g., GCSE results, age, sex) even in the absence of value-added analyses provide better estimates of schools' relative performance than would be possible using actual average GCSE results on their own and without any contextual analysis. It is clear that, in the absence of better (i.e., more reliable) intake and outcome data for all schools (preferably student-level data), the results of both OLS and multilevel [non-value-added] analysis should be interpreted cautiously. The evidence from the AMA analyses suggest that for most schools results are likely to be broadly similar. However, for a substantial minority (one-fifth) the danger of substantial inaccuracy in estimates is real. In other words false positives and false negatives will be encountered using the less sophisticated OLS methodology and aggregated school-level information. It is very important that this difficulty is recognised and taken into account in interpreting the results of non-value-added contextual analyses.

Multilevel analyses controlling for prior attainment only
A further multilevel model was tested using the AMA school data-base. This model is a simplified version of Model 4. It utilises *only* individual prior attainment data to account for variations in students' later total GCSE performance scores. It thus makes no use of any other student, school or ward-level intake information. By comparing the estimates of schools' effectiveness (the residuals) and their associated confidence limits obtained with this basic value-added model with those obtained from Model 4 (which controlled for additional information about student intakes) it is possible to establish whether such additional contextual information is necessary to make informed judgements about schools' GCSE performance. If little difference in the results of the two models (4 and 5) were identified, it could be concluded that the institution of a system of judging school performance involving the use of additional measures such as individual student age, sex, eligibility for school meals, ethnicity or the percentage of students taking free school meals in the school (an aggregate measure)

is not necessary and a simpler system using prior attainment data only could be adopted.

In fact the results reveal that the use of prior attainment data on its own provides a noticeably less good fit in terms of accounting for variation in students' total GCSE performance scores, than the use of prior attainment and the combination of student and school-based measures of intake characteristics. Overall, Model 5 (controlling for prior attainment only) accounted for 57.2 per cent of the school-level variation in GCSE performance scores. This compares with a figure of 79.1 per cent obtained with Model 4 (controlling for prior attainment, other background factors and the percentage of students taking free school meals). In terms of the total variation in students GCSE performance scores accounted for by the two models, the figures are 36.7 per cent (Model 5) versus 44.5 per cent (Model 4).

Given this, it is not surprising that marked differences in the estimates of schools effectiveness in promoting GCSE performance (residuals) are found between the two models. In all 49 (60.5 per cent) of the 81 schools' residuals were found to be statistically significant (2 plus SE) in the basic value-added model, compared with a figure of only 33 (40.7 per cent) for Model 4 (value-added and contextual). Thus just under 20 per cent of schools which appear to be significantly better or worse in terms of the basic value-added model are not so classified when account is taken of other student and school-level intake characteristics.

In all, 26 schools estimated as over- or under-performing in terms of GCSE performance were identified by both models. A further 23 were identified as significant by the basic Model 5, controlling for prior attainment only, but not by Model 4 which incorporated additional intake data (these can be viewed as false positives). In addition, eight schools were identified as significantly over- or under-performing by the basic Model 5 but were not identified by Model 4 (false negatives).

Discussion and conclusions

The need for better information about student intakes

The results of the OFSTED 'Assessing School Effectiveness' research project have been determined, to a large extent, by the necessity of using nationally available data to obtain measures of the characteristics of schools' student intakes so that schools' GCSE performance can be judged more appropriately on a 'like with like basis'.

The literature review presented in Section 2 provides strong evidence from a variety of research traditions (educational disadvantage/priority, urban deprivation, educational indicators and school effectiveness) of links between a relatively small number of selected measures of student background (including both measures of advantage and disadvantage) and academic performance. Such relationships have been identified both at the individual student-level and in analyses using cruder aggregate data at the school (and the LEA) level.

Developments in school effectiveness research have demonstrated the need to use appropriate data and statistical techniques in comparative studies of school performance. In particular, the adoption of value-added methodology enables the calculation of more reliable estimates of schools' relative effectiveness (residuals) in promoting student outcomes (e.g., GCSE attainment). Studies emphasise the importance of obtaining individual student-level information about personal, family and socio-economic background and, *of crucial importance*, prior attainment (preferably at entry to school, e.g., at secondary transfer for an analysis of GCSE performance) in analyses of schools' effects on cognitive achievement.

Unfortunately, no nationally available source of data concerning prior attainment exists as yet (although in some LEAs data are routinely collected at certain stages, such as secondary transfer). Until such data are available, therefore, it is not possible to conduct proper longitudinal value-added analyses of change in attainment (progress) over time. In addition, individual-level outcome data concerning students' GCSE examination results were not available from the DFE schools' file.[8] The NCER national 10 per cent sample of schools, however, does include individual GCSE results for students in the 15 plus age group (but does not include students who do not enter GCSEs). This file was used for many of the analyses conducted as part of the OFSTED project.

Nationally, only a relatively small number of variables are available from DFE school files. Of such measures five were found to be important in connection with the analysis of GCSE secondary school differences in performance. These were percentage of pupils with statements of special educational need, percentage of students taking free school meals (an indicator of low family income); total number of pupils on roll, percentage of the 15 plus age group who were female[9]; and school type (selective versus non-selective). Selective schools were excluded from the majority of analyses because, by their very nature such schools recruit academically more able pupils and do not have intakes which might be representative of their local area (as defined by ward in which the school is located) a necessary requirement for the use of census-based intake indicators discussed below. In the absence of adequate measures of prior attainment it is not appropriate to compare the performance of selective and non-selective schools because such comparisons inevitably cannot be made on a 'like with like' basis. Such comparisons would be possible, however, if adequate measures of prior attainment were available.

The utility of census-based measures
The project explored, in detail, the possibilities of using ward-level 1991 census information to provide estimates of the likely characteristics of student intakes to schools. Data concerning a wide variety of measures relating to the six groups

---

8 However, the NCER data base does include individual-level GCSE subject results.
9 However, this was not significant in analyses in which information about gender was available at the more detailed level of the individual student.

of background factors noted above were matched on the basis of ward in which a school was located (using the school's post code for identification purposes). This procedure inevitably provides only a crude indication of the likely characteristics of a schools' intake because school's actual catchment areas are not easily defined and will be affected by a number of factors. These include the location and popularity of other schools and transport facilities. Where a school is located near the ward boundary its catchment area is likely to cover other wards and thus intake measures derived from the ward in which it is located may be of limited value.

In order to address this issue it is possible to use census-based intake measures derived from a wider geographical area averaging figures for each measure across the ward in which a school is located and three neighbouring wards. However, whilst this procedure may provide more accurate information for some schools, for others the result is likely to be less accurate. Correlation analyses for the national school sample based on census measures derived from the ward in which a school is located compared with averages over three wards and schools' examination results were in fact very similar. Ideally, more accurate information about the likely characteristics of schools' intake could be obtained by using students' home addresses. Mapping students' post codes would provide a better picture of the characteristics of the neighbourhoods from which students are drawn. For the purpose of analysing GCSE results it would be appropriate to map addresses for the 15 plus age group. A pilot exercise concluded as part of the OFSTED project involving five schools showed that indicators obtained from using the ward in which a school is located could vary from those obtained from mapping post codes.

There are a number of serious disadvantages to the use of post code mapping, however. Not all schools hold post code data in computerised form, it would require a considerable input of resources by schools (and/or OFSTED) to produce a national data base of census measures extracted using a post-code mapping, and such a data base would need to be updated annually using post codes for the relevant age cohort. In addition, given the decennial nature of the census collection, there is an inevitable problem of ageing with census data. Changes in the characteristics of populations are likely to be particularly marked at the more detailed spatial scale of the enumeration district. Given the requirement of the OFSTED project to produce a methodology based on nationally available data for use as an *interim* measure prior to the introduction of a value-added approach when suitable prior attainment data are collected, we conclude that it is not appropriate to recommend investing considerable resources to obtain intake measures derived from pupil post-code matching.

The applicability of using census-based data
A number of caveats should be noted concerning the applicability of census-based measures. Whilst for most schools ward-level data may give a broadly representative picture of the likely characteristics of their actual pupil intakes, for schools in atypical circumstances census data are likely to prove less useful.

These include:

- those with many travellers or refugees in their intakes
- those in areas where many parents send their children to private schools
- those in some rural areas where pupils may be bussed long distances to school
- those in areas where the population is highly mobile
- those in areas where secondary schools are located close enough for parental choice to be a major issue and/or where selective schools 'cream' high ability children at secondary transfer
- those in disadvantaged inner city areas
- denominational schools.

The problems of under-recording of population in the census are greatest in inner city areas such as inner London. Such under-reporting was significantly higher in 1991 than in the 1981 Census (due to the individually-based Community Charge). Such under-reporting is likely to be higher amongst disadvantaged groups and communities (e.g., amongst the homeless). In addition, the issue of how valid it will be to use census data over the longer term needs to be considered because, inevitably, these data will become increasingly out-of-date due to the ten year time-lag between each census.

Despite these caveats concerning the validity of using census data to provide indicators of the likely characteristics of school intakes, fairly strong and highly significant correlations were identified between the various ward-level census measures covering the six groups of background variables identified through the literature review and schools' examination performance. Relationships were in the directions expected on the basis of previous research and those relating to socio-economic characteristics showed the closest associations with schools' examination results.

Methodology
Many of the census measures (and the school-based free school meals data available from the DFE school file) are highly associated, as might be expected. For example measures of housing tenure are correlated with certain of the socio-economic measures related to lone parenthood, unemployment and social class. It was considered important to examine the relationships between different intake indicators and GCSE performance to provide an appropriate theoretical basis for any grouping of schools. An approach involving testing the statistical relationships between census-based and school based data and schools' 1992 GCSE examination performance was therefore adopted. Measures relating to both socio-economic/educational advantage and disadvantage were included in the analysis, avoiding a focus on factors related to socio-economic deprivation alone.

Each of the six blocks of potential explanatory variables was tested using multilevel models and an overall 'best fit' model was identified. This demonstrated that, using a relatively small number of intake measures it is possible to account for a substantial proportion of the *school-level* variation in GCSE performance (62 per cent).[10] Of the eight intake indicators found to be important, three were school-based measures, three were census-based and two student-level characteristics.

Intake measures used in final model

- Total number of students on roll [school-level]
- Per cent of pupils with statements of Special Educational Needs [school-level]
- Per cent of pupils taking free school meals [school-level]
- Student sex (female versus male) [pupil-level]
- Student age in months (compared with average age of the 15 plus cohort) [pupil-level]
- Per cent of total persons in households with head in Registrar General's Group I or II (non-manual) [ward-level]
- Per cent of persons age 18 plus with higher educational qualifications [ward-level]
- Per cent of total persons of other ethnic origins (not Asian, not White) [ward-level].

Multilevel techniques were used to calculate expected (predicted) average total GCSE performance scores for the NCER non-selective school sample ($n = 388$). These performance scores represent the score expected given the characteristics of schools' intake. The differences between expected and actual scores (the residuals) enable comparisons of schools' GCSE performance to be conducted on a like-with-like basis. The overall statistical relationships between each intake measure (student, school or ward level) and students' GCSE performance scores thus form the basis for calculating expected scores for schools. In addition, because individual student-level data about GCSE performance and (limited) background characteristics were available for this sample, confidence limits associated with a measure of each school's relative effectiveness in promoting GCSE performance (residuals) could also be calculated. From these it is possible to identify schools for which actual GCSE performance was significantly different from that expected on the basis of the available intake data.

Because individual-level GCSE outcome data and information relating to student sex and age were not available on a *national* basis for the OFSTED project from DFE school files, further statistical analyses using only aggregate GCSE outcome data at the school-level and aggregated school- or ward-level intake measures (i.e., no individual pupil data about sex or age) were conducted. OLS regression techniques were employed to calculated schools' expected (predicted) GCSE performance scores and estimates of their relative effectiveness (residuals). The results were compared with those obtained using more appropriate multilevel models and, using the mainly aggregated data nationally available,

---

10 However, only around 12 per cent of the total variation in GCSE performance scores was accounted for by this model, a reflection of the limited amount of student-level background data available for analyses (only sex and age).

were found to be very similar. Given this, it is concluded that OLS techniques can be adopted to provide an *interim* method of putting secondary schools' GCSE performance in context. Nonetheless, it must be recognised that a major limitation of the OLS approach is the absence of confidence limits for individual schools' residuals. Using the OLS approach, it is not possible to identify for *which* schools' actual GCSE performance *differs significantly* from that expected on the basis of their intake.

In addition, as noted earlier, it is not as yet possible, using nationally available data, to calculate *value-added* estimates of schools' GCSE performance by taking account of prior attainment. However, for a sample of 81 inner London schools for which suitable student-level prior attainment and intake data were available (collected as part of a project conducted for the AMA) it was possible to compare the use of value-added models with the cruder OLS model for the same set of schools. The results of these comparisons (based on the AMA sample) give some indication of the validity of the models suggested for national use based on nationally available data.

The results indicate that the availability of more student-level background information and, in particular, prior attainment data affects the calculation of schools' expected GCSE performance scores and the estimates of their effectiveness and associated confidence limits. The overall correlation between results obtained using a value-added multilevel model (with detailed pupil-level background and prior attainment data, and the percentage of pupils eligible for free school meals) with those obtained using the OLS model based on aggregate data was 0.89. This is sufficiently close to suggest that, in overall terms, it is appropriate to use OLS techniques and the kinds of intake data currently available to obtain a better picture of schools' GCSE performance than is possible using raw examination data alone. In other words, as an *interim* solution, it is possible to go some way towards developing a usable method for putting schools' performance in context.

However, the comparisons also provided strong evidence that for some schools the use of the cruder OLS analysis and more limited aggregate intake information (particularly the absence of prior attainment data), does make a difference in an inner city context. The estimates of schools' effectiveness in promoting GCSE performance were very different for around one in five schools in the AMA sample. It can be concluded that although the OLS methodology is an improvement on looking at schools' raw GCSE results out of context, the results should still be regarded with caution because estimates are likely to be inaccurate for a significant minority of schools and it is not possible to establish for *which* schools estimates are inaccurate. It is crucial, therefore, that the results of such analyses are used only as a rough guide to assist in interpreting variations in schools' actual GCSE performance. The need for professional judgement and other information remains vital to test out and interpret the results of statistical analyses designed to take account of the influence of intake on examination performance.

Grouping schools
A major focus of the OFSTED project was to develop a method of grouping schools on the basis of relevant indicators of intake so that the actual GCSE performance of individual schools could be evaluated in the context of like schools serving similar kinds of intakes.

The possibilities of using cluster analysis techniques or some kind of index of educational advantage/disadvantage based on relevant indicators of intake were considered during the research. However, these techniques were rejected because it was felt important that any grouping of schools should be based on demonstrable statistical links between relevant intake indicators and GCSE outcomes. Given widespread acceptance that multilevel modelling techniques provide the most appropriate methodology for analyses of schools' effects on examination results, they were used to investigate the statistical relationships between different intake indicators and total GCSE performance scores at the level of the individual student and develop a model which provided a good fit to the GCSE data. The model was then used to calculate schools' *expected* GCSE performance scores, taking into account the characteristics of their intake. A cruder school-level OLS model which yielded broadly similar results was also developed.

The accuracy of these expected scores is improved where individual level data are available for their calculation. Schools' expected scores provide the best basis for grouping schools in terms of intake information relevant for the investigation of variations in GCSE performance.

In order to form school groupings the approach of *minimising* within group variation in expected scores was chosen. No 'correct' solution to the grouping of schools exists because any divisions are, to some extent, arbitrary. Given this, two possible grouping solutions were illustrated. The first used very fine divisions between schools, and comprised 23 categories. The second involved a smaller number of groups (10 in all) which were somewhat more varied in terms of schools' expected scores. With both grouping solutions, *actual* average GCSE outcomes varied widely.

In any comparisons of school performance between groups it is likely to be informative to consider individual schools' expected, actual and residual GCSE scores over several years, as well as variation in such scores amongst schools allocated to the same group.

Summary of results of research

(1) A set of key indicators of schools' intake characteristics was identified by the research after screening a much wider set of indicators derived from a detailed review of relevant literature. Despite the limitations of these indicators (especially those derived from the census) and the lack of prior attainment data, they were found to account for nearly 62 per cent of the school-level variance in students' total GCSE performance scores for a national sample of schools.

(2) The use of appropriate statistical techniques for the analysis of examination data has considerable theoretical and practical advantages. Outcome data are backward mapped to intake measures, thus the methodology proposed is based on demonstrable statistical relationships between intake characteristics and academic outcomes.

(3) Schools can be categorised into 'like' groups on the basis of expected (predicted) average total GCSE performance scores, and an OLS formula for calculating such scores using nationally available data has been derived. Schools with similar predicted scores have similar kinds of intake in terms of the indicators found to be significant for the explanation of variation in GCSE performance.

(4) There is no 'correct' solution to the question of the number of school groups adopted. Two possible groupings based on expected scores are illustrated. On balance, the use of a finer rather than a broader grouping is preferred. This ensures that schools in a given group are very similar in terms of the intake indicators taken into account.

(5) The advantages of using multilevel (rather than OLS techniques) are demonstrated. The ability to calculate confidence limits for the estimates of individual schools' effects on GCSE performance (the residuals) using multilevel methods is particularly useful. In order to employ multilevel techniques individual student-level outcome data (e.g., GCSE performance score) and certain intake measures (e.g., sex and age in months) are required.

(6) Considerable improvements in the reliability of estimates of schools' effectiveness (residuals) would result from the availability of more student-level intake information, particularly baseline prior attainment data at secondary transfer. This would allow the adoption of *value-added* multilevel models which investigate the crucial concept of progress over time.

(7) Several limitations exist in the use of census data (the only suitable nationally available source of intake information apart from school-based data collections), for putting school performance in context. Given this, the investment of considerable resources in mapping pupil post codes to the enumeration district-level to obtain a better picture of schools' actual catchment areas, is not considered appropriate. This is because the limitations of census data are likely to be greater at this finer spatial scale. Obtaining better student and school-based information about schools' *actual* intakes (so that the need to utilise crude census surrogates is removed) is likely to prove a more useful strategy in the long term. In particular, student-based measures of prior attainment would improve the quality of intake information markedly.

(8) Use of these indicators should provide a pointer to those schools whose results are unusually good or poor by comparison with similar schools. But the indicators on which such predictions are based *would need to be verified in terms of the actual intake of the school.*

The Assessing School Effectiveness research provides a valuable illustration of the possibilities and limitations of using nationally available information to put schools' performance in context. It points to the advantages of extending national secondary school data-bases to incorporate intake indicators identified as relevant for the investigation of variations in schools' GCSE performance.

The research also indicates that the inclusion of additional information about relevant characteristics of schools' intakes in PICSI reports would be valuable. It is possible to use the formula that has been derived from the analysis to calculate expected and actual average total GCSE performance scores and to produce estimates of a school's effectiveness. If such information is collected over a period of several years it will be possible to investigate, nationally, the stability of GCSE performance and to identify trends over time.

School effectiveness research emphasises the importance of taking account of variations in schools' intake. The project also demonstrates the importance of improving the quality of information available about schools' intakes (for example, to include more student-level information about the 15 plus age group such as sex, age, eligibility for free school meals and prior attainment). This extra information would allow the adoption of value-added, multilevel techniques for the analysis of schools' GCSE performance and would significantly improve the fit of the model and the precision of school groupings. This methodology and the availability of student-level data would enable the confidence limits attached to any school's residuals to be calculated. Confidence limits would indicate whether a school's actual GCSE results were *statistically significantly* different from those expected on the basis of its intake alone. If such information was calculated over several years it would provide inspectors with a much clearer indication of possible under- or over-performance through reference to the value-added by the school. Were a national system of putting secondary schools' examination results in context adopted, the extra information about performance would assist schools in the process of self-evaluation and review.

Further research needs and applications

The project contributes to the debate about ways of assessing secondary school effectiveness. Nonetheless it is clear that further research is still required. There is a need for more research to address the following tasks.

- Explore the further development of a national method of grouping schools to incorporate a value-added approach using appropriate pupil-level prior attainment data. Areas which could be usefully explored include the use of crude versus sensitive prior attainment measures, subject specific measures, general ability measures and differential school effects.
- Investigate the ways in which suitable contextual and value-added models for different age groups (covering both primary and secondary schools) could be developed (e.g., KS1 to KS3, KS3 to GCSE, GCSE to A-level).

- Investigate the extent to which this methodology could be applied to the study of subject differences in schools' performance at GCSE and could address the issue of secondary school departmental effectiveness.
- Investigate the applicability of contextual approaches to the analysis and interpretation of school performance in terms of non-academic (attendance/behaviour) outcomes.

Appendix 2.1

Groups of indicators identified as relevant for research

1 Personal characteristics

Age*
Sex*
Prior achievement — Best predictor and vital for value-added studies (not available nationally)

2 Family structure

Large family — (4 or more children)
One parent

3 Socio-economic

Parental unemployment
Low income — Eigibility for free school meals (FSM) probably better than receipt of free meals, receipt of other benefits
Car ownership — No car and two or more cars
Social class — Father's occupation a better predictor than mother's measures of both high and low social class are relevant – categories Registrar General's classification I II non-manual (high) and IV and V manual (low)
Housing — Measures of tenure e.g., Council versus owner-occupier (but sales of council housing in 1980s mean council and owner-occupier categories less distinct)
Measures of poor housing conditions (lack of amenities, overcrowding)

4 Educational

Parental educational qualifications
Parents' school leaving age

5 Ethnicity/language

Ethnic group - Various classifications have been used at individual-level
Language - Measures of English as second language and fluency in English have been used, incomplete fluency in English best predictor of attainment

6 Other

Mobility — Measures of population turnover and pupil mobility at school-level, refugees and travellers may be relevant
Population density — This has been included in the assessment of educational need for determining ssas and can help to distinguish rural/urban context
School characteristics — School status, size (pupil roll) and type (mixed, single status, county voluntary, grant maintained, CTC, number of people with special educational needs).

* available on OFSTED 10% NCER file
** a school-level data an uptake of fsm available on both OFSTED files

Appendix 2.2

Correlations between measures of GCSE performance and selected census indicators (NCER [10%] sample of schools)

	LARGE3D	LARGE4D	LARGE3T	LARGE4T	NONEARNT	NONEARND	CARO	CAR1	CAR2	CAR3	CAROD
AC5PCT	-.27	-.27	-.34	-.31	-.27	-.43	-.44	.44	.43	.41	-.34
EXAMSCOR	-.31	-.31	-.37	-.34	-.26	-.48	.44	.44	.43	.39	-.39

	OWNOCC	RENT	OWNOCCD	RENTD	OVERCR	OVERCRD	AMENIT	AMENITD	AMENHT	AMENHTD	WHITE
AC5PCT	.35	-.36	.34	-.36	-.33	-.31	.04 ns	-.04 ns	-.26	-.25	.15
EXAMSCOR	.39	-.40	.40	-.41	-.41	-.39	.00 ns	-.07 ns	-.27	-.27	.24

	BLACKC	BLACKA	BLACKO	INDIA	PAK	BANG	CHIN	OTHAS	OTHOTH	WHITEC	BLACKCC
AC5PCT	-.13	-.10	-.18	-.08 ns	-.14	-.10 ns	.02 ns	-.03 ns	-.06 ns	.15	-.11
EXAMSCOR	-.22	-.18	-.25	-.18	-.13	-.14	-.03	-.12	-.14	.23	-.20

	BLACKAC	BLACKOC	INDIAC	PAKC	BANGC	CHINC	OTHASC	OTHOTHC	NEWCC	OSIDEC	HHMOVE
AC5PCT	-.08 ns	-.16	-.08 ns	-.13	-.09 ns	.06 ns	-.02 ns	-.06 ns	-.13	-.10	.09 ns
EXAMSCOR	-.16	-.23	-.17	-.13	-.14	.03 ns	-.12	-.12	-.21	-.19	.07 ns

	PHHMOVE	PDHHMOVE	DHHMOVE	DENSITY	LONET	LONED	UNEMP	UNEMPM	UNEMPF	HHRGII	HHRGII
AC5PCT	.10	.11	.09 ns	-.13	-.31	-.28	-.40	-.40	-.32	.38	.44
EXAMSCOR	.08 ns	.07 ns	.04 ns	-.19	-.38	-.33	-.46	-.44	-.39	.39	.44

	HHRGIV	HHRGV	CHHRGI	CHHRGII	CHHRGIV	CHHRGV	HIGHED				
AC5PCT	-.34	-.35	.30	.44	-.30	-.29	.41				
EXAMSCOR	-.31	-.35	.31	.45	-.27	-.28	.38				

* All correlations statistically significant $p < 0.05$ unless marked ns. $N = 418$ schools except for DENSITY where $N = 408$.

Selection and calculation of census indicators

Errors of measurement in census data
It is widely accepted that there has been under-recording of the population in the 1991 Census. Forrest & Gordon (1993) note:

> 'Demographic checks indicate that there was an under-count of approximately one million people (2 per cent) in Great Britain as a whole (OPCS 1992). This was significantly higher than the 0.5 per cent estimated under-count in the 1981 Census. There has been speculation that this increased error resulted from people deliberately avoiding enumeration due to fear of the "poll tax".'

West, West & Pennell (1993) argue that

> 'There are good grounds for believing that under-reporting is greater among those in greater need through homelessness. For example, deliberate evasion may have occurred amongst those in the community who were the most disadvantaged' (p. 13).

Variable	Mean	Std Dev	Valid N	Label
LARGE3D‡	17.54	4.65	418	Households with 3 or more dependent children
LARGE4D‡	4.63	3.29	418	Households with 4 or more dependent children
LARGE3T†	4.79	2.03	418	Households with 3 or more persons 0–15
LARGE4T†	1.25	1.08	418	Households with 4 or more persons 1–15
NONEARNT†	35.44	8.04	418	Total households, no adults in employment
NONEARND‡	14.75	9.71	418	Households with dep children, no adults in employment
CAR0†	32.56	13.03	418	Total households with no car
CAR1†	67.44	13.03	418	Total households with one or more cars
CAR2†	22.98	10.38	418	Total households with two cars
CAR3†	3.85	2.21	418	Total households with 3+ cars
CAR0D‡	13.12	6.22	418	Total households with dependent children no car
OWNOCC†	68.75	15.50	418	Total households owner occupied
RENT†	29.44	14.91	418	Total households rented
OWNOCCD‡	71.92	17.36	418	Householders – Owner occupied with dependent children
RENTD‡	26.03	16.72	418	Householders – Rented, with dependent children
OVERCR†	2.03	1.71	418	Total households with over one person per room
OVERCRD‡	5.85	4.70	418	Households with over one person per room with dependent children
AMENIT†	1.10	1.11	418	Households lacking or sharing use of WC or B
AMENITD‡	0.40	0.65	418	Households lacking . . . with dependent children
AMENHT†	19.46	12.13	418	Households lacking... and no central heating
AMENHTD‡	15.03	12.86	418	Households lacking . . . no CH with dependent children
WHITE*	94.46	9.42	418	Total persons White
BLACKC*	0.80	1.78	418	Total persons Black Caribbean
BLACKA*	0.31	0.93	418	Total persons Black African
BLACKO*	0.28	0.44	418	Total persons Black other
INDIA*	1.66	3.80	418	Total persons Indian
PAK*	0.98	3.39	418	Total persons Pakistani
BANG*	0.32	2.02	418	Total persons Bangladeshi
CHIN*	0.28	0.30	418	Total persons Chinese
OTHAS*	0.38	0.81	418	Total persons other Asian
OTHOTH*	0.54	0.64	418	Total other persons other ethnic group
WHITEC**	91.29	14.10	418	White aged 0–15

Table continues

Table (continued)

BLACKCC**	0.80	1.78	418	Black Caribbean 0–15
BLACKAC**	0.45	1.31	418	Black African 0–15
BLACKOC**	0.69	1.07	418	Black other 0–15
INDIAC**	2.34	5.39	418	Indian aged 0–15
PAKC**	1.86	6.07	418	Pakistani aged 0–15
BANGC**	0.64	3.90	418	Bangladeshi aged 0–15
CHINC**	0.33	0.40	418	Chinese aged 0–15
OTHASC**	0.48	1.14	418	Other Asian 0–15
OTHOTHC**	2.34	1.07	418	Other other aged 0–15
NEWCC**	7.38	11.91	418	Persons aged 0–17 in hhold with head born in New Commonwealth
OSIDEC**	9.64	13.35	418	Persons aged 0–17 in hhold with head born outside the UK, Ireland, Old Commonwealth and USA
HHMOVE†	6.31	2.36	418	Mover households
PHHMOVE*	5.87	2.29	418	Persons in mover households
PDHHMOVE*	7.22	2.70	418	Persons in mover households with dependent children
DHHMOVE*	7.96	2.86	418	Mover households with dependent children
DENSITY	2763.42	2019.17	418	Residents per hectare
LONET†	13.14	5.85	418	Total lone parent families
LONED‡	18.69	9.12	418	Total lone parent families with dependent children
UNEMP*	5.48	2.83	418	Total persons unemployed
UNEMPM*	8.04	4.40	418	Total males unemployed
UNEMPF*	3.17	1.78	418	Total females unemployed
HHRGI†	6.25	4.23	418	Households with head RGI
HHRGII†	29.10	11.12	418	Households with head RGII
HHRGIV†	13.97	5.89	418	Households with head RGIV
HHRGV†	4.70	3.11	418	Households with head RGV
CHHRGI**	6.92	5.97	418	Persons 0–15 in household with head RGI
CHHRGII**	28.88	14.10	418	Persons 0–15 in household with head RGII
CHHRGIV**	13.55	7.98	418	Persons 0–15 in household with head RGIV
CHHRGV**	4.42	4.53	418	Persons 0–15 in household with head RGV
HIGHED*	11.09	6.55	418	Persons 18+ with HE qualifications

† calculated as percentages of total households
‡ calculated as percentages of total households with dependent children
* calculated as percentages of total persons/males/females
** calculated as percentages of total persons 0–15 or 0–17

Appendix 2.3

A finely differentiated school grouping based on OLS expected average total GCSE performance scores 1992 (NCER sample non-selective schools)

Group (N = 388)	N of schools	Range in predicted GCSE score	Range in actual GCSE Score	Average actual for group
G 1	4	12.40–16.44	17.59–20.96	18.58
G 2	3	17.16–18.69	16.86–20.19	18.42
G 3	10	19.28–20.85	15.90–25.54	22.26
G 4	14	21.02–21.96	15.76–32.88	24.05
G 5	13	22.07–22.92	15.73–27.68	20.91
G 6	10	23.01–23.91	15.32–26.86	19.46
G 7	13	24.01–24.97	13.56–44.82	26.24
G 8	18	25.00–25.93	17.09–26.71	23.13
G 9	25	26.03–26.99	19.57–35.94	27.69
G10	19	27.05–27.92	19.44–34.36	26.48
G11	19	28.02–28.84	21.59–38.18	29.54
G12	25	29.03–29.97	18.45–38.15	28.38
G13	25	30.16–30.96	22.73–39.88	31.04
G14	28	31.01–31.87	22.62–52.79	30.09
G15	22	32.08–32.97	24.67–35.43	30.31
G16	27	33.01–33.88	23.31–40.61	33.43
G17	22	34.08–34.99	21.44–52.56	34.32
G18	31	35.01–35.97	27.89–50.11	36.05
G19	19	36.00–36.72	31.63–51.02	38.28
G20	11	37.13–37.96	29.08–41.94	37.30
G21	15	38.10–38.99	35.36–51.83	39.35
G22	8	39.07–39.86	37.34–49.07	41.88
G23	7	40.05–43.01	36.26–47.40	41.35

3

Issues in School Effectiveness Research

Introduction

The first part of this volume stressed the origins of school effectiveness studies particularly through the assessment of the crucial impact of school intake, and presented two examples of policy-related research which demonstrated the need to contextualise schools' performance by explicitly measuring the impact of powerful external factors, such as students' socio-economic and family circumstances. In the UK, but also in Australia, parts of South East Asia and North America, the school effectiveness research tradition has assumed an increasingly high profile due to outside influences which have emphasised accountability mechanisms, the role of the consumer and which have attempted to develop national systems for the judgement of school quality. The latter trend was clearly demonstrated by the Assessing School Effectiveness research example described in the last chapter. There are, however, considerable dangers in the stress on judgement and accountability. In the UK the high profile policy of publicly naming schools identified as failing by inspectors – nicknamed 'naming and shaming' has proved highly controversial. Many in education believe it has been counter-productive, lowering teacher morale and leading to a culture of blame. Others, however, believe it is a necessary stimulus for action. The full impact of this approach to school improvement has yet to be evaluated, although it is a subject referred to in part 3 of this volume in relation to an evaluation of Raising School Standards project in Northern Ireland.

The work on inspection evidence by authors from the school effectiveness and improvement traditions has highlighted a number of concerns about the process of national inspection and the judgement of schools in the UK (see reflections by Earley, Fidler & Ouston, 1996; Gray & Wilcox, 1994, 1995; Wilcox & Gray, 1996). Goldstein (1997) has also provided a timely critique of the dangers of misapplying school effectiveness research for the purposes of accountability.

These concerns were at the forefront of my thinking after completing the Assessing School Effectiveness research and while working on a study of Departmental Differences in Secondary Schools' Academic Effectiveness (later published as Forging Links [Sammons, Thomas & Mortimore, 1997], when I was asked to prepare a keynote address for the Educational Effectiveness symposium of the 1995 Dutch Educational Research Association's annual conference in Groningen. The symposium organisers were keen to receive a contribution which would focus on examples of recent research. The paper Complexities in the Judgement of School Performance used this opportunity to review some of the most important (in my view) methodological considerations in current school effectiveness research and built on the important work by Dutch authors such as Scheerens (1992) and Creemers (1994a,b). The focus on complexity was deliberate, since the increasing use of multilevel techniques and analysis of large longitudinal data sets of student performance measures taken over several years during the late 1980s and early 1990s demonstrated the ways in which it was possible to create better models of student attainment and progress. The availability of data, rather than the limits of statistical methods, thus became the more important factor in restricting further advances in school effectiveness research. The development of multilevel methods enabled researchers to address the important but neglected question of the extent of internal variations in school effectiveness, as well as to obtain more precise estimates of the size of between school differences which had been a focus of earlier studies. Such analyses, it is argued, are vital to the investigation of equity issues in education, and to provide a better understanding of the complex world of schools. Indeed, the paper calls into question the applicability of the concept of school effectiveness, suggesting it is both time and outcome specific and can vary for different groups of students in the same school. The next chapter reproduces the Complexity keynote in the form in which it was later published in Educational Research and Evaluation. The paper uses a review of school effectiveness literature to expand current thinking about what constitutes an adequate school effectiveness research design.

It examines the aims and goals of school effectiveness research and definitions of effectiveness. Five issues in school effectiveness research which are relevant to the development of better methods of judging and measuring relative effectiveness, drawing on evidence from studies in different countries and contexts, are reviewed. These are (1) size and significance of effects, (2) consistency across outcomes, (3) stability over time, (4) the long term effects of schools, and (5) differential effects for different student groups.

It is concluded that effectiveness is best regarded as a relative concept dependent upon time period and age groups studied, and the choice of educational outcomes and intake measures. The emphasis on complexities was a deliberate response to attempts to develop crude systems for judging school performance as either 'good' or by implication, 'bad'. The paper concludes by arguing the case for making research results available to schools for the purposes of self-evaluation and review, a topic which receives further attention in Part 3 of this volume.

School effectiveness research is an academic field which has developed rapidly over the last 30 years (Creemers, 1994a). Empirical investigations of the theory that schools have a measurable impact upon their students' educational achievements have been conducted in a growing number of countries and the field is particularly strong in the UK, USA, the Netherlands and Australia, although school effectiveness research is becoming more common in Asia and the third world (Riddell, 1995). The central focus of school effectiveness research concerns the idea that "schools matter, that schools do have major effects upon children's development and that, to put it simply, schools do make a difference" (Reynolds & Creemers, 1990, p. 1).

A major impetus for the development of North American and British school effectiveness research is generally recognised to have been a reaction to the deterministic interpretation of findings by the US researchers Coleman et al. (1966) and Jencks et al. (1972) concerning the potential influence of schools and teachers on students' achievement (Rutter et al., 1979; Mortimore et al., 1988a; Reynolds & Creemers, 1990; Firestone, 1991 Mortimore, 1993b). Creemers, Reynolds and Swint (1994) have also pointed to different intellectual ancestries of school effectiveness studies in other national contexts. For example, in the Netherlands research traditions concerning matters such as teaching, instruction, curriculum and school organisation were influential, while in Australia the strong field of educational administration provided a stimulus.

Increasing concern about the standards of education achieved by school leavers, and of achievement in basic skills such as literacy and numeracy by younger age groups, is evident in many parts of the world today. This has led to pressure for the development of better systems for judging and monitoring the quality of students' education and the performance of schools and for greater accountability of teachers and schools to different consumer groups, policymakers, politicians, parents and students (Tomlinson, Mortimore & Sammons, 1988; Creemers, 1994a; Reynolds et al., 1994a; Scheerens, 1995).

In the UK school effectiveness research has become of increasing policy and political relevance during the 1990s due to the Government's explicit policy of applying market forces to education, with the intention of "raising standards" in schools. Reforms were intended to increase parental choice of schools via a policy of open enrolment, combined with the provision of additional information through the publication on a national basis of secondary schools' public examination results at GCSE (taken at age 16) and A level (taken at age 18) and attendance data, and the planned publication of national assessment results for primary schools. This so called "league table" policy was intended to inform parental choices and thus lead to the "withering away" of schools with poor results if they failed to improve (DFE, 1992). The national publication of league tables has been very contentious and its impact (if any) on parental choice has not been studied in any depth as yet. It has been heavily criticised by academics as well as practitioners because of the failure to utilise the knowledge gained from school effectiveness research about appropriate ways to compare and evaluate the impact of schools on their students' achievements (Mortimore, Sammons & Ecob, 1988; Fitz-Gibbon, 1991; McPherson, 1992; Goldstein et al., 1993; Sammons, Nuttall & Cuttance, 1993; Sammons et al., 1994b; Goldstein & Thomas, 1996).

The major flaw in the use of raw league tables to make judgements about performance is that they take no account of differences between schools in "the talents and motivations of individual pupils, the nature of their families and communities and even of what are sometimes seen as social determinants of achievement: gender, social class and race" (Mortimore, Sammons & Thomas, 1994, p. 316). As Nuttall argued "natural justice demands that schools are held accountable only for those things they can influence (for good or ill) and not for all the existing differences between their intakes" (Nuttall, 1990, p25).

This paper examines the aims and goals of school effectiveness research and definitions of effectiveness. Five methodological issues relevant to the development of better methods of judging and measuring relative effectiveness are identified and discussed through a review of past research. These are: (1) size of school effects; (2) consistency; (3) stability over time; (4) the long term effects of schools; and (5) differential effectiveness. Relevant examples of recent research are used to illustrate these issues.

It is beyond the scope of this article to review the important and related field of research into class effects, which can be seen as indicators of teacher effectiveness (see Luyten, 1995; Luyten & Snijers, 1996). Few studies have examined both school and class effects simultaneously, although this is becoming more common. Some such studies have suggested that class effects can be larger than school effects (e.g., Mortimore et al., 1988b, Rowe & Hill, 1994; Hill, Rowe & Holmes-Smith, 1995). Others have found greater variation between schools than classes (Bosker, 1991, reported in Luyten, 1995).

In order to investigate class effects such studies analyse student academic achievement over one school year. It is necessary to include schools with two or more classes per grade to allow separation of variance at the student, class, and school levels. Whilst there is undoubtedly a need for further research on class, teacher and school effects, focusing on the topic of effective instruction (Creemers 1994b, c), it is argued that studies which examine the effects on students of being a member of a particular school over a period of several years remain of considerable practical and theoretical relevance. This is because a student's educational experience commonly involves being taught by several different teachers in different years at the primary level, and a number of subject specialists at the secondary level in each year. Students and their parents may have (usually constrained) opportunities to choose schools, but not individual classes or teachers. Moreover, students spend several years in specific institutions and therefore the question of whether over several years the school attended has an impact on educational outcomes is important. Ideally, studies which investigate the differences in students' educational outcomes of being a member of a particular school over several years (preferably from entry to exit at the end of a particular phase such as primary or secondary) modelling class/teacher effects (at primary) and class/teacher or departmental effects (at secondary) are required.

Aims and goals of effectiveness research

In reviewing early school effectiveness studies in the US context, Firestone (1991) noted that the effective schools movement was committed to the belief that children of the urban poor could succeed in school (e.g., Edmonds, 1979; Goodlad et al., 1979). He recognised that "Effectiveness is not a neutral term... criteria of effectiveness will be the subject of political debate" (p. 2). Early school effectiveness research incorporated explicit aims or goals concerned with equity and excellence and focused on the achievement in basic skills of poor/ethnic minority children in elementary schools.

More recent research, especially in the UK, has moved away from an explicit equity definition towards a focus on the achievements of all students and a concern with the concept of progress over time rather than cross-sectional 'snapshots' of achievement at a given point in time. This broadens the clientele to include all students, not just the disadvantaged, and a wider range of outcomes (academic and social). As in the US, however, the majority of UK studies have also been conducted in inner city schools. The crucial importance of school intake has been increasingly recognised, and attempts to control statistically, for intake differences between schools before any comparisons of effectiveness are made (Mortimore, 1991b; Mortimore, Sammons & Thomas, 1994, Sammons, Mortimore & Thomas, 1996).

Definitions of effectiveness

Although in 1987 Reid, Hopkins and Holly concluded that, "while all reviews assume that effective schools can be differentiated from ineffective ones there is no consensus yet on just what constitutes an effective school" (p. 22), there is now greater agreement concerning appropriate methodology for such studies, about the need to focus explicitly on student outcomes and, in particular, on the concept of the 'value added' by the school in terms of relative progress (McPherson, 1992; Scheerens, 1992; Creemers, 1994a; Reynolds et al., 1994b; Sammons, 1994a, b). For example, Mortimore (1991a) has defined an effective school as one in which students progress further than might be expected from consideration of its intake. An effective school thus adds *extra value* to its students' outcomes in comparison with other schools serving similar intakes.

Although studies of variations between schools exist in both simple and more sophisticated forms, the majority attempt to take account of the differences in the characteristics of students entering and attending schools. Whilst the methodology for doing this has improved considerably over the last 20 years, most studies conducted before the mid-1980s are considered statistically weak (Creemers, 1994a; Scheerens & Bosker, 1995). The criteria for an adequate study of school effectiveness have been succinctly described by Scheerens (1992). These are:

- taps sufficient "natural" variance in school and instructional characteristics, so that there is a fair chance that they might be shown to explain differences in achievement between schools;
- uses adequate operationalizations and measures of the process and effect variables, preferably including direct observations of process variables, and a mixture of quantitative and qualitative measures;
- adequately adjusts effect measures for intake differences between schools (e.g., in previous achievement and socio-economic status of students);
- has units of analysis that allow for data analyses with sufficient discriminative power;
- uses adequate techniques for data analysis – in many cases multilevel models will be appropriate to do justice to the fact that we usually look at classes within schools, students within classes and perhaps even schools within specific types of environments;
- uses longitudinal data.

Definitions of school effectiveness are thus dependent upon a variety of factors. These include: the sample of schools; choice of outcome measures; intake controls used; methodology; and timescale (Sammons, 1994a). The need to interpret residual estimates of individual school's effects (as in outlier studies of highly effective or ineffective schools) by reference to the confidence limits associated with such estimates is also now widely recognised (McPherson, 1992; Goldstein et al., 1993; Sammons, Nuttall & Cuttance, 1993; Sammons et al., 1994a; Creemers, 1994a; Mortimore, Sammons & Thomas, 1994).

1   How large are school effects?

Evidence for the existence of school effects has been found across all phases of schooling and for a variety of usually academic educational outcomes. For example, in the UK context, Tizard et al.'s (1988) study of infant schooling demonstrated the existence of effects for reading and number work; Mortimore et al. (1988a, b) produced evidence for a variety of cognitive and non-cognitive outcomes concerning the junior phase. Studies by Cuttance (1987), Reynolds, Sullivan and Murgatroyd (1987), Rutter et al. (1979), Nuttall et al. (1989, 1990), Willms and Raudenbush (1989), Jesson and Gray (1991), Goldstein et al., (1993), Gray, Jesson and Sime, (1990, 1995), Sammons et al., (1994a), Thomas and Mortimore, (1996), demonstrate the existence of secondary school effects on examination results at age 16 years.

For the post–16 phase work on the existence of specific school effects is limited, although Thomas, Nuttall and Goldstein's (1992, 1993) work on A-level league tables for the Guardian project indicates the existence of significant school effects. Furthermore, Fitz-Gibbon's (1991, 1992) analyses at the departmental level indicates the importance of departmental effects for A-levels.

In considering the issue of the size of school effects it is important to examine the extent to which studies employ similar measures of students' educational outcomes and of intakes. Bosker and Scheerens (1989) discussed effect size as a function of the *choice* of dependent and independent variables. They note that specific characteristics of instructional processes show up more clearly when curriculum specific tests rather than general scholastic aptitude tests are used as the dependent variable (Madaus et al. 1979). Brandsma and Knuver (1989) concluded "schools can influence arithmetic progress more than language progress... This may be caused by the fact that arithmetic is a subject that is more uniquely learned at school, while language acquirement is more a joint operation of school and home environment" (p. 787). A recent study of departmental differences in secondary school effectiveness (Thomas, Sammons & Mortimore, 1995) also found that the percentage of variance attributed to the school varied for different measures of academic outcomes being lower for English (6 per cent) and higher for subjects primarily taught in school (e.g., mathematics 9 per cent, English literature 12 per cent, history 20 per cent).

Not only may choice of outcome measure affect the size of school effects identified, the choice of intake controls is important. Reynolds et al., 1994b's country report noted that UK research is particularly strong in terms of the multiple measures of student intake controlled for utilising multiple variables relating to prior pupil achievement as well as background factors relating to parental socio-economic status, education and ethnicity.

Such research demonstrates quite clearly that the magnitude of school effects identified varies according to the nature and extent of background factors controlled for and the method of analysis. The availability of appropriate (finely differentiated) measure(s) of prior attainment provide the best controls for intake differences, but other background factors exert a continuing impact (Willms, 1992; Sammons et al., 1994a, Thomas & Mortimore, 1996). For example, in a re-analysis of the School Matters (Mortimore et al., 1988a) data, Sammons, Nuttall and Cuttance (1993) demonstrated that the inclusion of a detailed set of individual student level background variables accounted for a significant proportion of variance in students' reading and mathematics attainment in Year 5 of junior schooling (age 10+) but their impact was greater for reading than for mathematics (a reduction of 20.6 per cent compared with 11.3 per cent of total variance in scores respectively). These results demonstrate that control for socio-economic and other background factors as well as an appropriate prior attainment measure is likely to be particularly important in analyses of effectiveness using language-based measures as outcomes.

Sammons, Nuttall and Cuttance's re-analysis revealed that, when prior attainment was controlled, background factors were much less important in accounting for relative progress than school attended. In terms of total variance the size of school effect was 8.7 per cent for mathematics and 7.3 per cent for reading. However, in terms of unexplained variance (i.e., that *not* accounted for by prior attainment and background factors) the school accounted for 17.2 and 18.7 per cent respectively for these two subjects.

Lack of control for relevant background factors also has an impact on the calculation of residuals (and their associated confidence limits) which give an estimate of the size and significance of individual school effects. Research conducted for the Office for Standards in Education (the body responsible for the national inspection of schools in the UK) to develop methods for putting secondary school performance in context (Sammons et al., 1994b) compared the residuals calculated using multilevel models which controlled for prior attainment only (basic model) with those which controlled for prior attainment and other background characteristics. For five out of 81 schools (6.2 per cent) in the sample the residuals calculated by the two models were statistically significantly different. By controlling only for prior attainment therefore a misleading picture of effectiveness was obtained for some individual institutions.

The OFSTED research indicated that the basic multilevel model controlling for prior attainment only accounted for 36.7 per cent of the total variance in students' GCSE scores, whereas the figure for a multilevel model controlling for prior attainment and individual-level background data (age, sex, ethnicity and eligibility for free school meals) and a contextual measure (per cent students eligible for free school meals) was substantially higher at 44.5 per cent. The school level variation (which represented 19.1 per cent of total variation when no controls were included) for this sample was reduced to 6.8 per cent. It was concluded that, in the absence of adequate controls for students' background characteristics as well as their prior attainment(s), estimates of schools' effectiveness in promoting their students' progress will reflect less favourably on the performance of schools in certain contexts (e.g., those with disadvantaged intakes in the inner city).

Importance of school effects
In terms of total variance in academic outcome measures Daly's (1991) review indicated that school effects are thought to represent between 8 and 10 per cent. Creemers (1994a) concluded that "About 12–18 per cent of the variance in student outcomes can be explained by school and classroom factors when we take account of the background of the students" (p20).

Expressed as percentages, school and classroom effects are not large, but in terms of differences between schools in students' outcomes they can be highly significant both educationally and statistically (Mortimore et al. 1988b; Gray, Jesson & Sime, 1990; Sammons, Mortimore & Thomas, 1993a; Sammons et al., 1995b). For example, Thomas & Mortimore (1996) report differences between schools' value added scores of between seven Grade E results and 7 Grade C results (over 14 points) at GCSE in a study of schools in Lancashire and similar results were identified in a study of inner London secondary schools (Thomas, Sammons & Mortimore 1995). Differences of this size have significant implications for students' subsequent education and employment prospects.

Furthermore, as noted earlier, in terms of relative student progress (value added) school effects are found to be more important than background factors such as age, gender and social class (see Mortimore et al. 1988a, b). In connection with equity differences, Mortimore et al.'s (1988a, b) research also demonstrated that, although no school removed social class differences in attainment, the absolute achievement in basic skills of working class students in the most effective schools was higher than those of middle class students in the least effective schools after three years of junior education. Again, such findings point to the *educational significance* of differences between schools in their effectiveness in adding value to student outcomes, and highlight the importance of using longitudinal designs which control for prior attainment.

Bosker and Witziers (1995) produced a major meta analysis of research concerning the size of school effects. They concluded that the "true" gross effect (no intake control) may be something like ten per cent, and the "true" net effect (with intake controls) about five per cent of total variance in student achievement. Bosker and Witziers (1995) went on to argue that, although the true net school effect may appear relatively insignificant, schools have an effect on all the students they serve. Taking this into account they claim that in the case of "stable performing secondary schools, that serve 1000 pupils over a period of five years each, the relative importance of the school is 50 times as high as the within school variation" (Bosker & Witziers, 1995, p. 9).

The extent to which models of school effects are generalizable across different national contexts and for different measures of students' educational outcomes clearly has implications for the debate about the size and significance of school for classroom/departmental effects. Scheerens and Bosker (1995) highlight the need for further comparative research into this issue.

2   Do schools perform consistently on different outcomes?

As noted in Section 1, a considerable body of evidence that schools can vary in their effectiveness in promoting students' academic achievement has accumulated. Much less evidence exists concerning schools' effects on other important aspects of student development (e.g., behaviour, attendance, attitudes, self-concept). The majority of studies have focused on only one or two measures of academic attainment, most commonly basic skills (reading/mathematics) at primary and total examination scores at secondary level. Only a minority of researchers have attempted to examine consistency in schools' effects on a range of different educational outcomes including social/affective and academic outcomes (examples of studies which have paid some attention to this issue include Reynolds, 1976; Rutter et al., 1979; Gray, McPherson & Raffe, 1983; Mandeville & Anderson, 1986; Cuttance, 1987; Mortimore et al., 1988a; Brandsma & Knuver, 1989; Knuver & Brandsma, 1993).

Different aspects of academic attainment

At the primary school level, Mortimore et al.'s (1988a) work provides evidence that some schools were more effective at promoting particular aspects of academic outcome than others. Modest positive correlations were reported between schools' effects on reading and mathematics ($r = 0.41$) and writing and mathematics ($r = 0.28$). Overall 19 schools were reported to have positive effects on all or three of the four cognitive outcomes examined. Around a quarter (12) were found to have positive effects on none or only one cognitive outcome out of a sample of 47 schools for which data on all outcomes were available.

Subsequent reanalysis of the data set suggested that the relationship between schools' effects on reading and mathematics may be somewhat closer ($r = 0.61$), Sammons, Nuttall and Cuttance, (1993). In addition to reporting an overall measure of association, this reanalysis examined the extent to which individual schools could be classified as highly effective or highly ineffective by reference to their residual (value added) estimates and attached confidence limits for the two basic skill areas. Only four out of 49 schools in the sample had significantly positive effects on students' progress in both outcomes and six had a markedly negative effect on both (at the 0.05 level). Thus around a fifth of the sample of schools were particularly effective or particularly ineffective for both cognitive areas but the majority of schools varied in effectiveness.

Bosker and Scheerens (1989) provided an extensive review of various aspects of stability in school effects. In connection with stability across effect criteria, they noted American studies by Mandeville and Anderson (1986) and Mandeville (1987) which, like the Mortimore et al. (1988a) study, investigated the correlation between effects on mathematics and reading. The results indicated fairly substantial consistency (r near 0.70), a figure in line with that reported by Bosker's (1989) study of elementary schools in the Netherlands ($r = 0.72$) and somewhat higher than that reported by Mortimore et al. (1988a).

A recent elementary study of reading and mathematics achievement by Yelton, Miller and Ruscoe (1994) in the USA demonstrated a fairly high level of consistency in school effects for these two outcomes within years. However, the study was based on aggregate data and therefore suffered from serious methodological weaknesses. An analysis of outliers revealed that "eight of the fifteen or sixteen outliers identified for each year were outliers in *both* reading and mathematics achievement" (pp. 9–10). Outliers were classified as residuals plus or minus one standard deviation larger than expected.

Relatively few studies have examined consistency in different measures of attainment at secondary school. Cuttance (1987) reported correlations for Scottish secondary schools of 0.47 and 0.47 for English and arithmetic respectively with an overall achievement indicator. Work by Willms and Raudenbush (1989) and Smith and Tomlinson (1989) also found some positive correlations between schools' effects in English and mathematics and overall exam score. Smith and Tomlinson concluded "There is a considerable tendency for schools that are successful in one subject to be successful in another and across all subjects, but there are also some important contrasts between the level of success achieved by the same schools with different subjects" (p. 276).

In the Netherlands, Luyten (1994) conducted a wide ranging analysis of the stability of school effects in secondary education examining performance in 17 subjects across five years. However, no student level prior attainment or background data were available, thus little control for intake variations between schools was possible (control for curriculum track was made). It was concluded that "differences between subjects within schools – appear to be of more importance than the general school differences" (p. 21). In total 40 per cent of the school level variance in student achievement was attributed to subject differences (where the school level represented 15 per cent of the total variance in student achievement).

Until recently little research had simultaneously investigated the issues of consistency in effects for different outcomes and across several years (see also section 3). As part of a study of departmental differences in secondary school effectiveness (Sammons et al., 1995a) the issue of consistency in schools' effects across a range of subjects (English, English literature, French, history, mathematics and science) and tscore a measure of overall attainment (total GCSE performance score) was explored. Consistency in effectiveness was examined over a three year period using three level models which provide a single value added measure of effectiveness (residuals) over three years for each outcome (Thomas, Sammons & Mortimore, 1995). Table 3.1 shows the correlations between the (shrunken) school level residuals for each outcome.

In all cases the correlations were positive but by no means perfect. Thus, although there was a general tendency for schools with positive effects in one area to have positive effects on others in some schools particular departments were relatively more effective than others. Thomas, Sammons and Mortimore's

Table 3.1 Correlations Between School Effects (Over Three Years) on Total GCSE Performance Score and Separate Subject Scores.[1]

	Tscore	Maths	English	English Literature	French	History	Science
Tscore	1.00	0.48	0.57	0.38	0.52	0.42	0.58
Maths		1.00	0.24	0.24	0.45	0.25	0.35
English			1.00	0.72	0.38	0.31	0.26
English Literature				1.00	0.25	0.25	0.25
French					1.00	0.33	0.34
History						1.00	0.37
Science							1.00

Note. N students = 17850; n schools = 94; n years = 3.
After Thomas, Sammons and Mortimore (1995).

---

1 It is important to note that any correlations between schools' value added scores (residuals) may be viewed as technically inflated estimates due to the "shrinkage" (towards the overall mean score) in calculating these scores, particularly for schools with small numbers of students and the 'true' correlations will be somewhat lower.

(1995) results demonstrate that at the secondary level departmental differences can be marked. School effects in English and science were found to relate more closely to tscore (total GCSE performance score) a composite measure of attainment than other subjects. Evidence concerning the existence of internal variations in effectiveness at the department level, indicate that a composite measure of achievement on its own provides only a partial indicator of effectiveness for some schools, although it remains a useful summary measure of overall achievement.

In addition to using correlations as a measure of consistency in effectiveness across different outcomes, Thomas, Sammons and Mortimore (1995) also examined the extent to which schools' actual GCSE performance differs significantly (at the 0.05 level) from that expected, taking into account their student intakes. Using a range of criteria (see Table 3.2) for grouping schools it was possible to identify in any year a group of outliers with highly significant positive effects in overall GCSE performance and several subjects (and no significant negative effects) and a group of schools with highly negative effects (and no significant positive effects) in terms of residuals derived from multilevel analyses. For example, in the 1991 analysis, 13 (14 per cent of the 94 schools) classified as broadly effective. A further 15 (16 per cent) were classified as broadly ineffective. Thus for 1991 just under a third of schools (30 per cent) could be classified as consistent outliers in terms of their effectiveness. However, for the majority (70 per cent) no clear cut picture emerged. Indeed around one in five (18 schools) were classified as highly mixed, recording significant negative effects in the same subjects and significant positive effects in others taking account of prior attainment and background. (Section 3 summarises the findings concerning stability over time).

Table 3.2 Consistency of School Effectiveness Across Different Measures at GCSE.

	Number of schools		
	1990 ($n = 69$)	1991 ($n = 94$)	1992 ($n = 77$)
Positive scores criteria [1]	10	13	12
Negative scores criteria [2]	10	15	7
Mixed scores criteria [3]	14	18	22
Other scores criteria [4]	28	41	27
No significant scores	7	7	3

Note. Criteria [1] significant +ve total score *and* 2 or more significant +ve subject scores (no significant −ve scores).
Criteria [2] significant −ve total score *and* 2 or more significant -ve subject scores (no significant +ve scores).
Criteria [3] 3 or more significant +ve *and* −ve subject scores.
Criteria [4] Any other combination of significant scores.
Residuals for 6 separate GCSE subjects and for total GCSE performance scores were included in the analysis.
After Thomas, Sammons and Mortimore (1995).

At the post-16 level in the UK, Fitz-Gibbon (1991, 1992) has also investigated departmental effectiveness at A-level. She argued that the results "do not lend strong credibility to the notion that schools doing well with students in one aspect will be effective in all aspects" (Fitz-Gibbon, 1991, p. 80) but no correlations were reported of departmental effects at A-level on different subjects and no overall measure of achievement was analysed.

The evidence concerning the extent of consistency in schools' performance in promoting different aspects of academic attainment or subjects is still fairly limited being confined to language/reading and mathematics/arithmetic at the primary level. Given the differences between secondary and primary schools (the greater use of subject specialist teachers and departmental organisation) it might be anticipated that the concept of overall school effectiveness would be less applicable than of departmental effectiveness and Luyten's (1994) and Thomas, Sammons and Mortimore's (1995) analyses provide some support for this view.

In different kinds of outcomes (academic and affective/social)
Early influential secondary school studies by Reynolds (1976) and Rutter et al. (1979) found fairly strong inter-correlations between schools' academic effectiveness and their social effectiveness (using attendance and delinquency rates). The Rutter et al. (1979) study concluded "On the whole, schools which have high levels of attendance and good behaviour also tend to have high levels of exam success" (p. 92). Substantial rank correlations were reported for delinquency and attendance (0.77); for delinquency and academic outcome (0.68); for delinquency and behaviour (0.72). However, later work by Gray, Jesson and Sime (1983) suggested that outcomes such as liking school and attendance were partially independent of schools' academic outcomes.

Fitz-Gibbon (1991) reported significant but modest positive correlations (ranging between $r = 0.26$ and $r = 0.53$) between effects on attitude to school, on attitude to subject and on aspirations and effects on A-level subject performance for chemistry and mathematics.

For primary schools, although Mortimore et al. (1988a) found some significant positive associations (and no negative ones) between particular cognitive and affective outcomes (e.g., attitude to mathematics and on self-concept with effects on mathematics progress), it was concluded that the two dimensions of school effects upon cognitive and upon non-cognitive areas were independent. Nonetheless, it was also noted that a number of schools had positive effects upon outcomes in both areas (14 out of 47 schools) whereas very few recorded positive effects for cognitive outcomes but were ineffective in non-cognitive outcomes (3) and six were broadly effective in promoting non-cognitive outcomes but were unsuccessful in most cognitive areas.

In the Netherlands, Knuver and Brandsma (1993) studied the relationships between schools' effects on a variety of affective measures (attitudes to language and arithmetic, achievement motivation, academic self-concept and school well being) and on language and arithmetic attainment. The correlations were very small but never negative. This study also found that "the strongest relationship between cognitive and affective outcomes can be found when subject specific affective measures are under study" (p. 201). In line with Mortimore et al.'s (1988a) work it was concluded that the two domains are relatively independent at the school level, but "not in any way contradictory to one another" (Knuver & Brandsma, 1993, p. 201).

Evidence on the extent of consistency of schools' effects on different kinds of educational outcomes appears to be mixed. At both secondary and primary level significant and positive, though far from perfect, correlations between schools' effects on different kinds of academic outcomes have been reported. However, few studies have examined both cognitive and academic/social outcomes. Of those that have, primary studies suggest that schools' effects on the two domains are positively weakly related and may be independent. At the secondary level results suggest that effects on academic and certain affective/social outcomes may be more closely linked, particularly for examination results and attendance and behaviour. Further research is needed to investigate the issue of consistency in schools' effects on different kinds of outcomes in greater depth and in a wider range of contexts before any firm conclusions can be drawn.

3 How stable are school effects over time?

Bosker and Scheerens (1989) reviewed evidence concerning two aspects of stability over time. They make the assumption that "since we might expect that organisational characteristics of schools are more or less stable over time, we must know if the rank order of schools on output remains the same no matter when we measure the effect" (p. 747). The extent to which schools' organisational characteristics do remain stable over time is, of course, debatable. Evidence from Mortimore et al. (1988a) has demonstrated that many inner city primary schools were subject to substantial change during a three year period. In the UK the implementation of the 1988 Education Reform Act has led to many far-reaching changes in curriculum, funding and organization. Rather similar market driven reforms are now affecting schools in countries such as Australia and New Zealand.

The evidence on whether the effects of schools vary over time is mixed. Early British secondary studies (Reynolds, 1976; Rutter et al., 1979) examined students' examination outcomes for different years and found in general consistency over time (rank correlations of 0.69 to 0.82 for the Rutter study). Work by Willms (1987) and Goldstein (1987) in the UK reveals correlations ranging between 0.80 to 0.60. In the Netherlands figures ranging between 0.75 to 0.96 (Bosker & Scheerens, 1989) have been reported.

However, Nuttall et al.'s (1989) research in inner London pointed to the existence of lack of stability in secondary schools' effects on student total examination scores over a three year period, although no indication was given of the correlation between years in the estimates of individual schools' effects. They concluded "This analysis nevertheless gives rise to a note of caution about any study of school effectiveness that relies on measures of outcome in just a single year, or of just a single cohort of students." (p. 775). In her A-level work, Fitz-

Gibbon (1991) also drew attention to the need to examine departmental residuals over several years.

Like Nuttall et al. (1989), Raudenbush (1989) also drew attention to the importance of longitudinal models for estimating the stability of school effects. Summarising results from a study of school effects over two time points covering four years (1980, 1984) in Scotland, it was concluded that the estimated true score correlation was 0.87. In the original analyses Willms and Raudenbush (1989) also looked at English and arithmetic results separately in addition to overall examination results and found that school effects on examination results in specific subjects were less stable than those on an overall attainment measure.

Research on stability in examination results for secondary schools by Sime and Gray (1991) also found considerable stability in schools' effects on students' mean examination results over a period of three years in one LEA and over two years in another (correlations ranging from 0.86 to 0.94).

In the Netherlands (Luyten, 1994) examined the stability of school effects in secondary education over a period of five years for 17 subjects, and the interaction effect of instability, across years and subjects. Unfortunately, because no measures of prior attainment were available it is possible that, with more adequate control for intake, the results might differ. Nonetheless, it was concluded that "schools produce fairly stable results per subject across years" (p. 20). It was further noted that "the interaction effect of subject by year is substantial" and "the general year effect turns out to be very modest" (p. 21).

The results of the study of departmental differences in effectiveness in inner London (summarised in section 2) utilised three-level models to examine the percentage of total variance in student scores attributable to fluctuations in results over time using an overall measure of examination attainment and results in individual subjects over a three year period (1990–1992). Table 3.3 (after Thomas, Sammons & Mortimore, 1995) shows that the percentage of variance due to school was higher than that due to year for all subjects and that the year effect is smallest for total GCSE performance score.

In addition to the three level models, two level models were used to calculate a separate intercept for each year at the school level and estimate the 'true' correlations between the school effects across the three years for each outcome measure (see Table 3.4).

The correlations over two years are all significant and positive and, as would be expected, are higher than those over three years. It is notable that the correlations for the overall measure of performance (Tscore) are higher than those for separate subjects, particularly over three years. This indicates that departmental effects are less stable over time and is in line with Smith and Tomlinson's (1989) analysis of English, mathematics and overall examination performance score. Gray et al. (1995) also recently concluded that, at the secondary level, school effects in terms of an overall measure of academic outcomes are fairly stable.

In addition to calculating correlations between schools' effects on a range of outcomes across years, the extent to which results were significantly better or

Table 3.3 Percentage of Variance in Students' Total GCSE Performance and Subject Performance Due To School and Year.

	1990–92*	
	Variance (%) due to school	Variance (%) due to year
Total GCSE performance score	6.2	1.1
English	4.1	1.8
English Literature	6.9	3.4
Mathematics	5.9	3.6
Science	6.1	4.7
French	7.8	7.8
History	15.3	3.6

Note. * 3-level model using 1990–92 data (1: candidate, 2: year, 3: school).
After Thomas, Sammons and Mortimore (1995).
N of schools = 94, N of students = 17850.

worse than predicted on the basis of intake measures across three years was examined. Thomas, Sammons and Mortimore (1995) found that, out of a sample of 69 schools for which data were available for each year, only three were consistently classified as broadly more effective in each year and three as less effective. Four schools consistently demonstrated mixed effects (significant positive effects in some subjects and significant negative effects in others). By analysing schools' effects for both consistency and stability over time simultaneously it was possible to identify a small group of ten outlier schools which could be clearly differentiated (as either positive, negative or mixed) in terms of stringent effectiveness criteria (noted in Section 2, Table 2).

Less attention has been paid to the stability of school effects at the primary than the secondary level, although Mandeville (1987) reports correlations in the

Table 3.4 Correlations Between Estimates of Secondary Schools' Effects on Different Outcomes a GCSE Over Three Years.

	1990 vs 1991	1990 vs 1992 Final Model	1991 vs 1992
Tscore	0.88	0.82	0.85
English	0.86	0.40	0.77
Mathematics	0.59	0.56	0.83
Science	0.52	0.41	0.59
History	0.92	0.71	0.83
English Literature	0.84	0.38	0.71
French	0.48	0.38	0.57

Note. Students = 17850; Schools = 94; Years = 3.
After Thomas, Sammons and Mortimore (1995).

USA ranging from 0.34 to 0.66. However, Bosker and Scheerens (1989) point out that these figures may be "deflated" because of the inadequacy of the statistical model used.

Although no correlations were reported, another example of USA research (the Louisiana School Effectiveness Study) examined selected schools which were consistent outliers either negative or positive terms in the state basic skills test (which focused on reading). Stringfield et al. (1992) concluded in their study of 16 outliers "most schools appear to have remained stable outliers, either positive or negative, for at least seven years" (p. 394).

Two recent studies of elementary schools in the USA have also pointed to the existence of substantial stability in school effects over time. Crone, Lang and Franklin (1994) investigated the issues of stability and consistency in classifications of schools as effective/ineffective at one point in time. They report correlations of residuals over the two years ranging between 0.49 and 0.78. It was concluded that stability was greater for composite measures of achievement and for mathematics than for reading or language. Unfortunately, the study utilised aggregate school level data in regression analyses, no prior-attainment measures were available and the intake data available for control was limited. This is likely to have had a substantial impact upon the estimate of schools residuals obtained and thus the results should be treated with caution.

Research by Yelton, Miller and Ruscoe (1994) likewise investigated the stability of school effectiveness for 55 elementary schools. Like the Crone, Lang and Franklin (1994) study, the research used aggregated data, no prior attainment data were included and the control for other intake characteristics was limited. It was found that of the 14 reading achievement outliers (defined as plus or minus 1 sd) identified for each year, nine were outliers in both years. For maths achievement 15 outliers were identified over two years with eight being outliers in both years. However, only three schools were identified as outliers in both cognitive areas and in each of the two years.

This review of the issue of stability in school effectiveness research suggests that there is a fair degree of stability in secondary schools' effects on overall measures of academic achievement (e.g., total examination performance scores) over time (correlations are fairly strong and all positive). The same trend is evident for basic skill areas in the primary sector, though correlations are lower. There is rather less evidence concerning stability for specific subjects at the secondary level, or concerning social/affective (non-cognitive) outcomes of education for any age groups.

On the basis of existing research it is apparent that estimates of schools' effectiveness based on one or two measures of students' educational outcomes in a single year are of limited value. Ideally, data for several years (three being the minimum to identify any trends over time) and several outcomes are needed. Further research which examines both stability and consistency simultaneously is required for a wider range of outcomes and in different sectors.

4 What are the long term effects of schools?
In contrast to the attention increasingly paid to the issues of stability and consistency in schools' effectiveness and the question of differential effects in multilevel school effectiveness studies, scant attention has been given to the continuity of school effects measured at different stages of a student's school career. In other words, what long term effects (if any) does previous institutional membership (e.g., primary school attended) have on later performance at secondary school?

Typically, models of secondary school effects utilise measures of students' prior attainment and background characteristics at intake (commonly at secondary transfer at age 11 in UK studies) in analyses of performance at a later date (e.g., in GCSE examinations taken at age 16, the end of compulsory schooling in the UK). Recent developments in UK school effectiveness research suggest that such models may be seriously mis-specified.

Sammons et al. (1995b) provided an example of the first attempt to address the issue of continuity in school effects using multilevel approaches. Their results suggested that primary schools exert a long term effect upon student performance, even after controlling for student attainment at secondary transfer. Their study was based on a nine year follow-up of the *School Matters* (Mortimore et al., 1988a) junior school cohort at secondary school. However, it was limited because it did not model junior and secondary school effects simultaneously by considering the full cross classification of individual students in terms of their secondary by junior school attendance using recently available techniques (Goldstein, 1995). A recent reanalysis of this data set by Goldstein and Sammons (1995) using such techniques provided a more detailed investigation estimating the joint contributions of primary and secondary schools. The total school (level 2) variance is the sum of a between-junior and a between-secondary variance. Table 5.5 provides details of the results of fitting three different multilevel models with total GCSE performance score as the explanatory variable.

The figures in Table 3.5 demonstrate that the variance in total GCSE performance scores attributable to junior school is much larger for Model C (no control for attainment measures at the end of junior education) than in Model A (control for attainment at secondary transfer and random variation at level 1). However, the junior school variance is larger in all cases. Further analyses were conducted controlling for English and mathematics attainment at entry to junior school (age 7 plus) but these made little difference to the results (and were not significant at the 0.05 level).

Goldstein and Sammons (1995) argued that the standard secondary school effectiveness model (where only secondary school attended is fitted at level 2) considerably over-estimates the secondary school effect. They concluded "the usual quantitative procedures for estimating school effectiveness need to be augmented with careful measurements of all relevant prior performances, including institutional membership. This applies to studies of value-added at A level (age 18 plus) where, in principle we can study the variation from primary, secondary and tertiary institution simultaneously" (p10).

Table 3.5 Variance Components Cross-Classified Model for Total GCSE Examination Performance Score As Response.[2]

	A	B	C
Fixed			
Intercept	0.51	0.50	0.25
Males	−0.21 (0.06)	−0.19 (0.06)	−0.34 (0.07)
Free school meal	−0.22 (0.06)	−0.23 (0.06)	−0.37 (0.08)
VR2 band	−0.39 (0.08)	−0.38 (0.08)	
VR3 band	−0.71 (0.13)	−0.71 (0.13)	
LRT score	0.31 (0.04)	0.32 (0.04)	
Random			
Level 2 (School)			
(Junior variance) σ^2_{u1}	0.025 (0.013)	0.036 (0.017)	0.054 (0.024)
(Secondary variance) σ^2_{u2}	0.016 (0.014)	0.014 (0.014)	0.019 (0.02)
Level 1 (Student)			
σ^2_{e0}	0.50 (0.06)	0.554 (0.06)	0.74 (0.05)
σ_{e01}	0.092 (0.03)	0.064 (0.03)	0.10 (0.05)
σ_{e02}	0.093 (0.018)		
σ^2_{e2}	0.033 (0.022		
−2Log likelihood	1848.8	1884.2	2130.3

Note. N of Junior Schools = 48; N of Secondary Schools = 116; N of Students = 785. After Goldstein and Sammons (1995).

The data set analysed by Goldstein and Sammons (1995) was relatively small (due to loss of sample over time) being based on only 758 students with 48 junior and 116 secondary schools included. Given the small sample size, caution should be exercised in interpreting the findings. Nonetheless, Goldstein (1995) has produced results broadly in line with those identified here in a larger study (but with less detailed background data) of Scottish schools.

Given the development of cross-classified techniques, there is a clear need for further investigations of the long term effects of schools using detailed longitudinal samples incorporating prior institutional membership and involving larger samples of schools drawn from a range of different socio-economic and geographical contexts. The need for collaborative research concerning the generalizability of multilevel educational effectiveness models across countries is becoming increasingly recognised (Scheerens & Bosker, 1995).

---

2 The exam score and LRT score have been transformed empirically to have N(0,1) distributions. FSM is a binary (yes, no) variable. At level 2 the subscript 1 refers to Junior and 2 to secondary school. At level 1 the subscript 0 refers to the intercept, 1 to males and 2 to LRT.

5 To what extent are schools differentially effective for different groups of students?

As noted in Section 1, the importance of controlling for prior attainment in studies of school effectiveness, so that the value-added by the school can be estimated, is now widely recognised as standard practice. Nonetheless, there is evidence that a number of pupil-level characteristics remain of statistical and theoretical importance (see Nuttall et al., 1989; Sammons, Nuttall & Cuttance, 1993; Sammons et al., 1994b). Willms (1992) argues that, of such factors, measures of socio-economic status are particularly important.

In considering whether schools perform consistently across differing school memberships it is important to distinguish contextual or compositional effects as they are some times labelled, from differential effects. Contextual effects are related to the overall composition of the student body (e.g., the percentage of high ability or of high SES students in a given year group or in the school's intake as a whole) and can be identified by between school analyses across a sample of schools. Such research has been a feature of UK studies of secondary schools in the main and has suggested that contextual effects related to concentrations of low SES, low ability and ethnic minority pupils can be important (e.g., Willms, 1986; Willms & Raudenbush, 1989; Nuttall, 1990; Sammons et al., 1994b).

Differential school effects

Differential school effects concern the existence of systematic differences in attainment between schools for different pupil groups (those with different levels of prior attainment or different background characteristics), once the *average* differences between these groups has been accounted for.

(i) Prior attainment

Although the study by Rutter et al. (1979) did not utilise multilevel techniques it did examine schools' examination results for the most and least able children and compared the results for children of different levels of prior ability (using a three category measure at intake – VR band). It was found that "the pattern of results for each school was broadly similar in all three bands" (p. 86). However, this study was based on a very small sample of 12 schools.

Smith and Tomlinson's (1989) study of multi-racial comprehensives produced some evidence of differential effectiveness for students with different levels of prior attainment (measured by second year reading test). In particular, differences in English exam results between schools were found to be greater for students with above-average than for students with below-average second year reading schools. The authors conclude that this is "largely because the exams are such that even the best school cannot achieve a result with a pupil having a below-average reading score" (p. 273). However, Smith and Tomlinson found little evidence that the slopes of schools' individual effects on examination results cross over. The same schools are most successful with the more *and* the less able students, "but a more able pupil gains a greater advantage than a less able one from going to a good school" (p. 273). The findings for mathematics were similar to those for English.

Nuttall et al.'s (1989) and Nuttall's (1990) secondary school analyses report evidence that schools' performance varies differentially, some schools narrowing the gap between students of high and low attainment on entry. The results suggest that variability in high ability students between schools is much larger than that of low ability students. These studies were limited, however, by inadequate statistical adjustment because the only prior attainment data available was the crude categorisation of three VR bands.

In the Scottish context, Willms and Raudenbush (1989) also report some evidence of differential school effectiveness for students of different prior attainment (VRQ) levels. However, in an earlier study of Scottish Secondary Schools Willms (1986) concluded "the *within* school relationships between outcomes and pupil characteristics did not vary much across schools" (p. 239).

Jesson and Gray (1991) investigated the issue of differential school effectiveness for students with different levels of prior achievement at the secondary level. These authors suggested that there is no conclusive evidence for the existence of differential slopes, "Pupils of different prior attainment levels did slightly better in some schools than in others... schools which were more effective for one group of pupils were generally speaking more effective for other groups as well" (p. 46). This conclusion is broadly in line with that of Smith and Tomlinson (1989). Jesson and Gray (1991) suggested a number of possible reasons for the difference between Nuttall et al.'s (1989) and their own results. They drew particular attention to the high degree of social differentiation in inner city areas and to the crude measure of prior attainment in Nuttall et al.'s (1989) research. They concluded that the use of a crude grouped measure rather than a finely differentiated measure of prior attainment may affect findings about the nature and extent of differential school effectiveness.

Most of the evidence concerning differential school effectiveness and prior attainment has been conducted at the secondary level. The original analyses for the School Matters (Mortimore et al., 1988a) study did not re-examine differential effectiveness for students with different levels of prior attainment. The subsequent reanalysis by Sammons, Nuttall and Cuttance (1993) found some evidence of differential school effectiveness for students with different levels of prior attainment, although this was less notable for reading than for mathematics. Nonetheless, the general conclusion was that although significant, differential effects were fairly modest and that schools which were effective for low attaining students also tended to be effective for those with high prior attainment also.

Research in the Netherlands by Brandsma and Knuver (1989) at the primary level also investigated the extent of differential effectiveness in language and arithmetic progress. No evidence of "equity differences", as these authors call such effects, were found in relation to pre-test scores for mathematics. However, for language "the effect of language pre-test on post-test differs slightly between schools" but "these differences are very small" (p. 787).

Gender

A few studies have pointed to the existence of differential school effects related to pupil gender (after taking account of the impact of gender at the level of the individual level). For example, Nuttall et al.'s (1989) study of examination results over three years in inner London points to the existence of such differential effects in terms of total examination scores "some schools narrowing the gap between boys and girls... and some widening the gap, relatively speaking" (p. 774). However, in the Scottish context Willms and Raudenbush (1989) who noted differential effects for prior attainment did not identify any differential effects for other background characteristics, including gender.

At the primary level the study by Mortimore et al. (1988a) produced no evidence of differential school effectiveness related to gender for reading or mathematics progress and the more detailed reanalysis supports the earlier conclusions (Sammons, Nuttall & Cuttance, 1993).

In the Netherlands, Brandsma and Knuver (1989) found no evidence of differential school effects related to gender for mathematics and only very small equity differences for the Dutch language. "The influence of gender, (overall positive for girls), does differ somewhat between schools" (p. 787) but the authors note that these differences are very small and cannot be explained by the school or classroom factors investigated in their study.

Ethnicity

Several studies at the secondary level point to the existence of differential school effects for students of different ethnic backgrounds. Nuttall et al. (1989) reported within school Caribbean-English, Scottish, Welsh (ESW) differences in effectiveness and comment that other ethnic differences vary across schools even more than the Caribbean-ESW differences "the Pakistani-ESW differences has a standard deviation of some 3 score points across schools" (p. 775). However, the authors draw attention to the lack of individual-level data about the socio-economic level of students' families which could confound ethnic differences with socio-economic differences.

Elsewhere in the UK, Smith and Tomlinson's (1989) study also produced evidence of differential school effectiveness for children of different ethnic groups although these differences were found to be "small compared with differences in overall performance between schools" (p. 268). The authors make a general conclusion about schools in their sample: "the ones that are good for White people tend to be about equally good for Black people" (p. 305).

At the primary level neither the original Mortimore et al. (1988a) analyses nor the reanalysis by Sammons, Nuttall and Cuttance (1993) found evidence of significant differential school effectiveness for specific ethnic groups. Brandsma and Knuver (1989) likewise found no indications of the existence of differential school effectiveness according to ethnic group in their study of Dutch primary schools.

Socio-economic indicators

The importance of taking into account relevant socio-economic factors in studies of school effectiveness has been noted earlier (Section 1). In addition to effects at the level of the individual pupil, compositional or contextual effects related to the proportion of students from particular social class groups or of low family income have been identified in some studies (see the discussion of contextual effects). Few studies have examined the extent of within-school differential effects related to socio-economic factors. Willms and Raudenbush (1989) report compositional effects related to SES, but no within-school differences related to such characteristics.

At the primary level Mortimore et al.'s (1988a) study found no evidence of differential effectiveness related to non-manual versus manual social class background. Sammons, Nuttall and Cuttance's (1993) reanalysis confirmed this earlier conclusion. These authors also tested for differential effects related to low family income (using the eligibility for free school meals indicator) but found no evidence to support their existence. Interestingly, the Mortimore et al. research found no case in their sample of schools where students from manual backgrounds performed markedly better on average than those from non-manual groups. Schools were unable to overcome the powerful effects of social class. However, it was found that students from manual groups in the most effective schools on average outperformed those from non-manual groups in the least effective schools. The school was the unit of change rather than the social class group within it. The Mortimore et al. (1988a) sample was fairly small (just over 1100 students and 49 schools) and it would be of interest to establish whether, with a larger sample, the negative findings concerning differential effects at primary level would be maintained.

The need to examine evidence for differential effectiveness using more complex models which focus on data for a number of years has been noted by Thomas et al. (1995). In a simultaneous analysis of differential effects as part of the study of departmental differences in secondary school effectiveness reported earlier Thomas et al. (1995) examined schools' effectiveness for different groups of students in different subject areas and in terms of total GCSE score over a three year period. This work illustrated the complex nature of differential effectiveness at the secondary level. Evidence of significant differential effects was identified both for total GCSE score and separate subjects. It was most notable for students of different levels of prior attainment at intake, and those of different ethnic backgrounds. Some evidence of differential effects was also found for socio-economic disadvantage and gender (see Thomas et al., 1995). Nonetheless overall all students in more effective schools and departments were likely to perform relatively well at GCSE, but some groups (those not socio-economically disadvantaged) performed especially well. In contrast, all students in ineffective schools tended to perform badly at GCSE but disadvantaged groups were relatively less adversely affected. Their results provided no evidence that more effective schools closed the gap in achievement between different student groups – they did not compensate for society. However, whilst disadvantaged groups did better in more effective than in less effective schools, the gap in achievement increased within the more effective schools. Sammons, Thomas and Mortimore (1996) highlight "the complex nature of the equity implications of our findings".

Overall, it appears that, at the secondary level, there is some evidence of important contextual effects related to schools' pupil composition in terms of SES, ethnicity and ability or prior attainment in UK and Scottish studies, but less evidence for such effects in studies of younger age groups. For differential school effects (within school differences) again, secondary studies suggest that gender, ethnicity and prior attainment may all be relevant. However, for prior attainment, it is important that the control measure adopted is adequate (finely differentiated). Less evidence for differential effects exists at the primary stage. These inconclusive results point to the need for further research into differential school effectiveness using large data sets which contain adequate individual-level socio-economic as well as ethnic data. Exploration of the possible reasons for the apparent primary-secondary school differences in differential effectiveness are also called for.

Discussion and conclusions

Methodological and conceptual issues
A number of methodological and conceptual issues which this review suggests are of particular importance in making judgements about school effectiveness are summarised below and are thus relevant to the increasing attention given to school accountability by policy-makers.

Absolute differences versus progress
There is now fairly general acceptance that studies of school effectiveness in cognitive areas require adequate control for prior attainment at the level of the individual student (McPherson, 1992; Reynolds & Cuttance, 1992; Scheerens, 1992; Creemers, 1994a, b). Ideally such measures should be collected at the point of entry to school at the beginning of a relevant phase (infant, junior or secondary). The use of baseline attainment or ability data collected after a period of years in the same school is likely to lead to a reduction in the estimate of school effects. For example, control of GCSE scores at 16 and for ability at age 17 is likely to lead to reduced estimates of departmental differences at A-level. Yet GCSE and arguably ability scores are themselves highly likely to have been influenced by earlier secondary school (or departmental) effects. Ideally, the measures of prior attainment used should be finely differentiated rather than crude categorisations such as VR band and related to the outcome areas under investigation.

Control for prior attainment may also be relevant for studies of schools' effects on non-cognitive outcomes such as self-concept, behaviour and attendance, especially at the secondary level, given the known relationships between such measures at the student level. The value of utilising a baseline measure of the relevant outcome (e.g., behaviour at entry for studies of behaviour) should also be considered.

Controlling for intake

In addition to control for prior attainment levels, there is evidence that control for other background characteristics of student intakes is also important for the accurate measurement of school effects. The availability of individual level data concerning socio-economic factors, sex, ethnicity and language fluency is necessary to ensure that comparisons do not unfairly reflect on schools serving disadvantaged communities. Such factors are likely to be of particular importance in studies using language-based measures of student outcomes and total measures of performance. Background data are also required for the exploration of possible *differential* within school effects for particular student groups. In addition, aggregate whole school or year group data concerning student composition especially related to ability and SES enable exploration of possible *contextual* or compositional effects.

The analysis of differential and contextual effects allow the topic of *equity* in educational effects to be addressed. This question of equity is increasingly relevant to those concerned with raising educational standards at both the national and international level.

Range of outcome measures

The results of school effectiveness studies are heavily dependent upon the choice of outcome measures used. In the main, research has focused on a narrow range of measures of academic outcomes (e.g., basic skill attainment in primary schools, total examination scores at secondary level). Further work on a wider variety of academic measures is required to examine stability and consistency over time in different aspects of academic attainment (e.g., writing, science, technology). Oracy at primary school, subject differences or attainment in "core skill" areas at secondary level might prove fruitful areas for further investigation.

In comparison with cognitive outcomes, relatively little attention has been paid to the affective/social outcomes of education (Ainley, 1994). The evidence from studies which have examined both affective/social and academic aspects suggests that schools' effects on the two domains can vary considerably, especially for younger age groups. It is also conceivable that school effects may be more marked for some of these outcomes (e.g., behaviour, attendance) whereas teacher (for primary) or departmental (for secondary) effects may be relatively more influential for academic outcomes.

Further investigations are required as Mortimore (1992) noted, "the adoption of a broad range of outcome measures is essential if studies are to address, adequately, the all-round development of students, and if they are to be used to judge the effectiveness of schools" (p. 156). Longitudinal, multilevel studies are required to examine the size, consistency and interrelationships between school and classroom/departmental effects for different kinds of outcomes and for different phases of schooling.

Long term effects of schools

Very little work has been conducted to establish the long term impact of schools across phases (e.g., primary to secondary or secondary to post school). Work by Rutter et al. (1979) suggested that the post-school work records of students from more effective schools were better than those of their counterparts from less effective schools. Sammons et al. (1995b); Goldstein & Sammons (1995) and Goldstein (1995) provide evidence for continuity of primary school effects on secondary school achievement at GCSE. They conclude that current models of secondary school effects may be seriously mis-specified because of the absence of control for prior institutional membership. Further long-term longitudinal research using cross-classified multilevel models is required using studies specially designed to investigate the overall impact of early as well as later school attended on students' subsequent educational attainments and employment patterns.

Sampling

A major limitation of many studies of school effectiveness is related to the nature of the sample of schools utilised for the analyses. Most studies have focused on inner city schools serving disadvantaged intakes (often with a substantial ethnic mix). Further research is required using more diverse samples of schools to establish to what extent existing findings of school effectiveness research (e.g., concerning the relative size of school/class/ departmental effects for different kinds of outcomes and different phases of schooling, and of stability of effects over time) apply to schools in other locations and serving different kinds of communities. In particular there is a need to examine the generalizability of multilevel models of educational effectiveness across different national contexts (Scheerens & Bosker, 1995).

Processes related to school effectiveness

The question of what factors and processes (school and classroom) relate to effectiveness is clearly of major interest. Less attention has been paid, particularly in multilevel studies, to testing process data. Despite methodological weakness in studies which have attempted to examine aspects of process, recent reviews of the relevant literature (e.g., NWREL, 1990; Levine, 1992; Scheerens, 1992; Reynolds, 1994) suggest that some consistency in major findings has emerged. Scheerens (1992) and Creemers (1994b) have also drawn attention to school and classroom process factors identified in a range of studies of school or teacher effectiveness. Eleven key factors have been identified in a recent review by Sammons, Hillman and Mortimore (1995).

Nonetheless, this is clearly an area which requires much further investigation. As Mortimore (1992) and Reynolds (1992) have noted research on *which* school and classroom processes are associated with greater effectiveness, *how* school organisational factors have an effect on student outcomes, and what mechanisms create the organisational factors is needed. For example, Scheerens (1992) and Creemers (1994) have suggested that school process factors may only be of importance in as much as they create advantageous conditions for learning at the classroom level.

Reynolds (1992) suggests that, in addition to large scale quantitative studies, in depth qualitative case studies of individual schools may have a key role to play in promoting our understanding of effective school and classroom practices and the processes of school change. Studies which utilise qualitative case study approaches to supplement multilevel investigations may prove particularly fruitful (Sammons et al., 1995a).

Theory development
This review of school effectiveness research has focused on methodological issues of relevance to questions of judging school performance and policy concerns related to accountability. Despite broad consistency in findings concerning school and classroom processes "The causal status of the relationships found between school characteristics and effect measures is relatively small because of the correlative nature of the research, a lack of theory and insufficiently sharp-edged conceptualisation" (Scheerens, 1992, p. 76). Creemers (1994b) has drawn attention to the urgent need for more coherent and developed theories of school effectiveness (Creemers, 1994b). It is to be hoped that the next generation of school effectiveness studies will be able to build on and test out existing findings concerning the processes of school effectiveness and to assist in the construction of a more coherent and developed theoretical body of knowledge concerning the ways schools influence their students' outcomes. In particular, research into the generalizability of models of educational effectiveness is likely to be of particular importance during the next decade (Scheerens & Bosker, 1995), and research which examines explicitly the links between class/teacher and school effects.

Overview

The review of evidence from school effectiveness research presented in this paper is not intended to be exhaustive. It does, however, demonstrate that the concept of school effectiveness requires careful investigation using appropriately designed, longitudinal studies in different phases of schooling. Judgements about school performance and relative effectiveness requires careful analysis and there is little justification for "league tables" approaches to the presentation of performance data about schools. On the basis of the evidence of school effectiveness research, attempts to make fine distinctions between schools such as those encouraged in the UK by the national publication of schools' raw examination results are not statistically valid. Moreover fine distinctions are not appropriate even when value-added models are used. Reference to the confidence limits attached to estimates of schools' effects (residuals) is needed to establish the statistical significance of any apparent differences in effectiveness identified. In addition to appropriate methodology, it is important to ensure that adequate control for differences between schools in the characteristics of their student intakes is made using relevant measures of prior attainment and socio-economic and other student background data (e.g., age, sex, ethnicity).

Extending the criteria for adequate studies of school effectiveness outlined by Scheerens (1992, 1995) recent research demonstrates that judgements about schools' effectiveness need to make specific reference to questions such as:

- Effective in promoting which outcomes?
- Effective over what time period?
- Effective for whom?

School effectiveness is perhaps best seen as a relative term which is dependent upon time, outcome and student group. Gray (1990) and Gray and Wilcox (1994) provide further discussions concerning the judgement of school quality.

As Creemers (1994a) has noted, technical developments in the school effectiveness field mean that the methodology for such analyses is fairly widely available. However, there is a need for more long term longitudinal studies with appropriate student intake and outcome measures across different phases of schooling (Goldstein, 1995). In addition, studies which collect detailed process data at the appropriate level (school, classroom, teacher) are required so that the relationships between such measures and school/departmental or class effects can be examined over several years. Comparative studies conducted in a variety of geographical and national contexts are also needed to enable the impact of such contexts to be established. Finally, research which explicitly examines the nature of any links between student intake measures and school or classroom process data would improve our understanding of the way both influence student outcomes.

The fact that measures of stability in effectiveness are not perfect over time should not be too surprising. It is important to recognise that change in school and departmental effectiveness is likely over time periods of more than one or two years due to changes in staff, in pupil intakes and in ethos. Schools may also change rapidly as a result of outside intervention (e.g., new principal, inspection or in the face of particular issues). Therefore the separate identification and measurement of time and school effects is required in order to evaluate schools more accurately and to enable trends over time in performance to be monitored.

The limited availability of appropriate data sets at present is a major constraint on the further development of school effectiveness research. Only through carefully designed, longitudinal (and therefore expensive) empirical studies which test out and build on the findings of existing studies will our understanding of the complexity of judgements about school effectiveness and the factors which influence effectiveness be extended. Such studies will also facilitate the development of better theories of educational effectiveness, and hopefully will be of relevance to those concerned with the processes of school change and the improvement of practice. In particular, I believe that the development of systems for giving schools better feedback about their relative effectiveness, using a range of measures over several years will provide a major boost to the development of schools' capacity for self-evaluation and review and act as a stimulus for improvement work.

4

DIFFERENTIAL SCHOOL EFFECTIVENESS

Introduction

The last chapter highlighted the concept of complexities in the judgement of school performance. One of the import aspects of internal variation alluded to was the question of differential effectiveness. By this school effectiveness researchers are concerned with the extent to which schools differ in their effectiveness (in measures of student progress or value added) for different subgroups of student. In other words whether schools are equally effective (or indeed ineffective) for all their students, or whether particular categories of student make greater gains or receive greater benefits than others. These questions are clearly relevant to the strong interest of early school effectiveness researchers in the achievement of disadvantaged groups.

Despite these equity concerns the question of differential effects received little attention in most early studies. My interest in the question was influenced by my involvement in the classic School Matters *(Junior School Project) research during the early 1980s. Our study of primary school children from age 7 to 11 years received considerable attention when published in 1988 and has been clearly described by Peter Mortimore in an earlier volume in this series. Due to its time scale, the range of outcome measures studied, and its use of multilevel approaches for the analysis, the research had a considerable impact on the rapidly growing school effectiveness research field.*

As part of that study my colleagues and I had made an attempt to address the question 'Are some schools more effective for particular groups of children?' (Mortimore et al., 1988a, p. 206). Our analyses, however, were fairly crude due to the limitations of the multilevel models then available, and we focused only on three factors by which we divided children into groups – namely gender (boys versus girls), social class (non-manual versus manual) and ethnic background (Caribbean background versus other ethnic backgrounds).

The results suggested that

> 'in general, schools which were effective in promoting progress for one group of pupils (whether those of a particular social class, sex or ethnic group) were usually also effective for children of other groups. Similarly, those schools which were ineffective for one group tended to be ineffective for other groups' (Mortimore et al., 1988a, p. 217).

The research indicated that even the powerful impact of social disadvantage could be mitigated, although not eliminated by effective primary schools. It was concluded that

> 'Even though overall differences in patterns of pupil attainment are not removed in the most effective schools, the performance of all children is raised and . . . disadvantaged children in the most effective schools can end up with higher achievements than their advantaged peers in the less effective schools' (Mortimore et al., 1988a, p. 217).

The claim that effective schools tend to 'jack up' the performance of all their students is a bold one and one which required further testing as multilevel methods were further refined and more commonly used. In 1990 the Economic and Social Research Council (ESRC) awarded a grant to enable a further follow up of the Junior School Project cohort to investigate the long term effects of schools (the subject of Chapter 5) and to reanalyse the existing data to test out the findings reported in School Matters using more detailed multilevel models. The Models of Effective Schools project involved both Peter Cuttance and Desmond Nuttall. Our concern was to investigate internal variations in primary school effects (both in terms of consistency across different outcomes such as reading and mathematics and differential effects for different groups of students) as well as to enable a follow-up of the primary school students at age 16, the end of compulsory education in the UK.

Interest in differential effects was further stimulated by the publication of an important paper by Desmond Nuttall and colleagues (Nuttall et al., 1989) on secondary school examination data which pointed to the existence of statistically significant differential effects. A subsequent paper by David Jesson and John Gray (Jesson & Gray, 1991), however, suggested differential effects were fairly modest, and a debate about the measurement of such effects ensued.

In our paper 'Differential School Effectiveness' (written as part of the Models of Effective Schools Project) we focused on primary schools and sought to examine possible internal variations for a range of different student groups. We also extended the School Matters research by exploring differential effectiveness for students with different levels of prior attainment, a subject not addressed in *the original research. The published paper is an extended version of one originally presented at the British Educational Research Association annual conference in Stirling in 1992.*

Given the political context of the time, the paper also sought to make a contribution to the ongoing debate about the publication of school league tables by demonstrating the difference in school ranks apparent if raw results were compared with those based on school effectiveness models. This section of the paper highlighted the need to consider the statistical significance of the individual estimates of school effects (residuals) and pointed to the inappropriateness of making fine (i.e., ranked) distinctions between individual institutions as occurred in league table presentations. The research also drew attention to internal variations in school effects on reading and mathematics progress again pointing to the need to use several measures of school performance rather than one crude indicator.

Background

The demand for information about the functioning of educational systems from politicians, managers and consumers of the service has increased markedly during the last two decades. This demand reflects, in large part, the desire for greater accountability of the providers of the service, and general concern about 'standards' and 'value for money' in education (see Kogan, 1986; Tomlinson, Mortimore & Sammons, 1988; Marks, 1991) and has led to a debate about the development of appropriate performance indicators for schools.

The issue of appropriate and valid ways of measuring and reporting on schools' performance (as measured by pupils' examination or test results and the construction of performance indicators) has become increasingly relevant following the changes to schools' responsibilities and finance resulting from the introduction of the Education Reform Act 1988. The Government's recent decision (for the Parents' Charter) requiring the publication of 'league tables' of schools' raw examination and National Curriculum Assessment results, and its considerable doubts over the use of statistics adjusted in some way to take account of factors concerning differences between schools in their intakes in terms of pupils' prior attainments and the incidence of socio-economic disadvantage, have added fuel to the controversy. The 1991 Education (Schools) Act proposed new rules for the provision of information about school performance (Section 16). It requires league tables of examination results, National Curriculum Assessment results and truancy rates. Although it was originally intended that presentation of national curriculum results should include descriptions of the context of the school, it has now been stated that factual information should be published 'without footnotes or academic commentaries'. (For further discussion of the issue of valid comparisons of schools see Cuttance, 1986, 1988; Mortimore, Sammons & Ecob, 1988; Fitz-Gibbon, 1991; Jesson & Gray, 1991; Mortimore, Mortimore & Sammons, 1991; Nuttall, 1991; Goldstein et al., 1992; McPherson, 1992).

Academic interest in the fields of school effectiveness and school improvement research has expanded rapidly during the last two decades (see Creemers, Peters & Reynolds, 1989; Creemers & Scheerens, 1989; Reynolds, Creemers & Peters, 1989; Mortimore, 1990; Mortimore, Mortimore & Sammons, 1991; Reynolds, 1991). The International Congress for School Effectiveness and Improvement was inaugurated in 1988 and now publishes a journal devoted to these topics. Methodological advances, particularly the availability of appropriate statistical software for the analysis of multilevel data using models such as the ESRC's Multilevel Models' Project (Goldstein, 1987; Paterson & Goldstein, 1991), a package of which is now widely available (Prosser, Rasbash & Goldstein, 1991), have enabled more efficient estimates of school differences in pupil achievement (especially progress over time) to be obtained. Despite some controversy (see Preece, 1988, 1989) there is now substantial academic agreement as to appropriate methods of estimating school differences/effects and the kinds of data required for valid comparisons to be made. Indeed, the major limitation upon such studies is now the availability of appropriate longitudinal data sets, with data collected over time at the appropriate levels (individual pupil, class, school, LEA), particularly data sets which include measures of pupils' prior achievement and selected background characteristics, and process data related to school and classroom policy and practices.

To date, multilevel research on school effectiveness has provided strong evidence of the existence of differences between schools (both primary and secondary) in their overall effectiveness in promoting pupils' academic attainments (see Mortimore et al., 1988a, b; Bondi, 1991 on primary schools, and Willms, 1986; Cuttance, 1986, 1988; Smith & Tomlinson, 1989; Nuttall, 1990; Daly, 1991; Nuttall, Thomas & Goldstein, 1992; Jesson & Gray, 1991 on secondary schools). In addition, there is some evidence of variations between schools in their effectiveness in promoting different kinds of academic outcomes (Mortimore et al., 1988 a, b; Smith & Tomlinson, 1989). Indeed, analyses at A-level have indicated that departmental differences in effectiveness may be a more relevant concept than overall school differences in effectiveness (Fitz-Gibbon, 1991; Tymms, 1992).

However, there is less consistency in findings concerning differential school effectiveness (differential slopes) for particular groups of pupils (divided by sex, ethnic or social class group) and for different levels of prior attainment. Nuttall's (1990) work on the examination performance of inner London secondary schools found evidence of differential school effectiveness for pupils of different levels of prior achievement and for different ethnic groups. Smith and Tomlinson (1989) also found some evidence of differential effectiveness for pupils of different levels of prior achievement but concluded that, in general, the same schools do well and badly with the above-average and with the below-average pupils. Jesson and Gray's (1991) investigation of differential school effectiveness for pupils of differing levels of prior attainment argues that there is no conclusive evidence for the existence of differential slopes. They suggest that the use of a crude rather than a finely differentiated measure of prior attainment may affect estimates of variation in slopes between schools. Goldstein et al.'s (1992) recent work on secondary school examination results, however, supports the existence of differential slopes.

The issue of differential school effectiveness is clearly of importance in considering how schools' results should be compared. If schools differ markedly in their effectiveness for particular pupil groups the use of an overall measure could prove misleading. However, if there is little evidence of differential slopes an overall measure can provide a valuable indicator of school performance and a more valid basis for comparisons than raw results.

The main focus of the ILEA's Junior School Project research (Mortimore et al., 1988a, b) was the analysis of overall school effectiveness, and an early version of Goldstein's (1987) model was used to establish the proportion of variance in reading and mathematics attainment after three years in junior school attributable to the school level, whilst controlling for prior attainment and a variety of pupils' background characteristics (Mortimore et al. 1986, 1988a). School-level residuals were calculated and interpreted as indicators of school effectiveness for each outcome.

The original JSP research, however, did not address the issue of differential school effectiveness in any depth and the original analysis of this issue (which examined school effects for different sex, social class and ethnic groups) suffered from a number of limitations. In particular, it did not examine differential effectiveness (slopes) for pupils of different levels of prior achievement.

Aims

This paper presents results from a multilevel reanalysis of JSP data for pupils' mathematics and reading progress over three years from entry to junior school (autumn year 3) to the end of the third year in junior school (summer year 5). The main focus of the reanalysis is the extent of differential school effectiveness for i] prior attainment and ii] for specific pupil groups (divided by sex, social class, ethnic group and eligibility for free school meals).

In addition, the paper examines the presentation of schools' test results in reading and mathematics. Raw test averages are compared with school-level residuals from multilevel analyses which give an indication of the value-added by the school.

Readers who are unfamiliar with multilevel methods may wish to concentrate on the summary of results and conclusions.

For reading the analyses are based upon all pupils with valid Edinburgh Reading Test (ERT) scores on these two occasions ($N = 1115$). For mathematics, analyses are based upon all pupils with valid NFER Basic Mathematics Test (BMT) scores on these two occasions ($N = 1240$). Pupils were drawn from a total of 49 out of the 50 schools in the original JSP sample, one school having dropped out of the Project before the third year (see Mortimore et al., 1986 Part C).

A useful introduction to multilevel modelling is given by Paterson & Goldstein (1991). The analyses presented in this paper utilise the ML3E extended version of the ESRC Multilevel Models program (Prosser, Rasbash & Goldstein, 1991) based on Goldstein's (1987) multilevel model. They differ from the original JSP analyses being based on a slightly larger number of cases, because the JSP analyses focused on pupils with both valid reading and valid mathematics scores in years 3 and 5 ($N = 1101$), and the results are therefore not directly comparable with the original JSP findings (although the estimates of overall school effectiveness remain similar).

The design of the analyses of mathematics is identical to that adopted in the analyses of reading progress to enable comparisons of the results. However, the findings for reading are based on a slightly smaller number of cases (1115 versus 1240 for mathematics). Therefore the estimates for the effects of individual schools for reading and mathematics are based on slightly different pupil samples and this should be remembered in comparing the results for the two cognitive areas.

Estimation of variance of school means

The intercept explanatory variable 'CONS' was used to establish the extent of variance of school means for i) year 3 raw scores, and ii) year 5 raw scores and the intra-school correlations were calculated. The results are shown in Table 4.1. The results indicate the existence of significant differences between schools for both cognitive areas. It appears that substantially more of the variance in raw year 5 mathematics scores (summer) is attributable to differences between schools than of the variance in raw mathematics scores at entry (autumn year 3) – nearly 15 per cent compared with just over seven per cent. This is in line with the findings for reading attainment at these two points in time but the trend is more marked. It appears that differences between junior schools increase over time for both cognitive areas.

The analysis of progress

Schools differ markedly in the prior attainments and background characteristics of their pupil intakes. The original JSP analyses demonstrated strong relationships at the pupil level between factors such as age, sex, low family income, social class, ethnic background and mathematics attainment (Mortimore et al., 1988a). It was also shown that raw reading and raw mathematics scores in years 3 and 5 were strongly correlated ($r = 0.762$, $p < 0.0001$ for reading, $r = 0.688$, $p < 0.0001$ for mathematics).

The multilevel models used for the analysis of later reading and mathematics outcomes were therefore elaborated to take account of relevant background factors and prior attainment measured at entry to junior school. In these models, therefore, what is meant by the term 'progress' refers to analyses of an educational outcome relative to (adjusted for) a previous measure of attainment at entry.

In order to better specify the model, the variable pupils' raw score year 3 was centred around its mean and added to the fixed part of the model at the pupil

Table 4.1 Estimation of Variance of School Means (a) Year 3 (b) Year 5.

a) Attainment in year 5

Parameter		Raw BMT* score Summer year 5		Raw ERT* score Summer year 5	
		Level 2 Estimate	Level 2 S Error	Estimate	S Error
School σ^2	CONS/CONS	17.92	4.54	73.05	19.22
		Level 1 Estimate	Level 1 S Error	Estimate	S Error
Pupil σ^2	CONS/CONS	102.6	4.204	443	19.19
Intra-school correlation		0.1487		0.1416	

b) Attainment at entry year 3

Parameter		Raw BMT score Autumn year 3		Raw ERT score Autumn year 3	
		Level 2 Estimate	Level 2 S Error	Estimate	S Error
School σ^2	CONS/CONS	3.585	1.135	71.68	19.77
		Level 1 Estimate	Level 1 S Error	Estimate	S Error
Pupil σ^2	CONS/CONS	47.13	1.93	529.1	22.91
Intra-school correlation		0.0707		0.1193	

level. This process also aids interpretation of the results for the fixed part of the model. A number of background factors were included in the fixed part of the model. Two age-at-test measures for year 3 and year 5 were available and were tested, neither was significant. The two measures of age were highly correlated at the pupil level and the inclusion of prior attainment (which was significantly related to age at test in year 3) appears to remove any separate age effect. Because of this only the variable age-at-test year 5 was included in the model. This variable was centred around its mean.

Reading progress and background effects
Although marked sex differences in reading attainment (in favour of girls) were evident on each occasion reading was assessed, the effect of sex just failed to reach significance when prior reading attainment and other background factors were included in the model.

Dummy variables were created contrasting effects for different groups based on father's occupation (skilled manual, semi and unskilled manual, long term unemployed and father absent, father's occupation not known) with those of non-manual backgrounds. It was considered important to include the father's occupation in the not known group because treating this as missing would lead to a further reduction in sample size. Effects related to these social class variables were all significant, as were the effects for incomplete fluency in English, and eligibility for free school meals. The variable related to Asian background was not found to be significant when fluency in English was included in the model. The results for the fixed effects are shown in Table 4.2.

The effect for the Caribbean group is statistically significant in this model. Two other ethnic categories (Southern Irish and Other) were also tested but were not found to be statistically significant (effectively, therefore, these became part of the 'CONS' group). The ethnic categories are 'dummy' variables contrasted with the category English, Scottish, Welsh and Northern Irish (ESWI), the most numerous of the ethnic groups.

It should be noted that, in the fixed effects results (Table 4.2), the intercept variable 'CONS' represents the year 5 reading attainment of girls of average age and prior achievement level, from non-manual backgrounds, who are fully

Table 4.2 Effects of Background Factors and Prior Attainment on Year 5 Reading Attainment (Fixed Effects).

Parameter	Models including prior attainment	
	Estimate	S Error
CONS	59.85	1.352
* ERTRWY3	0.68	0.0185
AGEERT5	−0.163	0.116
* CARIBBEAN	−2.673	1.216
ASIAN	−2.728	1.879
SEX	−1.014	0.809
* FREEMLS	−2.569	0.999
* STAGE	−4.209	1.767
* SKILLED	−4.158	1.095
* SEMIUNSK	−5.977	1.834
* LTUFABS	−3.459	1.391
* FNK	−6.457	1.536

* $p < 0.05$.

Where:
ERTRWY3 = Raw ERT reading score year 3
AGEERT = Age at test in months year 5
CARIBBEAN = Caribbean ethnic background 1 versus the group English, Scottish, Welsh, Irish 0
ASIAN = Asian ethnic background 1 versus the group English, Scottish, Welsh, Irish 0
SEX = Male 1, female 0
FREEMLS = Eligible for free meals 1 versus not eligible 0
STAGE = Incomplete fluency in English 1 versus fluent 0
SKILLED = Father in skilled manual 1 versus non-manual work 0
SEMIUNSK = Father in semi or unskilled manual 1 versus non-manual work 0
LTUFABS = Father long term unemployed, economically inactive or absent 1 versus non- manual work 0
FNK = Father's occupation not known 1 versus non-manual work 0

fluent in English, and not eligible for free school meals, and of English, Scottish, Welsh or Irish or Other ethnic backgrounds. The interpretation of the estimates of the effects of pupils with different background characteristics shows how the reading attainment of such groups differ from that of the 'CONS' group.

Mathematics progress and background effects
Although sex differences in mathematics attainment (in favour of girls) were evident at each testing, the sex differences were much smaller than those found for reading. Nonetheless, when progress over time was examined the effects of sex were found to be significant (with boys making less progress than girls over the three years) taking into account prior mathematics attainment and background factors in the fixed effects part of the model (see Table 4.3).
Effects related to prior attainment were (as would be expected) highly significant. Only sex, eligibility for free school meals and the ethnic group OTHERNK (those from all other ethnic groups [except Asian or Caribbean] and those whose ethnicity is not known versus English, Scottish, Welsh and Irish) were significant. In contrast to the findings for reading none of the social class variables were significant when prior attainment was included in the model. Also in contrast to reading, the estimates of the effects for the Caribbean group and those not fully fluent in English were not significant. These factors are not important in predicting mathematics progress although they are important for reading. Only the measure of poverty (eligibility for free school meals) was found to be significant for both progress in mathematics and in reading, with those eligible for free meals making poorer progress than others.

Pupils in the ethnic category OTHERNK appear to attain more highly in mathematics than the ESWI group. The interpretation of this finding is not easy because the group consists of an amalgam of various other ethnic groups (e.g., Italian, Turkish, Greek, Chinese) and those for whom no ethnic data were available. It was not found to be significant in the equivalent analyses of reading.

In the fixed effects results (Table 4.3), 'CONS' represents the year 5 raw mathematics attainment of girls of average age and prior achievement level, from non-manual backgrounds, who are fully fluent in English, not eligible for free school meals, and of English, Scottish, Welsh or Northern Irish ethnic backgrounds. The interpretation of the estimates of the effects for pupils with different background characteristics shows how the mathematics attainment of such groups differs from that of the 'CONS' group.

The extent to which two models (A including background factors only, B including background factors and prior attainment) account for the variance in pupils' year 5 reading and mathematics attainment was also examined. The results are shown in Table 4.4.

The overall inclusion of background factors only (model A) accounts for around 11.3 per cent of the overall variance in pupils' year 5 mathematics attainment. This is substantially less than the equivalent proportion for reading attainment (20.6 per cent) indicating that the set of background factors included in the model are poorer predictors of mathematics than of reading attainment.

Table 4.3 Effects of Background Factors and Prior Attainment on Year 5 Mathematics Attainment (Fixed Effects).

Parameter	Models including prior attainment	
	Estimate	S Error
CONS	28.75	0.732
* MTHSRWY3	1.032	0.0326
* SEX	−0.878	0.417
AGEMAT5	−0.0767	0.0613
ASIAN	1.5	1.038
CARIBBEAN	−0.14	0.649
* OTHERNK	1.556	0.63
STAGE	0.136	1.015
* FREEMLS	−1.454	0.523
SKILLED	−0.982	0.581
SEMIUNSK	−1.445	0.959
LTUFABS	−0.569	0.737
FNK	−0.849	0.825

* $p < 0.05$.

Where:
MTHSRWY3 = Raw BMT mathematics score year 3
AGEMAT5 = Age at test in months year 5
ASIAN = Asian ethnic background 1 versus the group English, Scottish, Welsh, Irish (ESWI) 0
CARIBBEAN = Caribbean ethnic background 1 versus the Group ESWI 0
OTHERNK = All other ethnic backgrounds 1 and ethnic group NK versus the group ESWI 0
SEX = Male 1, female 0
FREEMLS = Eligible for free meals 1 versus not eligible 0
STAGE = Incomplete fluency in English 1 versus fluent 0
SKILLED = Father in skilled manual 1 versus non-manual work 0
SEMIUNSK = Father in semi or unskilled manual 1 versus non-manual work 0
LTUFABS = Father long term unemployed, economically inactive or absent 1 versus non- manual work 0
FNK = Father's occupation not known 1 versus non-manual work 0

For model A the intra-school correlation indicates that around 14 per cent of the variance in final mathematics attainment not accounted for by the set of background factors is attributable to schools (compared with only 9.8 per cent for reading). The inclusion of prior attainment and background factors (model B) causes a marked reduction in the amount of unexplained variance in year 5 attainment. The reduction is greater for reading (60.9 per cent) than for mathematics (49.3 per cent). In addition, there is an increase in the intra-school correlation showing that 17.2 per cent of the variance in year 5 mathematics attainment not attributable to the impact of pupils' background characteristics

Table 4.4 Estimation of Variance of School Means in Year 5 Attainment Using Different Models (A) Background Factors only, (B) Background Factors and Prior Attainment.

a) background factors only

		Mathematics Level 2 Estimate	S Error	Reading Level 2 Estimate	S Error
School σ^2	CONS/CONS	14.92	3.84	40.04	11.76
		Level 1 Estimate	S Error	Level 1 Estimate	S Error
Pupil σ^2	CONS/CONS	92.01	3.77	369.5	16.0
Intra-school correlation		0.1395		0.0978	

b) background factors and prior attainment

		Mathematics Level 2 Estimate	S Error	Reading Level 2 Estimate	S Error
School σ^2	CONS/CONS	10.52	2.58	37.67	9.287
		Level 1 Estimate	S Error	Level 1 Estimate	S Error
Pupil σ^2	CONS/CONS	50.58	2.073	164.1	7.105
Intra-school correlation		0.1722		0.1867	

and initial attainment (pupils' mathematics progress over three years in junior school) is attributable to differences between schools rather than differences between individuals.

These results are in line with those found for reading (intra-school correlation 0.187) and demonstrate the existence of significant school effects and the importance of utilising multilevel techniques for the analysis of differences between schools in pupils' cognitive attainments at any point in time, and the need for a longitudinal perspective. However, they also demonstrate significant differences in the importance of particular background factors in accounting for attainment and progress in the two cognitive areas.

It should also be noted that the fixed effects for all the pupil-level background factors shown in Tables 4.2 and 4.3 are much larger and are highly statistically significant when no measure of prior attainment is included in the model (see Table 4.5).

Comparing the results in Table 4.5 with those in Tables 4.2 and 4.3 demonstrates that the inclusion of a measure of prior attainment has a marked effect in reducing estimates of the impact of background factors because the model is longitudinal rather than cross-sectional, focusing on *change* in pupils' attainment over time. When attainment (rather than progress) is considered it is found that at any given point in time, background factors are powerful predictors of attainment. All factors, with the exception of membership of the Asian (significant for reading) and the OTHERNK (those from all other ethnic groups

Table 4.5 Effects of Background Factors on Year 5 Attainment (Fixed Effects).

Parameter	Mathematics		Reading	
	Estimate	S Error	Estimate	S Error
CONS	32.8	0.925	70.5	1.795
SEX	−1.332	0.562	−7.2	1.184
AGEMAT5 or AGEERT5	0.359*	0.0804	0.729*	0.169
ASIAN	−0.174	1.394	−6.552*	2.906
CARIBBEAN	−3.352*	0.862	−6.33*	1.843
OTHERNK	823	0.847	−0.79	1.769
STAGE	−6.081*	1.34	−15.96*	2.779
FREEMLS	−2.852*	0.701	−7.342*	1.48
SKILLED	−3.711*	0.774	−10.7*	1.618
SEMIUNSK	−5.785*	1.279	−16.11*	2.715
LTUFABS	−3.236*	0.985	−9.4*	2.062
FNK	−4.546*	1.095	−10.71*	2.253

* $p < 0.05$.

[except Asian or Caribbean] and those whose ethnicity is not known versus the group English, Scottish, Welsh and Irish) ethnic groups, had a statistically significant impact for both cognitive areas. Age-at-test had a strong positive effect, but all other factors were negatively related to attainment.

It is because background factors are highly related to pupils' initial attainment levels at entry to junior school that their impact is reduced when initial attainment is added to the model. The results of the analysis of change in attainment over time demonstrate that, in this statistical model, background factors are of less importance in accounting for progress (the 'value-added' approach) than are school differences, but that school differences seem to be greater for mathematics. The measures of prior attainment available in the JSP data set are finely differentiated. Where no or only crude measures of prior attainment are available, the estimates of the impact of background factors upon pupils' later attainment are likely to remain large. It is, of course, acknowledged that because of the correlation between background factors and levels of initial attainment, the inclusion of initial attainment in the model may remove variance caused by the former.

Differential school effectiveness

Prior attainment

The measure of prior reading attainment was allowed to vary randomly at both the pupil and school levels. The results for the estimates of random effects are shown in Table 4.6.

The intra-school correlation was recalculated for average attaining pupils (r = 0.1859) for the model which included prior attainment and all the back-

Table 4.6 Estimates of Random Effects for Prior Attainment on Year 5 Raw Reading Scores.

Parameter		Level 2	
		Estimate	S Error
School σ^2	CONS/CONS	36.77	9.081
	ERTRWY3/CONS	−0.501	0.160
	ERTRWY3/ERTRWY3	0.00767	0.004
		Level 1	
Pupil σ^2	CONS/CONS	161.0	7.318
	ERTRWY3/CONS	−0.897	0.141

ground factors shown in Table 4.2. This indicates that, for the pupil with average prior reading attainment, nearly 19 per cent of the *unexplained* variance in year 5 raw reading scores is attributable to differences between schools. (In terms of *total variation*, this represents 7.3 per cent of the total variation in pupils' year 5 raw reading scores.)

The loglikelihood function was calculated to test the hypothesis that adding prior reading attainment to i) levels 1, and ii) levels 1 and 2 of the random parameter matrix led to an improved fit of the model to the data. The result showed that allowing prior reading attainment to vary randomly at level 2 as well as level 1 led to a highly significant improvement in fit ($p < 0.00002$). There is thus evidence of the existence of differential slopes for individual schools.

The plot of slopes for the 49 individual schools is shown as Figure 4.1. This is based on the predicted raw reading scores (summer year 5) for individual pupils, controlling for initial attainment and background factors compared with their raw year 3 reading scores. The graph reveals marked differences between schools in their overall effectiveness (in terms of raising reading scores above predicted levels). It is also notable that individual schools' slopes do not tend to cross. The interpretation of this result is that effective schools appear to raise later reading attainment scores for pupils irrespective of initial attainment level. Conversely, less effective schools seem to depress later attainment scores for all.

However, Figure 4.1 shows that the slopes for schools are closer for pupils of high initial reading attainment (year 3) than for those of low initial attainment. The interpretation of this result is not clear. It may be due to 'ceiling effects' on the tests. Alternatively, it is possible that there are indeed greater variations in effectiveness between schools for pupils of low initial attainment. An analysis of the intra-school correlations by raw initial (year 3) reading score was made to examine the relative importance of school-level variation in comparison to pupil-level variation. It was found that the intra-school correlation is higher for pupils with low than for those with high initial reading attainments.

In order to allow for possible ceiling effects a quadratic function of initial reading attainment was included in the model. The results also suggest the existence of modest differential school effects for pupils with different levels of prior attainment. The overall pattern of marked differences between schools in their effects, and of greater differences between schools for pupils of low initial reading attainment remains unchanged.

The measure of prior mathematics attainment was similarly allowed to vary randomly at both the pupil and school levels. This was done to establish whether there was any evidence for differential slopes for individual schools, and to establish whether allowing effects to vary randomly improved the fit of the model to the data. The results are shown in Table 4.7.

The intra-school correlation was recalculated for average attaining pupils ($r = 0.1847$) for the model which included prior attainment and the background factors. This indicates that for the pupil with average prior mathematics attainment 18.5 per cent of the variance in year 5 raw mathematics scores is due to differences between schools. This estimate is close to that found for reading.

The loglikelihood function indicated that the inclusion of prior attainment at levels 1 and 2 led to a significant improvement in model fit. The estimate for differential effectiveness for pupils of differing prior attainments was also found to be highly statistically significant using the loglikelihood function. This is in line with the findings for reading and provides some evidence for the existence of modest differential slopes. This means that, although in general terms schools which are effective in promoting progress in mathematics are successful for pupils of low and high initial attainments, and those which are ineffective tend to be ineffective for all pupils, some schools are more effective for pupils with low attainments than for those with high initial attainments and vice versa.

The plot of slopes for the 49 individual schools is shown as Figure 4.2. This is based on the predicted raw mathematics scores (summer year 5) for individual pupils, controlling for initial attainment and background factors compared with their raw year 3 mathematics scores. The graph reveals marked differences between schools in their overall effectiveness (in terms of raising mathematics scores above predicted levels). In contrast to the results for reading there is some evidence that individual schools' slopes cross. Thus the evidence for differential school effectiveness appears to be stronger for mathematics than for reading progress over time. Schools with flatter slopes seem to be more effective for low

Table 4.7 Estimates of Random Effects for Prior Attainment on Year 5 Raw Mathematics.

Parameter		Level 2	
		Estimate	S Error
School σ^2	CONS/CONS	11.07	2.7
	MTHSRWY3/CONS	0.036	0.152
	MTHSRWY3/MTHSRWY3	0.0385	0.0169
		Level 1	
Pupil σ^2	CONS/CONS	48.85	2.052
	MTHSRWY3/CONS	−0.355	0.148

attaining pupils, while those with steeper slopes seem to be more effective for those with high initial attainments.

An analysis of the intra-school correlation by raw initial (year 3) mathematics score was made. This indicates that the intra-school correlation is higher for pupils with low and for those with high initial attainments than for those of average initial attainment. This suggests that, for mathematics progress, school attended has a greater impact for pupils with low or high attainment at entry than for those of average attainment at entry.

Sex

At the individual level there was some evidence of greater variation in reading attainment of boys than of girls. Given this, the model was elaborated to allow the variable SEX to vary randomly at level 1. In addition, in order to test whether the sample schools were differentially effective in promoting reading progress for the two sexes, SEX was allowed to vary randomly at level 2. The results demonstrated that the effect of SEX were only significant at the pupil level (level 1).

Previous analyses have shown that the effects of prior attainment were near significance at level 2 (the school). The variable SEX was found to be significant at level 1 (the pupil) in a model where prior attainment was allowed to vary randomly at level 2 (see Table 4.8).

As with reading there was evidence of some variation in mathematics attainment for the two sexes. The model was elaborated to allow the variable SEX to vary randomly at level 1. In addition, in order to test whether the sample schools were differentially effective in promoting mathematics progress for the two sexes, SEX was allowed to vary randomly at level 2.

The estimates of the effects of SEX did not reach significance at the school level but were significant at the pupil level. These results provide no indication of significant differential school effectiveness in promoting mathematics progress for the two sexes and are in line with the findings for reading and the conclusions of the original JSP analysis.

Table 4.8   Estimates or Random Effects for Sex and Prior Attainment on Year 5 Raw Reading Scores.

Parameter		Level 2	
		Estimate	S Error
School σ^2	CONS/CONS	36.82	9.089
	ERTRWY3/CONS	−0.51	0.163
	ERTRWY3/ERTRWY3	0.007815	0.00409
		Level 1	
Pupil σ^2	CONS/CONS	148.1	9.018
	ERTRWY3/CONS	−0.921	0.142
	SEX/CONS	14.51	6.619

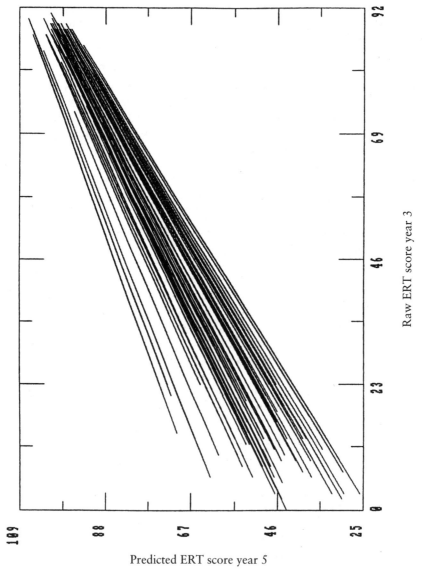

Fig. 4.1 Plot of school slopes showing predicted reading score year 5 (ERT = Edinburgh Reading Test).

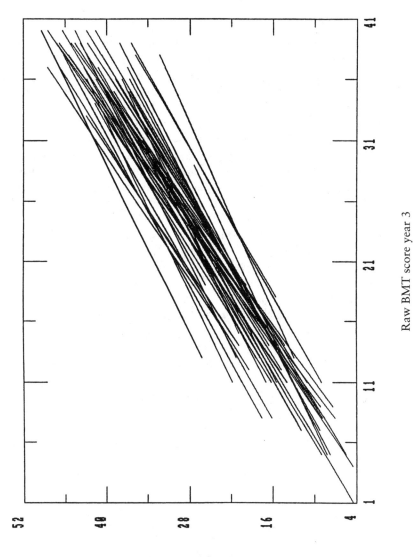

Fig. 4.2 Plot of school slopes showing predicted mathematics score year 5 (BMT = Basic Mathematics Test).

Previous analyses have shown that the effects of prior mathematics attainment were significant at level 2 (the school). The loglikelihood function indicates that including the variable SEX improved the fit of the model (p.,0.02) at level 1 (the pupil) when prior attainment was allowed to vary randomly at the school and pupil level (see Table 4.9).

Social class
The fixed effects (shown earlier in Table 4.2) demonstrate marked differences in the effects of membership of different social class groups on pupils' later reading attainment. In addition, there was evidence of differences between the social class groups in the variance of the measure of initial and of later reading attainment.

In order to establish whether schools were differentially effective for pupils of different social class backgrounds the model was reset to allow the different social class categories to vary randomly at levels 1 and 2. The categories chosen were 'non-manual', 'manual', and 'any other'. It did not prove possible for all three of these social class groups to be included in the model random at levels 1 and 2, therefore each group was tested separately. The same set of background factors were included in the fixed part of the model. The findings indicate that the estimates of the effects for the 'non-manual' group do not reach statistical significance either at level 2 or level 1. It appears, therefore, that the sample schools are not differentially effective for pupils of non-manual backgrounds in comparison with other pupils.

Similar models were tested for the 'manual' groups, and for the 'any other'. As with the non-manual group the estimates did not reach statistical significance.

The results of the fixed effects analysis (shown earlier in Table 4.3) demonstrated no significant social class differences in the effects of social class background on pupils' later (year 5) mathematics attainment, after controlling for initial mathematics attainment at entry to junior school. There was evidence, however, of some differences in the variance of mathematics attainment for the various social classes.

Table 4.9 Estimates of Random Effects for Sex and Prior Attainment on Year 5 Raw Mathematics Scores.

Parameter		Level 2	
		Estimate	S Error
School σ^2	CONS/CONS	10.99	2.684
	MTHSRWY3/CONS	0.0363	0.151
	MTHSRWY3/MTHSRWY3	0.0375	0.0167
		Level 1	
Pupil σ^2	CONS/CONS	44.9	2.659
	MTHSRWY3/CONS	−0.342	0.147
	SEX/CONS	4.017	2.047

Table 4.10 Estimates of Random Effects for Different Social Class Groups on Year 5 Raw Mathematics Scores.

Parameter	Level 2	
	Estimate	S Error
FMANUAL/FMANUAL	14.24	3.881
FNONMAN/MANUAL	10.31	2.957
FNONMAN/FNONMAN	7.897	3.189
FANYOTH/FMANUAL	10.69	2.998
FANYOTH/FNONMAN	7.857	2.651
FANYOTH/FANYOTH	9.586	3.334
	Level 1	
CONS/CONS	50.22	4.369
MTHSRWY3/CONS	−0.358	0.15
SEX/CONS	3.298	2.048
FMANUAL/CONS	−1.684	2.496
FNONMAN/CONS	−4.461	2.83

In order to establish whether schools were differentially effective for pupils of different social class backgrounds the model was reset to allow the different social class categories to vary randomly at levels 1 and 2. The categories chosen were non-manual, manual, and any other (see Table 4.10).

The results indicate that, when prior attainment and sex are included in the random parameter matrix at level 1, there is no evidence of significant random variations at the pupil level by social class. At level 2 (the school) the results do suggest the existence of significant differential effectiveness for the three social class groups. However, given evidence of significant differential school effectiveness for pupils' of different levels of prior attainment, and the strong link between attainment and social class group at any one point in time, it was considered important to allow for differential slopes for pupils of different prior attainment levels in the model.

It did not prove possible to obtain estimates for a model which included the three social class groups and prior attainment random at the school level (it is possible that this is a refection of the relatively limited sample size available for the analysis, $N = 1240$). Therefore each group was tested separately. When a model with the group (father in non-manual work) was included random at level 2 with prior attainment also random at level 2, convergence could not be achieved. However, it was possible to achieve convergence with the variable (father in manual employment) and prior attainment random at the school level (see Table 4.11).

These estimates of differential school effects show that the social class category (father in manual work) is not significant when the measure of prior attainment is allowed to vary randomly at the school level. The estimate of the effects of prior attainment (differential slopes) remains statistically significant in this model.

Table 4.11 Estimates of Random Effects for Manual Social Class on Year 5 Raw Mathematics Scores.

Parameter		Level 2	
		Estimate	S Error
School σ^2	CONS/CONS	9.671	2.748
	MTHSRWY3/CONS	−0.0671	0.156
	MTHSRWY3/MTHSRWY3	0.0416	0.0175
	FMANUAL/CONS	1.139	1.752
	FMANUAL/MTHSRWY3	0.255	0.143
	FMANUAL/FMANUAL	2.067	2.057
		Level 1	
Pupil σ^2	CONS/CONS	44.71	2.675
	MTHSRWY3/CONS	−0.354	0.146
	SEX/CONS	3.707	1.018

For both cognitive areas the results of the analysis of level 2 variation for different social class groups provide little indication of the existence of differential school effectiveness when a prior attainment measure is included at level 2.

Ethnicity

From the fixed effects analysis of reading progress it was found that when prior attainment was allowed to vary randomly at level 1, the negative effect for the CARIBBEAN group no longer reached statistical significance. There was evidence of differences between the CARIBBEAN group and all other pupils in the variance of reading attainment. In order to establish whether some schools in the sample were differentially more or less effective in promoting the progress of pupils of this ethnic group a model which allows the variable Caribbean to vary randomly at levels 1 and 2 was tested.

The results provide no evidence of differential school effectiveness in promoting later reading attainment for pupils of CARIBBEAN background compared with those of other ethnic origins. In addition, there was no evidence of significant random variation at level 1.

Similar analyses were conducted for the ASIAN and for the ESWI (English, Scottish, Welsh or Northern Irish) groups which were not examined separately in the original JSP study. The results for neither the Asian nor the ESWI group were statistically significant.

No evidence of significant ethnic differences in progress in mathematics over time were found from the fixed effects analysis (see Table 4.3) when a measure of prior mathematics attainment was included in the model. However, there was some evidence of differences in the variance of attainment by ethnic group. Separate models were specified to test the hypothesis that schools were differentially effective for particular ethnic groups (CARIBBEAN, ASIAN, ESWI). The results provided no evidence to support the view that some schools were more or less effective in promoting mathematics progress for specific ethnic groups.

These findings are in line with those found for reading and support the general conclusions of the original JSP study which were less detailed (only examining the Caribbean group versus all others). The results of the reanalysis support the conclusions of Smith & Tomlinson's (1989) research on secondary schools which found no evidence of differential school effectiveness for different ethnic groups in examination results, but contrast somewhat with those of Nuttall's (1990) analysis of examination results in inner London schools.

Low family income
The variable eligibility for free school meals provides an indicator of poverty or low family income. This variable had a significant negative effect on later reading attainment in the fixed effects analysis. There was also evidence of differences in the variance of later reading attainment between pupils eligible for free school meals and other pupils.

The original JSP analyses did not examine the issue of differential school effectiveness for pupils eligible for free school meals. To test its impact the variable FREEMLS was included in the random effects part of the model. It was not found to be significant at either the school or the pupil level. This indicates that, for reading attainment, there is no evidence of differential school effectiveness for pupils on low family incomes.

Eligibility for free school meals was the only socio-economic indicator which had a significant effect on later mathematics attainment in the fixed effects analysis (see Table 4.3). There was also some evidence of differences in the variance of prior and later mathematics attainment between pupils eligible for free school meals and other pupils.

As with reading, the variable free school meals was included in the random effects model for mathematics, but was not found to be significant for either the school or the pupil level. In line with findings for reading, no evidence of differential school effectiveness for pupils eligible for free school meals was found for mathematics progress.

Size of school effects on later attainment

The original JSP analyses provided estimates of the overall effects of individual schools (the level 2 raw residuals) on pupils' later (year 5 summer) mathematics and reading attainment. However, no comparisons were made of the size of these estimates of schools' effects on pupils' attainment and other descriptions of school performance in terms of raw results unadjusted for prior attainment or pupils' background characteristics. This is a matter of increasing public and policy interest given the Government's decision requiring the publication of 'league tables' of schools' raw examination and National Curriculum Assessment results, and expression of doubts about the use of statistics adjusted in some way to take account of differences between schools in their intakes.

Analyses were undertaken to examine differences between schools in estimates of their performance based upon pupils' raw results at a given point in time (a cross-sectional approach) and estimates of schools' effects upon pupils' later attainment adjusted to take account of the impact of prior attainment and background (a 'value-added' or longitudinal approach). The average of pupils' raw scores (year 5) for each of the 49 schools in the multilevel analyses was ranked separately (from 1 lowest to 49 highest) for each of the two cognitive areas. Estimates of schools' effects on year 5 reading and on year 5 mathematics scores were obtained from the level 2 (school) residuals from the multilevel analyses controlling for prior attainment (year 3) at entry to junior school and background factors. Comparisons of schools' rank positions on the 'raw' and 'value-added' estimates of performance were made. The results for reading demonstrate that taking account of the impact of pupils' prior attainment and background has a dramatic effect on many schools' rank positions.

For reading, at one extreme a school with a low average year 5 score but with a positive residual from the 'value-added' analysis moved down 38 rank places when ranked in terms of its 'raw' results alone. In all six schools moved down 20 or more places. In contrast, at the other extreme, four schools moved up 20 or more places if raw results rather than 'value-added' residuals were used. Table 4.12 gives details of the extent of change in rank positions.

A fifth of schools moved up or down 20 or more places when their ranked positions in terms of adjusted or 'value added' results were compared in terms of average raw scores. Fifty-three per cent moved by between five and 19 places, and around a quarter (26 per cent) moved up or down by fewer than five places.

Table 4.12 Changes in Schools' Rank Positions Comparing 'Raw' With 'Value-Added' Results in Terms of Raw Year 5 Reading Scores.

	Schools*	
	N	%
−35 or more	2	4.1
−25 to −34	1	2.0
−20 to −24	3	6.1
−15 to −19	4	8.2
−10 to −14	5	10.2
−5 to −9	6	12.3
0 to ± 4	13	26.5
+5 to +9	3	6.1
+10 to +14	6	12.3
+15 to +19	2	4.1
+20 to +24	2	4.1
+25 or more	2	4.1

Note. Changes in rank positions (rank of raw score − rank of school residual)
* $N = 49$.

The results of the equivalent analysis for mathematics are shown in Table 4.13.

For mathematics, at one extreme a school with a low average year 5 score but with a positive residual from the 'value-added' analysis moved down 24 rank places when ranked in terms of its raw results. In all six schools moved down 15 or more places. In contrast, at the other extreme, three schools moved up 15 or more places if raw results rather than 'value-added' residuals were used.

Just over 12 per cent of schools moved up or down 15 or more places when their ranked positions in terms of adjusted or 'value-added' results were compared with their rank average raw score positions. Nearly 39 per cent moved by between five and 14 places, and half moved up or down by fewer than five places.

The analysis of schools' rank positions in terms of 'raw' and 'value-added' results provided evidence of less movement than the equivalent analysis for reading (only a quarter of schools move fewer than five places in the reading analysis). Nonetheless, the results confirm the conclusions of the reading analysis that schools' positions in league tables comparing 'raw' and 'value-added' results can vary markedly. Given the strength of relationships between prior attainment and, to a lesser extent, background factors and later attainment it is clearly misleading to use 'raw' results as measures of schools' effectiveness in promoting pupils' academic achievement. Multilevel analyses of pupil attainment provide more valid estimates of effectiveness. Even these should be used with caution. Only those residuals greater than two standard errors in size differ significantly from the overall estimate and can be used as indicators of better or poorer performance than expected.

Furthermore, the JSP multilevel reanalysis indicates only a modest positive correlation ($r = 0.616$) between schools' effectiveness in different cognitive

Table 4.13 Changes in Schools' Rank Positions Comparing 'Raw' With 'Value-Added' Results in Terms of Raw Year 5 Mathematics Scores.

	Schools*	
	N	%
−20 to −24	1	2.0
−15 to −19	2	4.1
−10 to −14	5	10.2
−5 to −9	3	6.1
0 to ±4	24	50.0
+5 to +9	8	16.3
+10 to +14	3	6.1
+15 to +19	1	2.0
+20 to +24	2	4.1

Note. Changes in rank positions (rank of raw score − rank of school residual)
* $N = 49$.

areas, even in the so- called 'basic skills' of reading and mathematics. However, it should be noted that this correlation is higher than that found in the original JSP analysis ($r = 0.41$) which may be a reflection of improved model specification in the reanalysis. In terms of markedly positive residuals (estimates of school effects) only four of the 49 schools had a markedly positive effect on both areas (estimates greater than 2 standard errors in size) and six had a markedly negative effect on both (estimates greater than 2 standard errors). Thus around a fifth of the sample of schools appear to be particularly effective or particularly ineffective in promoting these two cognitive outcomes, but the majority of schools varied in their effects in different areas.

Summary of results

- The reanalysis of JSP data has contributed to current knowledge about the dimensionality of school performance. For two cognitive areas (reading and mathematics) detailed multilevel analyses of differential effectiveness produced the following results.
- Evidence of increases over time in the variance of pupils' attainment attributable to differences between schools, particularly for mathematics.
- Evidence that background factors and initial attainment account for more of the variance in reading than in mathematics attainment and that school differences may be greater for mathematics.
- Evidence of marked differences between individual schools in terms of their effects on later attainment (level 2 residuals) after controlling for prior attainment and background factors.
- Some evidence of differential effectiveness (differential slopes) of individual schools for pupils with different prior attainment levels. For reading, school differences were found to be greatest for pupils with low initial attainment. For mathematics, school differences were greater both for those with low and those with high initial attainment, and lower for those with average initial attainment. In general, effective schools raised the performance of all pupils, although there was greater indication of differential effectiveness for mathematics than for reading.
- No evidence of differential school effectiveness for different pupil groups (divided by sex, ethnic group, social class group, eligibility for free school meals).
- Although schools' effects (estimates of the level 2 residuals for mathematics and reading) were positively correlated, the relationship was not perfect ($r = 0.62$). Only four schools had significantly positive residuals for both areas and six had markedly negative residuals for both areas. The overall concept of effective versus ineffective schools therefore appears to be too simplistic to describe the dimensionality of schools' effects.
- Comparisons ranking schools in terms of their raw results (average raw reading scores and average raw mathematics scores in year 5) with rank-

ings based on residuals derived from multilevel analyses controlling for prior attainment and background factors (a 'value-added' approach), revealed marked differences in 'league table' positions. The results support the contention that simplistic league tables of schools' raw results do not (on their own) provide valid indicators of school performance and can be misleading. They point to the value of utilising multilevel models in comparisons of school performance, and the need to look at a variety of measures of educational outcomes.

- The findings also demonstrate the importance of utilising a good (finely differentiated) measure of prior attainment at entry in studies of school differences in effectiveness. When such a measure is included, little evidence of differential school effectiveness for particular groups (divided by sex, social class, ethnic group or low family income) is found. The results of the reanalysis support recent conclusions by Jesson and Gray (1991) on the existence of moderate differential slopes for pupils of different levels of prior attainment. In general, the findings support the conclusions of the original JSP research. Effective schools tend to be effective for all pupil groups and, conversely, ineffective schools tend to be ineffective for all groups.

The multilevel reanalysis of JSP data presented here did not take into account the issue of adjusting for measurement unreliability. Current work by Yang et al., (1992) has indicated that adjustment for different estimates of reliability in a measure of prior attainment can affect both fixed and random parameter estimates and standard errors. In particular it is shown that if adjustment is made at both level 1 and level 2 there is an increase in the intra-school correlation (due to a reduction in estimates of level 1 variation). The estimates of the intra-school correlation shown in this paper are therefore likely to be smaller than those which would be found if adjustment for test-reliability were made. It is hoped to conduct further analyses of this issue when a modified version of the ML3E becomes available.

Conclusions

The original JSP analyses concluded that there was little evidence of differential school effectiveness according to sex, social class or ethnic group, and the reanalysis of reading and mathematics attainment supports this conclusion. The reanalysis has gone further than the original JSP by looking at two further ethnic groups (ESWI and Asian) and another socio-economic indicator of low family income. However, the reanalysis does indicate that there is some evidence of differential school effectiveness for pupils of differing prior attainment levels. This area was not examined in the original JSP analyses. Nonetheless, the plots of individual slopes of school performance suggest that overall, 'good' schools boost the later attainment of students of differing levels of prior attainment, whereas in less effective schools later attainment is lower than predicted for all groups. This conclusion is in line with recent work on secondary schools' examination results (Jesson & Gray, 1991). This investigated the extent of differential effectiveness for pupils of different prior attainment levels and found that the use of a finely-differentiated measure of pupils' prior attainment reduced estimates of the extent of differential school effectiveness. It was concluded that

> 'there is some evidence of differential effectiveness. Pupils of different prior attainment levels did slightly better in some schools than in others. . .'.

However, it was also noted that

> 'schools which were more effective for one group of pupils were, generally speaking, more effective for other groups as well' (p. 246).

The JSP reanalysis utilised a finely differentiated measure of prior attainment, and the results seem to support Jesson and Gray's main conclusions.

The results of the reanalysis also have significant policy implications for the publication of schools' examination and test results. Comparing schools' rank positions in terms of 'raw' rather than 'value-added' results has a marked impact upon schools' 'league table' positions for the majority of schools. Given the strength of relationships between prior attainment, background factors and pupils' later reading and mathematics attainments, it is clearly misleading to use 'raw' results as measures of schools' effectiveness in promoting pupils' academic achievement. Multilevel analyses of pupil attainment provide more valid estimates of effectiveness. Schools with very large positive or negative residuals appear to be performing better or worse than would be predicted on the basis of their pupil intakes.

Even such estimates, however, should be treated cautiously. Variation from year-to-year and for different subjects/outcomes mean that no single measure of school effectiveness should be given undue emphasis. Thus Goldstein et al., (1992) have recently argued

> 'It is clear that the uncertainty attached to individual school estimates, at least based upon a single year's data, is such that fine distinctions and detailed rank orderings are statistically invalid' (p. 5).

Comparisons of the estimates (residuals) of schools' effects upon pupils' later reading and mathematics performance indicate that the relationship, although positive and significant, is not very strong. Some schools were more effective in raising pupils' performance in one cognitive area rather than the other. Only a few schools had a marked positive effect on both reading and mathematics (4 schools had positive residuals greater than two standard errors above that predicted for both areas) and only a few a marked negative effect on both areas (6 schools had residuals greater than two standard errors below that predicted for both areas).

The issue of differential school effectiveness for different subjects or kinds of educational outcome is clearly of importance in an era where the publication of school league tables is to become mandatory. The project's findings indicate that no simplistic division of schools into 'good' or 'bad' is possible, even on the basis of results in 'basic' subjects such as reading and mathematics. The results point to the need to be cautious in comparing schools on only one measure of educational outcomes at one point in time and the value of looking at performance in different areas and over time. As Goldstein et al., (1992) have argued that 'the best use of residuals is as a screening device and as feed back to individual schools about potential problems' (p. 6). The results of multilevel analyses can help to identify schools where pupil performance appears to differ markedly from that predicted on the basis of their pupil intakes and points to the need for further investigation. They can thus assist in the development of good indicator systems to measure schools' performance (McPherson, 1992).

5

Continuity of School Effects

Introduction

This chapter illustrates another important but as yet barely researched issue in school effectiveness studies, that of the long term impact of schools. A brief reference was made to the question of continuity in school effects in the earlier chapter on Complexities in Judging School Performance. Here, the first example of research which explicitly sought to investigate the concept of continuity in school effects using multilevel methods is presented. The research was conducted in 1992 as part of the Economic and Social Research Council, (ESRC), funded Models of Effective Schools Project during my time at the Centre for Educational Research as the London School of Economics. As with the focus on Differential Effectiveness (described in the last chapter), the study was planned with Desmond Nuttall and Peter Cuttance and used information about the earlier School Matters cohort.

  On the whole, previous school effectiveness research had not sought to conduct long-term follow ups to establish whether the particular school attended continues to influence student achievement at a later stage in their educational career. On the face of it, this omission is surprising. Cohort studies involving the tracking of national samples of children had been a feature of a number of major health and education projects such as the 1958 National Child Development Study (Hutchison, Prosser & Wedge, 1979) and the 1970 British Birth Cohort Study (Osborn & Millbank, 1987) in the UK. These studies have traced the continuing influences of personal, family and social background factors on children's long term educational outcomes and later life chances in terms of employment, criminal records, physical and mental health etc. However, little attempt was made to establish the continuing effects of individual schools, although the Rutter et al. (1979) Fifteen Thousand Hours *research did attempt to follow up young people's post-school outcomes to some extent and suggested that attending an effective secondary school had a positive impact on employability and reduced the likelihood of criminal behaviour. In the USA the investigation of children involved in the Headstart anti-poverty programmes which sought to use pre-school education and care to improve the long term educational prospects for disadvantaged children generally pointed to the positive impact of pre-school provision on students' later educational and social outcomes. The emphasis here, however, was on type of provision rather than the effects of particular institutions (e.g., see Lazar & Darlington, 1982).*

In school effectiveness research it was generally assumed that students from more effective schools would be at an advantage because of their higher attainment levels at any given stage in their school career, whether measured in terms of achievement at the end of primary education in basic skills (as in Mortimore et al., 1988a School Matters *study of junior schools) or in terms of public examination results at secondary level (as in Smith & Tomlinson's 1989 study of multiracial comprehensives). The possibility that, in addition, to their effects on student progress over the particular period of a research study (measured over one or more years), the particular school attended might have a measurable impact upon students' subsequent performance several years later in a different institution was not discussed in the literature.*

The paper which forms the basis of this chapter therefore broke new ground by highlighting this potentially influential issue. The Models of Effective Schools project was able to capitalise on the existence of the ILEA's Junior School Project data base (used in the School Matters study) which had followed primary children's attainment, progress and development over four years and during secondary transfer from 1980 to 1984 (Mortimore et al., 1986, 1988a). Using the ILEA's transfer and public examinations files it was possible to add details about students' performance in reading and verbal reasoning at age 11, secondary school attended and public examination results at age 16 to the primary school data base.

This enabled the analysis of students' achievement over nine years across the two phases of schooling, primary and secondary. At the time of the analysis cross-classified multilevel models had not been developed. Therefore, the research attempted to control for student achievement at secondary transfer in order to take account of primary school influences on progress from age 7 to 11. The research was important in suggesting potentially influential long term effects of primary schools on public examination results at age 16 over and above their influence on students' level of attainment at the end of primary schooling. The implications of the finding were far reaching, in that they suggested that current models used in secondary school effectiveness research worldwide might well be inadequate. Controlling for attainment at intake to a particular phase (e.g., at age 11 in the UK context or indeed at age 16 for stud-

ies of A-level results) is insufficient if the previous institution attended has a continuing effect on students' later outcomes. The results also suggested that school effectiveness studies which examine progress over only a couple of years, particularly that which only controls for attainment measured after several years in a given school, is likely to miss important effects of earlier schools attended. For example, models which followed student progress from say age 14 to 16 years at secondary school could miss important influences of secondary schools on progress from 11 to 14 and primary schools on progress over a longer term period. My colleague at the Institute, Harvey Goldstein, developed cross-classified models which facilitated a better approach to the analysis of student progress across phases through the separation of primary and secondary school effects. His work in Scotland (Goldstein, 1995) and a reanalysis of the Models of Effective School data (Goldstein & Sammons, 1997) supported and extended the findings reported in the Continuity of School Effects paper. In particular we commented on the simple approach to value added analysis then advocated in the UK by the School Curriculum Assessment Authority (now the Qualifications and Curriculum Authority). We argued that our analysis had profound implications for the future development of the school effectiveness field, noting that from our data junior schools appeared to have a stronger influence on students' public examination results at age 16 than secondary school attended. The results we suggested

> 'raise interesting and potentially important questions about the use of data relating only to the secondary school period in order to study variation between secondary schools' (Goldstein & Sammons, 1997, pp. 228–229).

We concluded that the need for school effectiveness research to become involved in very long term studies of schooling, rather than being restricted to a single phase was demonstrated by the findings. We also argued that further work involving larger samples and a range of geographical and social contexts, was called for.

The publication of the Continuity of School Effects paper received fairly wide press coverage during 1994–95 and the implications of the findings concerning the relative importance of the primary phase of schooling seemed to have some influence on the thinking of policymakers about the need to pay greater attention to the implications of school effectiveness and improvement research for this age group. Increased awareness of the association between reading performance at ages 7 and 11 in particular was relevant to the increased emphasis of a National Literacy strategy to raise levels of attainment at Key Stage 2, prior to secondary transfer.

A number of multilevel studies have investigated the issue of differences in secondary school effectiveness using students' public examination results as outcome measures (see, for example, Cuttance, 1986; Willms, 1986; Nuttall et al., 1989; Smith & Tomlinson, 1989; Willms & Raudenbush, 1989; Thomas,

Nuttall & Goldstein, 1993). These studies have provided evidence of significant variation in students' examination performance at the school-level. Fewer studies have investigated primary school effects but work by Mortimore et al., (1988a), Tizard et al., (1988), Brandsma & Knuver (1989), Bondi (1991) and Thomas & Nuttall (1992) have also provided evidence of significant school-level variation in primary pupils' academic outcomes.

Relatively little research has been conducted to examine the stability of individual school effects over time (as measured by school-level residuals), although reviews of this issue have been provided by Bosker and Scheerens (1989), and more recently by Gray et al., 1993 and Sammons, Mortimore and Thomas (1993). Sammons, Mortimore and Thomas (1993) have also reviewed the issue of consistency in schools' effects across different kinds of outcomes (both academic and social or non-cognitive) as well as stability over time. Research (by Willms & Raudenbush, 1989; Gray & Simes, 1991) provides evidence of considerable stability in individual schools' effects on examination results between years. However, Goldstein et al., (1992) rightly point to the need for caution in interpreting individual school effects at any one point in time.

> 'It is clear that the uncertainty attached to individual school estimates, at least based upon a single year's data, is such that fine distinctions and detailed rank orderings, are statistically invalid.'

Nonetheless, these authors concluded that

> 'a study of residuals differentiated by intake achievement and by subject, can suffice as a screening device and a feedback to individual schools about potential problems'.

A study by Thomas, Sammons and Mortimore (1994) indicates broad but by no means perfect stability in secondary schools' effects on total performance in General Certificate of Secondary Education (GCSE) examinations at age 16 over time, and supports the views that results should be looked at over more than one year.

Another important issue which has, as yet, received very little attention in the growing field of school effectiveness research is the continuity of effects at different stages in a student's school career. Thus, although progress over time at primary school or at secondary school has been examined, to our knowledge only Entwistle and Hayduk (1988) have investigated whether elementary school attended continues to have an effect upon a student's later academic performance at secondary school.

Entwistle and Hayduk (1988) found that the experiences of children in their early (junior) school years had a substantial impact on their achievements in the later secondary school. They attribute this to three sets of factors: the influence of significant others (parents and teachers); effective school practices; and the adaptation of pupils to the school environment (socialisation). They comment that

'If children's early social experience has such long-lasting effects, much of the "home background" influence measured in models of educational attainment in the secondary school may actually represent influences that were exerted much earlier in the schooling process . . . If so, taking into account children's early school experiences could substantially alter the interpretation of the secondary school models' (1988, p. 158).

However, Entwistle and Hayduk's research was based on a very small number of schools and thus did not employ multilevel methodology to investigate the percentage of variance in students' attainment at secondary school attributable to primary school.

Aims

This paper presents results of multilevel analyses of GCSE examination results undertaken to examine the issue of continuity in primary school effects and the existence of secondary school effects in the UK. The principal aim was:

to establish whether there is any evidence that primary school attended has a long-term effect upon students' later academic performance as measured by performance in public examinations at age 16, the end of compulsory schooling.

In addition, the size and significance of secondary school effects on GCSE performance were also investigated to enable comparison with the results of the analysis of continuing primary school effects.

Methodology

Follow-up data for the Inner London Education Authority's Junior School Project (JSP) sample were used for the analysis of primary and secondary school effects. The JSP was a longitudinal study of approximately 2000 pupils who entered junior classes in September 1980 and transferred to secondary school in September 1984 (see Mortimore et al., 1986, 1988a). The JSP data set contained a rich set of pupil-level background data, as well as measures of attainment at entry to junior school and annually over a three-year period. The structure of the data is well suited to multilevel modelling techniques and sub-sets of the data have been used to illustrate the working of the ML3 model (Prosser, Rasbash & Goldstein, 1991; Woodhouse et al., 1992).

The results of the original JSP provided evidence of significant primary school effects for a variety of cognitive and non-cognitive outcomes (Mortimore et al., 1986, 1988a). Recent multilevel reanalysis of the JSP data set broadly confirmed the original findings but extended them in a variety of ways, in particular, by investigating the extent of differential school effectiveness (Sammons, Nuttall & Cuttance, 1993).

The JSP student cohort took GCSE examinations in 1989. In order to investigate the continuing effects (if any) of primary school attended on students' later secondary school performance, data concerning students' total GCSE examination scores were extracted from the 1989 Inner London Education Authority examination files. Data were matched for 1116 students (48.8 per cent) of the full JSP cohort (2287 pupils). Outcome variables for attainment at age 16 include total GCSE examination performance score (the calculation of which is described in Greenhill & Chumun, 1990), English Language and English Literature scores, Mathematics, Mathematics SMP and Computer Studies. The extended data set thus enables the longitudinal analysis of pupil attainment over a nine year period.

The analyses focuses on total GCSE performance score because this provides a useful summary measure of overall achievement at the end of compulsory education and is a measure used to judge school performance in the UK by bodies such as the Office for Standards in Education (OFSTED) and the Department for Education (DFE). Whilst performance in specific subjects is also of interest and may provide valuable indications of departmental differences in effectiveness (see Sammons et al., 1994c) total GCSE scores are most likely to provide evidence relevant to the existence of any overall school effects. There is also evidence that total GCSE performance score is a better measure for identifying highly ineffective and highly effective schools and that schools' effects on total GCSE performance tend to be more stable over time than effects on specific subjects (see Thomas, Sammons & Mortimore, 1994).

In terms of the total JSP sample for whom a valid measure of attainment was obtained at secondary transfer (a total of 1623 pupils), overall examination performance scores were matched for 995 pupils representing 61.3 per cent of those for whom attainment data at entry to secondary school were available. The original pupil sample was depleted for various reasons. Firstly, due to pupil mobility (outward) from the sample schools during the four years of the Project. Secondly, no examination data were available for pupils who transferred to non-ILEA secondary schools (state or private). Thirdly, pupils who did not take any GCSE examinations (whether because they were not entered for examinations because of low ability or who did not take examinations due to early leaving or non-attendance) would not obtain a valid performance score at age 16. Finally, any pupils who changed surnames after moving to secondary school could not be traced.

Although the junior school sample for whom total examination scores could be matched represents just under half the original age cohort, a substantial reduction is to be expected over a nine year period where resources were not available to trace individual pupils at age 16 years. Nonetheless, the reduction in the sample is unlikely to have affected the validity of the findings although, given the relatively small size of the sample at GCSE, the possibilities of finding statistically significant differences at the school level, in particular, are reduced. Nonetheless, the number in the sample is sufficient to examine the size of school effects in terms of the percentage of variance attributed to the school compared with the pupil level.

The variation in the GCSE scores of sample matched is probably considerably restricted through non-random attrition and the selection effects noted above.

Because of this the estimates of the between-school variance obtained from the multilevel analyses should be interpreted as a *minimum* level. Had data been available for pupils who transferred to non-ILEA secondary schools and for those who took no GCSE examinations (i.e., who effectively scored zero) a higher percentage of the variance would be expected to be between schools.

Hypotheses

Two multilevel models were developed for the analysis of the GCSE data. One was designed to examine the possible impact of secondary school attended and the effects of prior attainment and background factors on examination performance at age 16 years. The second was designed to test the continuing impact (if any) of primary school attended on students' later attainment in public examinations. In conducting the investigation *it was hypothesised that any continuing primary school effect upon students' later secondary school examination performance would be eliminated if account was taken of the level of attainment students had reached at the end of primary education*. It was expected that any primary school effect would operate by affecting the progress students made during their primary schooling, and thus the level of attainment reached at the end of primary education prior to secondary transfer at age 11.

School effects

Secondary school analysis

The JSP data set includes detailed information at the pupil-level concerning attainment at entry to junior school (Year 3), and at later points in time (Year 5), as well as two measures of attainment at transfer to secondary school. These were London Reading Test [LRT] scores and Verbal Reasoning [VR] band (a three category measure of verbal reasoning ability based on primary teachers'/Headteachers' judgements). In addition, information concerning pupils' characteristics and family background at primary school (including sex, age, social class of father's occupation, low family income, ethnic group and measures of pupils' fluency in English) is available. This provides a particularly rich data set for examining secondary school effects using multilevel models.

Listwise deletion of pupils for whom information about LRT score, VR band, examination score or secondary school attended was missing was conducted. As a result, the secondary school analysis was based on a total of 943 students. These were drawn from 120 secondary schools. It should be remembered that the JSP pupils were not distributed randomly between secondary schools. The numbers of students varied between only one and 43 at the secondary school-level.

A variance components multilevel model was fitted to examine the random variation at the student (level 1) and school levels (level 2). The results provided evidence of the existence of significant random variation between secondary schools in overall examination performance scores (intra-school correlation

0.1148) accounting for about 11.5 per cent of the overall variation. This is slightly larger than estimates of effects obtained in other analyses of secondary schools (see the review by Daly (1991) which reports figures of between 8 per cent and 10 per cent). However, this figure is rather lower than that reported in the multilevel reanalysis of primary school effects on reading and mathematics attainment in Year 5 of junior school for the JSP pupil sample (Sammons, Nuttall & Cuttance, 1993). Thus, at age 16 rather *less* of the total variation in students' total examination performance scores is attributable to secondary school attended, than of the variation in reading or mathematics scores for the same students at primary school (intra-school correlation 0.1416 and 0.1487 in Year 5 of junior school). Differences amongst primary schools in basic skills attainment thus appear to be rather larger than those amongst secondary schools in total GCSE attainment. It may be that school effects are more important for pupils of primary age than for older secondary students (a point discussed later in the paper under 2.1). Alternatively, departmental differences or differences for particular groups of pupils (such as those of high attainment on entry) may be more apparent at the secondary stage.

It is well recognised amongst academics concerned with school effectiveness research that any analysis of school differences in students' attainments should take account of students' prior attainment at entry (Goldstein, 1987, 1992; Sammons, 1989a; Sammons, Nuttall & Cuttance, 1993; Thomas, Nuttall & Goldstein, 1993). As noted above, two measures of attainment at transfer to secondary school were available for the secondary school analysis, LRT score and VR band. These provided good baseline measures for control of attainment at entry.

Including a continuous measure of prior attainment (LRT) at secondary transfer in the fixed part of the model reduced the unexplained variance by 25.8 per cent. In addition to LRT score, VR band was added to the fixed part of the model. The inclusion of VR band reduced the unexplained variance by 30.6 per cent in total (see Table 5.1). The variable LRT score was centred to ensure efficient specification of the model and to aid interpretation of the results. The categorical variable VR band was incorporated as two dummy variables band 2 (average ability) and band 3 (low ability) contrasted with band 1 (high ability) students. In all 217 (23.0 per cent) of the 943 students included in this analysis were allocated to band 1, 534 (56.5 per cent) to band 2 and 192 (20.4 per cent) to band 3.

In the fixed effects results the comparison ("CONS") group represents the examination performance of band 1 (high ability) students with average LRT scores at transfer. These results demonstrate that membership of band 2 or band 3 has a significant negative relationship with students' later total examination performance scores, even when account is taken of LRT attainment at age 11 years. It is likely that the negative impact of low VR band is a reflection of a variety of influences. Primary teachers'/headteachers' assessments of ability may be based on more general knowledge of children's performance, attitudes, motivation and approach to work. In addition, however, being placed in a low VR

Table 5.1 Random and Fixed Parameter Estimates for Analysis of Students' Total GCSE Examination Performance Scores Controlling for Prior Attainment (Secondary School Analysis).

(a) *Random effects*

	Level 2	
	Estimate	S Error
Secondary school σ^2 intercept	14.07	4.522
	Level 1	
Pupil σ^2 intercept	146	7.088
Intra-school correlation	0.0879	

(b) *Fixed effects*

	Estimate	S Error
Intercept CONS	30.18	1.084
LRT	0.406	0.0527
Band 2	−8.936	1.116
Band 3	−12.68	1.817

Where:
CONS   Average examination score of Band 1 students of average LRT attainment.
LRT   London Reading Test raw score.
Band 2 Dummy variables contrasted with Band 1 (high ability students).
Band 3 Band 2 = average ability students, Band 3 = low ability students.

band may itself have a negative impact upon secondary teachers' and parents' expectations of such pupils, and upon pupils' own self-esteem and attitude to school work. This may also affect later examination performance. These findings concerning VR band are in line with those reported by Goldstein et al., (1992) in an analysis of examination results for inner London schools.

After controlling for the impact of the two measures of prior attainment the intra-school correlation was reduced to 0.0879, indicating that, for this model, about nine per cent of the unexplained variation in total examination performance scores was attributable to differences between secondary schools.

Background effects
Measures of students' background characteristics used in the earlier primary school reanalyses (Sammons, Nuttall & Cuttance, 1993) were included in the fixed effects part of the secondary school model. The results demonstrated that, even when account is taken of attainment at entry to secondary school at age 11 years, background factors (sex, social class, ethnic group and low family income) continue to have a highly significant impact upon students' overall examination performance at age 16 years. Relative to the comparison ("CONS") group (who represent band 1, English, Scottish, Welsh or Irish, female students of average age and LRT performance, with fathers in work, and not eligible for free school meals) the estimates of the impact of the following

categories: male; skilled manual; semi or unskilled manual; father in long-term unemployment or absent; father's occupation not known[1]; eligible for free school meals were significant and negative (see Table 5.2). The effect for father in semi or unskilled manual work was particularly marked indicating that social class remains a very important predictor of later academic achievement.

In contrast to results from the primary school analysis of reading and mathematics progress, estimates for membership of specific minority ethnic groups were found to be significant and positive in terms of overall GCSE examination performance. The estimate for the Asian sub-group was markedly positive, that for the Caribbean and the Other/not known groups smaller but also positive and significant. These findings are generally in line with those reported by Nuttall et al., (1989) and Nuttall (1990b) in a much larger multilevel analysis of students' examination results for all ILEA secondary schools. Nuttall's work also found that relative to those of English, Scottish or Welsh backgrounds, students of other ethnic groups tend to obtain better examination results, although his estimates for the Caribbean group were not significant or positive. It is possible that the more detailed data concerning prior attainment (LRT and VR band) and the availability of social class data and individual-level free meal data for the JSP reanalysis may be responsible for the different estimate obtained for the Caribbean group in the present analysis. In the absence of adequate data concerning social class and low income it is highly likely that ethnicity and socio-economic effects will be confounded.

Measures of students' fluency in English at primary school were not found to be related to their later examination performance at secondary school. It is likely that the majority of the JSP sample of pupils had become fluent in English by the age of 16 (the fluency data were collected in Year 4 of junior school). Thus, incomplete fluency in English at a relatively early stage in pupils' school career does not appear to hinder pupil progress and attainment over the long term, whereas at primary school it was found to have a negative impact on reading progress during Year 3 to Year 5 (Sammons, Nuttall & Cuttance, 1993).

Age in months was also not found to be significant, a reflection of its relationship with the two prior attainment measures LRT ($r = 0.13$) and VR band ($r = -0.12$).

It is notable that the estimates for the impact of low VR band remain negative and highly significant, even when controlling for a wide range of background factors in the fixed effects analyses. The inclusion of a full set of background variables and prior attainment measures in the fixed effects part of the model accounted for 36.6 per cent of the total variance in GCSE performance. In comparing the proportion of the unexplained variance attributable to differences between schools, rather than to differences between pupils, it was found that nearly nine per cent of the unexplained variance was attributable to the school (intra-school correlation = 0.0873). This estimate is markedly lower than that obtained in the equivalent multilevel analysis of primary school effects described in Sammons, Nuttall and Cuttance, 1993 (where estimates were in the region of 18 to 19 per cent for reading and mathematics).

---

1 This category was included to avoid any further reduction in sample size.

Table 5.2 Random and Fixed Parameter Estimates for Analysis of Students' Total GCSE Examination Performance Scores Controlling for Prior Attainment and Background Factors (Secondary School Analysis).

(a) *Random effects*

	Level 2	
	Estimate	S Error
School σ² intercept	12.8	4.123
	Level 1	
Pupil σ² intercept	133.6	6.487
Intra-school correlation	0.0873	

(b) *Fixed effects*

	Estimate	S Error
INTERCEPT	32.57	1.43
LRT	0.391	0.0514
BAND 2	−8.375	1.085
BAND 3	−12.51	1.776
SEX	−2.184	0.894
AGEMTHS NS	0.0657	0.116
SKILLED	−2.971	1.171
SEMIUNSK	−6.612	1.881
LTUFABS	−3.448	1.521
FNK	−4.623	1.515
ASIAN	9.313	1.766
CARIBBEAN	2.807	1.201
OTHERNK	5.339	1.163
FREEMLS	−2.983	1.050
NKFMLS NS	−1.02	3.388
STAGE3 NS	−0.752	3.714

Note. NS = Not statistically significant ($p > 0.05$).

Where:
LRT	London Reading Test raw score.
Band 2	Dummy variables contrasted with Band 1 (high
Band 3	ability students), Band 2 = average ability, Band 3 = below average students.
SEX	Males contrasted with females.
AGEMTHS	Student age in months.
SKILLED	Set of dummy variables contrasted with father in non-manual work where SKILLED = father in skilled manual work
SEMIUNSK	= father in semi or unskilled manual
LTUFABS	= father long term unemployed or absent
FNK	= father's occupation not known
ASIAN	Dummy variable students of Asian family background contrasted with those of English, Scottish, Welsh or Irish (ESWI) background.

Table continues

Table 5.2 (continued)

OTHERNK	Dummy variable students of other ethnic backgrounds (excluding Asian or Caribbean) and those whose ethnic background is not known contrasted with those of ESWI background.
FREEMLS	Dummy variables contrasting those students eligible for free school meals with those not eligible.
FREENK	Dummy variables contrasting students whose eligibility was not known with those not eligible for free school meals.
STAGE 3	Dummy variable contrasting students assessed (at primary school) as having incomplete fluency in English with those fully fluent in English.

There are likely to be a number of reasons for this difference in the size of the intra-school correlation. It is important to remember, however, that because only those pupils with a valid examination performance score were included in the analysis, the estimates of secondary school effects may have been reduced. Some schools may not have entered pupils of low ability for examinations whereas others may have been more likely to enter such pupils. In addition, because of the longitudinal nature of the follow-up survey, the sample of students for whom examination performance scores were obtained was not random and for many secondary schools only one or two students were included.

It is also possible, however, that school differences in overall examination performance may be smaller at the secondary than at the primary level. Overall examination performance may be affected by departmental and subject differences within schools (see Fitz-Gibbon, 1985; Smith & Tomlinson, 1989; Thomas, Sammons & Mortimore, 1994). Departmental differences may be more marked than overall differences between schools in total examination performance scores (because the former may counterbalance each other) and the concept of overall school effectiveness may be rather less relevant to secondary than to primary schools. Indeed, multilevel analyses of the 1991 GCSE results in English language and mathematics in a large sample of 116 schools have indicated relatively low correlation (0.48) between these two subjects although the correlation remains positive and significant (see Thomas, Nuttall & Goldstein, 1993). Nonetheless, recent evidence (see Thomas, Sammons & Mortimore, 1994) does suggest that total GCSE performance score tends to be a more stable measure over time than subject performance, and that it is useful in identifying outlier schools (those highly effective or highly ineffective in terms of overall academic performance).

The estimates of school effects for primary schools were larger for both reading and mathematics during the junior years, although individual schools were not necessarily equally effective in both areas. It is likely that overall examination performance score masks important subject differences in students' GCSE attainment. Further research into departmental differences with a larger sample is required to investigate this issue in depth. (Sammons et al., (1994c) have reported interim results of such a study which is currently under way in inner London.)

It is also possible that school effects are relatively less important for older than for younger age groups as suggested by Entwistle and Hayduk (1988). Students' overall examination performance may be more affected by cultural and social pressures (e.g., the propensity for early leaving, parental expectations, peer pressures, subject choices etc.) than performance in the basic skills of reading and mathematics at a younger age.

Two findings of the secondary school analysis are of importance. The addition of background factors (which are socio-cultural measures), does not impact upon the between school variance. The estimates change from 8.8 to 8.7 per cent of the unexplained variance attributable to the school. Background factors nevertheless have a significant effect on the amount of overall variance explained (reducing the total unexplained variance by six per cent after controlling for the two measures of attainment at transfer).

Any analyses of schools effects upon students' attainment which do not make adequate control for the impact of such socio-cultural factors in addition to prior attainment are likely to favour schools with more socially advantaged intakes whilst making schools with disadvantaged intakes appear to be less effective than they are in reality (Willms, 1992). This conclusion is clearly of relevance to the ongoing debate in the UK and elsewhere about the merits of the publication of league tables of raw and value-added tables of schools' examination and test results as a means of promoting greater accountability of schools to consumers (parents and pupils).

Differential school effectiveness
A variety of multilevel analyses were conducted to examine the impact of differential secondary school effectiveness. The inclusion of the variable raw LRT score in the random parameter matrix at level 2 was statistically significant (at approximately 5 per cent), providing some suggestion that secondary schools may have varied in their impact upon the examination scores of students of different levels of prior reading attainment. However, when sex, VR band and LRT score were allowed to vary randomly at the pupil level (i.e., different level 1 variances were fitted for these categories), the estimates of differential school effectiveness for the variable LRT score were reduced. Allowing for random variation at the pupil level (level 1), therefore, reduced the estimate of differential effectiveness for schools (level 2) which no longer reached significance. However, the very small numbers of students in some secondary schools in this analysis should be remembered in interpreting findings on differential secondary school effectiveness.

Primary school analysis

Listwise deletion of pupils from whom information about LRT score, VR band, examination score, reading or mathematics attainment at entry to primary school (age 7, Year 3) was missing was conducted. The analysis of the continuing effects of primary schools was therefore based on a total of 785 students (representing over 83 per cent of the pupils included in the secondary school analysis). These were drawn from a total of 48 of the original JSP sample of 50 primary schools. The numbers of students varied from 1 to 46 at the school-level.

Unfortunately, given the nature of the student sample, the data set was not fully nested (i.e., all pupils from one primary school transferring to a particular secondary school) and therefore it was not possible to conduct a three level analysis of the separate impact of primary and secondary schools on GCSE performance. However, given that all primary schools fed several secondary schools it seems unlikely that these any primary school effects on GCSE performance found in this analysis would be confounded by secondary school effects. Recent unpublished research by Goldstein (personal communication) using cross-classified multilevel models supports this interpretation.

A variance components multilevel model was fitted to examine the random variation at the student (level 1) and school levels (level 2) (see Table 5.3). The results provided evidence of the existence of significant variation in students' total examination performance scores for primary schools (intra-school correlation 0.0663). This is somewhat lower than the intra-school correlation found in the equivalent analysis of secondary schools (intra-school correlation 0.1148), indicating, as might be expected, that differences between secondary schools in students' examination performance at age 16 years were larger than those found between primary schools on the same performance measure at age 16 years.

Including two measures of prior attainment at entry to primary school (Edinburgh Reading Test score and Basic Mathematics Test score at age 7 years) accounted for over a fifth (22.8 per cent) of the total variation in students' final GCSE examination performance scores at age 16. Thus, even at the relatively young age of seven years (Year 3) important and long-lasting patterns of differences in pupil attainment are fairly well established.

The intra-school correlation (0.0599) for this model indicated that around six per cent of the unexplained variance in total GCSE performance was attributable to differences between primary schools, suggesting that primary school attended may have some long term impact on performance at a later stage (over a nine year period).

Table 5.3 Random Parameter Estimates for Analysis of Students' Total GCSE Examination Scores (Primary School Analysis).

	Level 2	
	Estimate	S Error
Primary school σ^2 intercept	15.27	6.016
	Level 1	
Pupil σ^2 intercept	215.2	11.19
Intra-school correlation	0.0663	

Background effects

Measures of students' background characteristics used in the earlier primary school analyses (Sammons, Nuttall & Cuttance, 1993) and the secondary school analysis reported earlier in Section 1 were included in the fixed effects part of the primary school model. The variables – incomplete fluency in English, age in months and eligibility for free school meals not known – were omitted because these were not found to be significant in the secondary school analysis. The results again confirm that, even when account is taken of attainment of entry to primary school, background factors (sex, social class, ethnic group and low family income) continue to have a significant impact upon students' overall examination performance at age 16. The inclusion of attainment at entry and background factors accounted for 31.4 per cent of the variance in final examination performance scores (background factors accounting for an additional 8.6 per cent of the overall variance). When background factors were added to the model the intra-school correlation was further reduced (0.0406).

It was hypothesised that any continuing primary school effect on students' later examination performance would be removed if measures of attainment at the end of primary schooling were added to the model. (It was expected that the primary school effect would have operated by raising attainment *prior* to secondary transfer.) In order to test this hypothesis, LRT score and VR band were added to the fixed effects part of the model. This increased the percentage of overall variance accounted by a further six per cent (reaching 38.0 per cent in total). For this model, the intra-school correlation was little changed at 0.0424 indicating that differences between primary schools, although small, remained relatively stable.

The model was further elaborated to allow for random variation in the impact of prior attainment and sex at levels 1 and 2. The results provided evidence of the existence of small but significant differences between primary schools in students' overall examination performance scores at age 16 years even when account is taken of attainment at secondary transfer.

In addition, the results also provided some indication of the existence of small but significant differential primary school effects for students with different levels of prior attainment at transfer, as measured by LRT score (see Table 5.4). This may mean that long-term primary school effects on students' later secondary school attainment can vary for students with different levels of attainment at transfer to secondary school. The results are also in line with the multilevel reanalysis of progress in reading and mathematics in junior school reported by Sammons, Nuttall and Cuttance (1993) which also revealed small but significant differential effects for pupils of different prior attainment levels for this cohort.

It was also found that the fixed effects estimates of the impact of mathematics and reading attainment at age 7 years were no longer statistically significant when reading attainment (LRT score) and VR band at transfer were included in the model. As might be expected, attainment at age 11 is a better predictor of performance at age 16 than attainment at age seven. The estimates of background effects for this model are shown in Table 5.5. As with

Table 5.4 Random Parameter Estimates for Analysis of Students' Total GCSE Examination Performance Scores Controlling for Prior Attainment and Background Factors (Primary School Analysis).

		Level 2 Estimate	S Error
Primary school σ^2 intercept		15.27	6.016
	LRT/Intercept	0.357	0.159
	LRT/LRT	0.263	0.115
		Level 1	
Pupil σ^2 intercept		215.2	11.19
	SEX/Intercept	17.17	5.624
	LRT/Intercept	1.45	0.118
Intra-school correlation		0.0566	

the findings for the secondary school analysis reported earlier, the results demonstrate the continued negative impact of certain social class categories and of membership of VR bands 2 and 3, and the positive impact of membership of certain ethnic groups. These results on the impact of background factors again point to the need to include pupil-level details of such characteristics as well as data about prior attainment in multilevel analyses of school performance.

The model fit was improved by allowing the variable sex to vary randomly at level 1, and LRT score to vary randomly at levels 1 and 2. For this model, the intra-school correlation was 0.0566 (see Table 5.4). This indicates that, for the pupil with average LRT score at transfer, primary school attended still accounted for nearly 5.7 per cent of the unexplained variance in total examination performance scores at age 16 years. Interestingly, allowing for the random structure of the data at both the pupil and school level increases the intra-school correlation.

The hypothesis that primary school effects on students' secondary school examination performance would be removed if account was taken of students' attainment at secondary transfer was not confirmed. The results of the multilevel analyses provide evidence of the existence of small but significant differences between primary schools in students' overall examination performance scores at age 16 years.

Discussion and conclusions

The follow-up of the JSP cohort's GCSE examination performance at age 16 provides a valuable data set for the longitudinal analysis of attainment at different time points over a period of nine years.

Table 5.5 Fixed Parameter Estimates for Analysis of Students' Total GCSE Examination Performance Scores Controlling for Prior Attainment and Background Factors (Primary School Analysis).

Fixed effects

	Estimate	S Error
INTERCEPT	31.17	1.51
MTHSRWY3	0.114	0.0819
ERTRWY3	0.043	0.0293
SEX	−2.733	0.82
FREEMLS	−2.750	1.013
SKILLED	−2.299	1.141
SEMIUNSK	−4.616	1.768
LTUFABS	−1.678	1.497
FNK	−1.678	1.463
ASIAN	8.833	1.623
CARIBBEAN	3.115	1.131
OTHERNK	4.766	1.135
LRT	0.286	0.0566
BAND 2	−7.214	1.337
BAND 3	−11.55	2.073

Where:
MTHSRWY3 Raw Basic Maths Test score entry to junior school Year3
ERTRWY3 Raw Edinburgh Reading Test score entry to junior school Year3

The results point to the existence of significant secondary school effects on students' overall examination performance. They also demonstrate the continued impact of background factors after controlling for prior attainment at transfer (see also Sammons, 1994b). The latter indicates the importance of adequate control for both prior attainment and socio-cultural factors in analyses of secondary schools' examination results to ensure that estimates of the "value-added" by schools do not reflect unfairly upon schools receiving disadvantaged intakes. Such findings are important for the ongoing debate about the value of publishing league tables of schools' raw examination results as measures of performance and to improve accountability.

The results also demonstrate that attainment in basic skills (Reading and Mathematics) at age seven are good predictors of later performance in public examinations at age 16. The finding that the VR band measure has an additional impact upon later GCSE performance even when controlling for reading attainment and background factors is also of interest. It may be picking up some non-school or only partially school-related variance such as student aspirations and attitudes or teacher response to student social/cultural background and behaviour. This would suggest overlap in the variance component attributable to socio/cultural background and this school cognitive measure. If this is the case then the variance between schools in fact may be higher than estimated. The same argument can be applied to the other background measures in the models. Given this it seems appropriate to interpret the between school variance as a lower estimate – it is the variance that can be *uniquely* attributed to a school.

The results of the analysis of primary school effects point to the existence of small but significant continuing effects of primary schools. They are in line with the conclusions of Entwistle and Hayduk's (1988) examination of the continuing effects of elementary schools. The latter's research, however, was limited because of the very small number of schools (3 in all) examined. Because of this Entwistle and Hayduk were not able to identify school effects using multilevel techniques. Whilst our analysis has focused upon academic effectiveness, it is interesting to note that, in a small scale study of a Scottish primary cohort which investigated truancy Gerrard (1989) found evidence of long-term effects of primary schooling. Gerrard (1989) reports on attendance patterns at secondary school in relation to primary school attended. His findings suggest that primary school attended may also have long term impact on social outcomes (such as attendance).

> 'It seems that children are being prepared by different primary school to cope in different ways and with different measured success with the demands of Scottish secondary school education' (Gerrard, 1989, p. 110).

Unfortunately, because of the nature of the sample, the data set was not uniquely nested between primary and secondary schools. Therefore, it was not possible to undertake a 3-level analysis to examine the impact of primary and secondary schools simultaneously.

However, recent advances in multilevel modelling techniques developed by Goldstein will allow the use of cross-classified analyses which enable the effects of secondary and primary schools to be modelled simultaneously. Early work by Goldstein (unpublished personal communication) supports the broad conclusions of the present study on the continuing effects of primary schools. However, further work in this area using cross-classified analyses and a larger sample of schools (and preferably a broader range of educational outcomes) would be of considerable theoretical interest and practical relevance.

Because of the restricted range of GCSE attainment among the JSP cohort which could be followed up at age 16 (it is likely that some more able students transferred to non-ILEA private or state schools and those at the lower end of the ability range would not usually be entered for GCSEs) it is argued that the 4.2–5.7 per cent of variance showing the continuing long term "primary school effect" should be treated as a lower estimate. Given this, and the findings of substantially higher estimates of the percentage of variance attributable to primary school attended in the "value-added" analyses of reading and mathematics progress at primary school (reported by Sammons, Nuttall and Cuttance, 1993), compared with the equivalent "value-added" analyses of total GCSE scores at secondary school, the results also suggest the importance of primary schools in determining students' later school performance may have been under-rated. The primary school effect appears to be two fold: firstly affecting the rate of progress made during primary school in basic skills such as reading and mathematics and thus raising (or lowering) the level of attainment reached by secondary transfer; and, secondly also affecting attainment at age 16 directly in some way.

The interpretation of *how* primary school attended affects later performance is not obvious. It is possible that such differences reflect the impact of primary schools upon their pupils' long-term attendance, attitudes to school, study skills, motivation or self-esteem which may affect later performance in secondary school.

It is also possible that more effective primary schools may have helped to raise pupil achievement by contributing to their sense of self-efficacy (their beliefs about their capabilities to exercise control over their own level of functioning and other events that affect their lives). Bandura's (1992) work suggests that

> 'Learning environments that construe ability as an acquirable skill, de-emphasise competitive social comparison, and highlight self-comparison of progress and personal accomplishments are well suited for building a sense of efficacy that promotes academic achievement' (p. 5).

The fact that differential primary school effects were not identified for pupils of different background groups (sex, social class, low income, ethnicity) and were very modest for those of different levels of prior attainment (Sammons, Nuttall & Cuttance, 1993) indicates that effective schools were broadly effective for all and likewise ineffective schools were generally ineffective for all. This suggests that effective school processes are likely to operate in similar ways for different pupil groups. We can hypothesise that the continued effects of primary schools on later examination performance at secondary school may reflect the impact upon pupils' beliefs about their own abilities and thus their learning behaviour at secondary school. Further research in this area would help to improve understanding about the different ways primary schools may affect later outcomes. In addition, further research using larger samples of schools and cross-classified models would enable the separation of primary and secondary school variance and thus enable better estimation of the size and significance of both secondary and primary school effects.

We believe that our results concerning continuity of effects are of theoretical significance because, to date, very little attention has been paid to the question of the long term impact of primary schooling. The results are also clearly of educational significance for policy makers and practitioners alike. In the UK primary education can be regarded (in resource terms and perhaps in Government or popular esteem) as the poor relation of secondary education. Yet it appears that primary school effectiveness is important not only for pupils during their period of primary education but also in later years. Given this perhaps greater attention should be paid to ways of promoting effectiveness at the primary level and in investing in this crucial phase of schooling. The results are of relevance also to primary teachers and headteachers. They point to the importance of their role in promoting pupils' educational outcomes and the long term impact they can have on their former students' continuing school careers.

Part 2

Understanding School Effectiveness

In this section a number of articles and chapters are drawn together to consider what school effectiveness research tells us about the ways school and classroom processes seem to influence student outcomes. To what extent can we illuminate the black box of how school and classroom experiences combine to foster or inhibit students' progress and their social and affective development? This forms the crucial link between the input/output models which were discussed in Section 1 where the question of measuring school effectiveness was addressed.

Far fewer school effectiveness studies have sought to explore the way processes can influence outcomes than have tried to measure and model the relationships between prior attainment and background factors and student outcomes in order to investigate the question of 'what makes a difference in school effectiveness?' Yet, for both policy makers and especially practitioners the most interesting and valuable aspect of school effectiveness research is when it helps to inform and improve our understanding of the ways schools and teachers influence student learning, and thus their attainment, progress and development. Section 2 of this book draws on the results of both primary and secondary school research in an attempt to explore this crucial issue.

6

The Question of Education Quality

Introduction

One of the major influences on my career in school effectiveness was involvement as Project Coordinator in the School Matters research from 1981 to 1986. The ILEA's Junior School Project, directed by Peter Mortimore, was a major longitudinal study which broke new ground by building on the earlier Fifteen Thousand Hours *study both methodologically and conceptually in examining the impact of primary schools. It remains remarkable even today in the range of social and affective, as well as the more usual academic outcomes it explored, and in the rich variety of information about classroom processes and school organisation it analysed, including detailed observational data collected over three separate school years from the same student age cohort. Much has already been written about the study whose strength reflects the foresight and inspiration of Peter Mortimore. Full details of the research design and technical aspects are provided in the four volumes of the Junior School Project final report, while* School Matters *(Mortimore et al., 1998a) provides an account of the main findings and their interpretation. Given the familiarity and accessibility of this material to those engaged in the school effectiveness field, I have not included extracts from these earlier publications in this book. Rather, I have chosen a chapter which forms part of a volume on* Improving Education: Promoting Quality in Schools *(Ribbins & Burridge, 1994). The chapter was originally written in 1991 as part of a series of seven* seminars which focused on different aspects of quality in education. The series was organised as a joint venture between the Education Department of the City Council in Birmingham and the School of Education of the University of Birmingham. It thus provides an unusual (for the time) example of a partnership between policy makers (LEA) and academics seeking to link theory, research and practice in order to

> 'test, refine and develop the ideas that underpin Birmingham's quality initiative in the light of these theories and practices' (Ribbens & Burridge, 1994, p. xii).

The series was also intended to help establish the reputation of the city in partnership with the university as a leading national centre in the study and practice of all aspects of quality in educational contexts; to generate a network of authorities interested in sharing ideas and experiences and to enable and encourage staff in Birmingham to contribute to the city's quality development initiative. It is interesting to note that during the 1990s, Birmingham became in many ways a 'flagship' Labour authority under the influence of its Chief Education Officer, Sir Tim Brighouse and had a significant impact on the development of the Labour opposition's (now Government) thinking on education over this period.

When asked to contribute to the seminar series I sought to consider findings from the school effectiveness research with which I had been involved during the 1980s and to explore some of the implications of the School Matters' results for improving the quality of schools. In order to do this, it was important to put the results in the context of UK education in 1991 which was experiencing the many changes and upheavals of the national policy developments post–1988 and the implementation of the far reaching Education Reform Act. The chapter also sought to look beyond the UK context in examining the concept of educational quality and increasing interest in the potential use of performance indicators in education. Indeed, at the time I was involved in a study commissioned by EUROSTAT for the European Union on the development of educational indicators for members states (see West et al., 1993, 1995). The chapter thus provided a considered reflection on the School Matters research several years after its publication and in the light of discussion of its findings with practitioners in professional development sessions across the UK and in North America. In addition, it investigated the links between the School Matters' findings based on junior school students and those of the earlier secondary school Fifteen Thousand Hours research conducted also in the inner London context but focusing on older students. In this way it attempted to analyse the extent of consistency in findings and their generalisabilty across sectors.

This chapter examines the methods and some of the main findings from a recent research study into school effectiveness and their implications for improving the quality of schools. The first section provides some background context, focusing on the increasing interest in measuring quality in education, evident among policy-makers and practitioners alike. Recent changes in legislation are noted and various definitions of quality are outlined. The second section describes some of the findings from studies of school effectiveness, and the way such findings can assist those concerned with developing systems to measure quality and monitor the performance of schools.

In the UK, schools have been subjected to a period of rapid and radical change as a result of the implementation of the provisions of the 1988 Education Reform Act (ERA). The introduction of a National Curriculum and associated teacher assessments (TAs) and externally monitored standard assessment tasks (SATS) for pupils at four Key Stages (at age 7, 11, 14 and 16 years) are beginning to have a major impact upon the content and processes of teaching and learning – what is taught, how it is taught and how it is assessed.

In addition, as with other public services, local education authorities (LEAs) and schools have been increasingly affected by public and government demands for greater accountability and freedom of information during the past decade (Tomlinson, Mortimore & Sammons, 1988). For example, the 1982 Local Government Finance Act requires the auditors appointed by the Audit Commission to satisfy themselves that authorities have made proper arrangements to secure 'the three Es, economy, efficiency and effectiveness in the use of resources'. However, in commenting on these 'three Es', Young (1990, p. 7) has noted that their interpretation, particularly in education, is not always easy.

> '*Desirable*, and indeed simple as the criteria may appear to be, their meanings are complex and their use for auditing or managerial purposes is often problematic, indefinite and capable of different and conflicting interpretations ... in the provision of services like education, the last of the "three Es", effectiveness, can be far more difficult to judge, since one cannot measure effectiveness with any precision against some sort of numerical scale.'

More recently, Local Management of Schools (LMS) has been introduced in an attempt to ensure greater financial independence and accountability for individual schools. As with the other changes resulting from the ERA, such as the National Curriculum and national assessment, the Government's stated aim is to improve the quality of education in schools. It is intended that LMS should be evaluated nationally and locally in order to assess 'the success of LEAs in implementing schemes' and of 'local management in raising the quality of education in schools' (DES, 1988, para.226).

Levacic (1990) has commented that, in evaluating the impact of LMS, predetermined goals can be investigated because the Government has made explicit its aims for this initiative:

> To improve the quality of teaching and learning in schools ... Effective schemes of local management will enable governing bodies and headteachers to plan their use of resources ... to maximum

effect in accordance with their own needs and priorities, and to make schools more responsive to their clients, parents, pupils, and the local community and employers.' (DES, 1988, paras 9 & 10)

Recent legislation has given LEAs some specific responsibilities for the monitoring of the education service, particularly of schools (DES, 1988). The favoured term to describe the measures to be used to evaluate the performance of schools is performance indicators (PIs).' A PI can be defined as an item of information collected at regular intervals to track the performance of a system' (Fitz-Gibbon, 1990, p. 1). Much discussion has occurred among policy-makers, researchers and educational practitioners alike concerning the choice of PIs, their benefits, limitations and possible abuses or negative side-effects.

The collection and use of performance indicator information is intended to promote accountability, and to help to identify areas where improvement is required at a variety of levels: LEA, individual school or classroom. In order to be useful for planning and policy-making purposes indicators will need to be collected on a regular basis to allow the monitoring of changes over time. In addition, a consistent approach to PIs would need to be adopted by LEAs and individual schools if valid comparisons are to be made between different authorities or institutions.

Uncertainty over the future of LEAs arising out of proposals in the recent White Paper *Choice and Diversity* (DfE, 1992) will limit a coherent approach to the monitoring of school quality at the local level.

Quality and schools

The stated intention of the recent legislative changes is to improve the quality of schools. Although this aim may receive broad support, there is no consensus as to how the quality of schooling should be defined, and whether it can be measured. Even more controversy abounds as to how the quality of schooling may be improved. As Mortimore and Stone (1990, p. 69) commented, the question of how to measure educational quality is not new and is 'intimately bound up with more fundamental questions about the nature of education itself'. The concept of education as 'an essentially instrumental activity designed to bring about the achievement of specifiable and uncontroversial goals' was contrasted with a more 'Aristotelian' view of education practice' as an essentially ethical activity guided by values which are open to continual debate and refinement by practitioners and others'.

Mortimore and Stone drew attention to the normative or comparative element implied by the term quality, noting that the term has been used as: an attribute or defining essence; a degree of relative worth; a description of something good or excellent; and a non-quantified trait. While recognizing the need to employ caution, Mortimore and Stone nevertheless argued that it is possible to discuss the educational quality of different components of the education system and that such discussion is necessary to ensure the important goal of accountability of the education service.

An influential OECD report entitled *Schools and Quality* (1989) discussed some of the issues involved in attempts to measure and improve quality. It concluded that

> 'The assessment of quality is thus complex and value laden. There is no simple uni-dimensional measure of quality. In the same way as the definition of what constitutes high quality in education is multi-dimensional, so there is no simple prescription of the ingredients necessary to achieve high quality education; many factors interact – students and their backgrounds; staff and their skills; schools and their structure and ethos; curricular; and societal expectations.' (OECD, 1989, p. 27)

School effectiveness research can help to disentangle and clarify such interactions, and because of this, I believe it has an important role to play in analysing the constituents of quality in education.

Likewise, Lagerweij and Voogt (1990) have stated that the concept of quality of education cannot be easily defined in a clear and exact manner. They noted that 'any definition of quality should be expected to change over time, because it necessarily reflects a society's interpretation of educational needs and the intensity of its moral and financial commitment to fulfilling them' (p. 100). In addition, they commented that

> 'Judgements about quality can refer to different, but closely related aspects, for example: the goals or functions, the contents, the processes, the effect and the conditions or means' (p. 100)

and discussed some of the ways educational improvement may be directed to one or more of these aspects.

Increased variability in school quality

Reynolds (1991) predicted that the 1990s would see an increase in the influence that schools have over the development of young people. He suggested that a variety of factors would be responsible for this phenomenon (e.g., the integration of children with special educational needs, concern to keep troublesome, delinquent or disturbed children in ordinary schools and the policy of decentralization of power within the education system down to the level of the individual school). Reynolds argued that the result of such changes, in the short term at least, is likely to be

> 'a substantial increase in the variation in their quality between individual schools ... the huge additional range of powers, roles and responsibilities that will fall upon schools and particularly upon their principals or headteachers, will also increase school variability substantially. Schools will also differ markedly in their ability to cope with rapid externally induced change.' (Reynolds, 1991, p. 5).

He pointed out that the Government's intention in Britain is that the major mechanism of quality control will be locally determined market mechanisms of parental choice:

> 'schools judged by parents to be ineffective will rapidly lose pupil numbers and will eventually shut' (p. 5).

He argued that in the long term this may reduce the variation between schools in their quality. Such a view may well prove to be unduly optimistic however, since the extent to which parents in reality are able to exercise choice will vary much in different geographical localities (e.g., depending upon the existence of alternative schools and their accessibility) and in the availability of appropriate information upon which to make informed choices.

There is growing evidence from research into choice of secondary schools that parents vary in the way they choose schools, and in the extent of the child's involvement in the decision. For example, in a survey of parents West and Varlaam (1991) found that the four factors most frequently mentioned as important (without prompting) when choosing a secondary school were: the child wanting to go there, good discipline, emphasis on good examination results and ease of access. Three-quarters of parents said there were particular schools to which they did not want their child to go – the predominant reason given was 'its bad reputation'.

The extent to which parents' knowledge and judgements of quality and effectiveness are based on up-to-date and accurate information is open to question. School reputations may rest upon past circumstances and bear little relation to the current situation, for example. Particular difficulties surround the use of information about pupils' academic achievement (as demonstrated through examination or test results) as a measure of the quality of schools. Unless information about schools' examination or test results is presented in context and takes account of the 'value added' by the school (by taking account of differences between schools in the pupils taught, in terms of background factors and prior attainment at intake) it is likely to be misleading (see Mortimore et al., 1988a; Gray, 1990; Nuttall, 1990; Sammons, Nuttall & Cuttance, 1993). In the absence of a strong commitment to provision of useful and usable information for parents (Tomlinson, Mortimore & Sammons, 1988), it seems unlikely that reliance on the market mechanism of parental choice will lead to improvements in the quality of schools.

Secondary schools have been required to publish their 'raw' (i.e., not adjusted for intake and prior attainment) public examination results since 1981. Under the ERA, LEAs are required to publish National Curriculum assessment results. There is no requirement to take account of intake, although schools may provide details about their intakes as contextual information.

In commenting on parental choice of schools and possible problems arising from the intention to publish National Curriculum assessment results, Nuttall (1990, p. 25) argued that 'insofar as the initiative is to increase the accountability of schools, natural justice demands that schools are held accountable only for those things that they can influence (for good or ill), and not for all the preexisting differences between their intakes. The investigation of differential school effectiveness, concentrating on the progress students make while at that school, therefore has a major role to play in future'.

School effectiveness research
During the past two decades much research has been conducted into the field of school effects or school effectiveness in the search for ways of creating effective schools (see reviews by Gray, 1981; Purkey & Smith, 1983; US Department of Education, 1986; Reynolds, 1989; Mortimore, 1991a). The terms school effects and school effectiveness are sometimes used interchangeably. However, most commonly school effects refers to the impact particular schools have on their pupils' educational outcomes, taking account of differences in intake, whereas school effectiveness refers to studies of the factors and processes related to positive or negative effects on such outcomes. An effective school is one that has a positive effect upon its pupils' educational outcomes, when account is taken of intake.

Research into ways of measuring the quality of schools has a number of advantages, as Mortimore and Stone (1991) have argued. As a way of identifying components of educational quality, research is superior to other methods. This is because its methods should be public and can be examined for bias, and its scope is very broad. While the idea of perfectly value-free research is now recognized as something of a myth, there are well-tried methods of limiting bias and of guarding against systematic distortion of evidence-gathering techniques. Perhaps the most potent aspect of research, as opposed to other ways of gathering information, is that its methods are stated publicly and are open to critical scrutiny.

Although the cost of research and the time needed to conduct the necessary longitudinal surveys of pupil progress mean that it is not a practical way to assess individual schools on a regular basis, the results of research, its techniques and instruments have much to offer those concerned with school improvement, school self-evaluation and the development of performance indicators for schools.

Educational goals for students – criteria for measuring quality
Ron Edmonds (Edmonds 1979a, b, 1982) was much concerned with school effectiveness research and improving the quality of schooling in the USA. He has stated that the major aim of schools should be *educational excellence*, which he defined as meaning that students become independent, creative thinkers and learn to work co-operatively. Edmonds (1982, p. 14) made a particular point of emphasizing student mastery of basic skills. John Goodlad (1984) has also pro-

posed a variety of different goals for education. These are much broader than a narrow focus on basic skills, and include:

- academic development;
- intellectual development;
- vocational goals;
- social, civic and cultural goals;
- personal goals.

If the need to strive for educational excellence and the importance of the different goals noted above are accepted, then the findings of research that attempts to increase our understanding of how schools influence their students, and what promotes their progress and development, need careful consideration.

As noted earlier, several extensive reviews of school effectiveness literature exist. In this chapter the main focus of attention is on the findings of the Inner London Education Authority's Junior School Project (JSP) (Mortimore et al., 1986, 1988b). There are several reasons for looking at this study in some depth:

- it is the largest piece of school effectiveness research undertaken to date either in the UK or in North America;
- it examines a much broader range of pupils' educational outcomes than previous studies, avoiding a narrow focus on basic skills alone;
- the results support and confirm many earlier findings about school effectiveness;
- the results also extend our knowledge of what contributes to effectiveness in several important ways;
- in contrast to most previous studies, it examines both factors to do with the school as a whole and those to do with the teachers' classroom practices, and shows how these relate to student outcomes;
- the study demonstrates the methodology of school effectiveness research clearly, utilizing data about individual pupils, their teachers, classes and schools at the appropriate level;
- it makes clear the need to ensure that judgements of effectiveness of schools are based on appropriate comparisons of 'like with like', by taking full account of intake.

Similarities between the JSP findings and those produced by other research are discussed later in this chapter.

The ILEA junior school project

The project began in September 1980 when an age cohort of nearly 2000 seven-year- olds entered their junior classes, and concluded four years later when the students transferred to secondary school. The students attended a random sample of fifty schools in inner London.

Aims

The project was designed to produce a detailed description of the students, teachers, curriculum and organization of schools in an inner city area. It attempted to answer three major questions.

1. Are some schools more effective than others in promoting students' learning and development, when account is taken of variations in the students' backgrounds?
2. Are some schools more effective than others for particular groups of children (for girls or boys, for those of different social class origins, or different racial backgrounds)?
3. If some schools are more effective than others, what factors contribute to such positive effects?

Information about students, schools and teachers

In order to answer the above questions information was collected on three major topics: students' background characteristics; students' learning and development; and school characteristics. These provided measures of the student intakes to schools; measures of students' educational outcomes; and measures of school and classroom *processes*.

Intakes

Considerable evidence exists, from a strong tradition of educational research, of the importance of socio-economic and family background factors as influences upon pupils' educational achievements at all stages of their school careers (see, for example, Douglas, 1964; Davie et al., 1972; Rutter & Madge, 1976; Hutchison, Prosser & Wedge, 1979; Essen & Wedge, 1982; Mortimore & Blackstone, 1982; Sammons, Kysel & Mortimore, 1983). Detailed information was obtained for each child about: sex, age, social, ethnic, language and family background; pre-school and infant school experiences; and initial attainments at entry to junior school. These data were needed so that the study could: establish the impact of such background factors on students' attainments, progress and development; take into account differences between schools in intakes; quantify the relative importance of school compared with background as influences upon students; and explore the effectiveness of schooling for different groups.

Outcome

The second set of information is related to students' learning and development. Reynolds and Creemers (1990, p. 2) have noted that 'In certain countries the school effectiveness movement has already become associated with a narrow, back-to-basics orientation for the teaching of basic skills'. Such a narrow focus is inappropriate because the goals of education, teachers' aims and objectives for pupils' learning and the curriculum are much broader than a concern with the 'three Rs' alone.

The JSP study focused on more than attainment in 'basic skill' areas, important though these are. To take account of the diversity of aims and curricula of primary education a wide range of outcomes was examined. In addition to reading and written maths, practical maths skills and visuo-spatial skills were assessed. Creative writing was assessed using measures that included the quality of language and ideas, as well as more technical aspects. To broaden the assessments of language, students' speaking skills were also studied. Oracy assessments focused on the ability to communicate effectively, and children were not penalized for using non-standard English.

Of equal interest to those concerned with the quality of education are non-cognitive or social outcomes of education, which have tended to be neglected in previous studies of school differences and effects. Information was obtained about students' attendance, their behaviour in school, their attitudes to school and to different types of school activities, and their self-concepts (including their perceptions of themselves as learners) in school.

Processes
The third set of information is related to the characteristics of the schools, their organization and numerous aspects of the learning environment experienced by students. A wide variety of information was collected and the areas covered included the following: aims and philosophy of schools, aims and philosophy of class teacher, organization of school and policies, rewards and punishments (school), organization of classrooms, curriculum, structure of teaching, rewards and punishment (class), teacher-student communication, support services, books and resources, school appearance and classroom appearance.

Interviews (with headteachers and their deputies, class teachers, special needs teachers and parents), observations (of classrooms over a three-year period, and of the school environment) and questionnaires (with headteachers, deputy headteachers, teachers and students) were used to collect these data. Field officers also made detailed observations and kept extensive field notes of teachers and students in the classroom setting.

Measuring school effects
The analysis of all these data was complex, and various statistical techniques, particularly multilevel modelling (see Goldstein, 1987) were employed. The intention was to find out what impact schools had on their students' progress and development, once account was taken of their attainment at entry to junior school, and of the influences of age, sex and other background factors. It was possible to analyse student progress because the Project was longitudinal. By studying the progress of individual students account could be taken of the very different levels of skill possessed by children at entry to junior school. For each student, therefore, her or his initial attainment at entry was the baseline against which progress during later years was measured (Sammons, 1989a).

Table 6.1 Difference Between Schools in Students' Attainment at Entry.

	Reading skills	Maths skills	Visuo-spatial skills
Average score all schools	45.5	24.1	24.4
School with highest average score at entry	62.6	29.3	31.1
School with lowest average score at entry	17.3	18.3	19.4
Maximum possible score	19	42	40

Note. $N = 50$.

The influence of background factors and school differences in intake
Strong relationships between background factors (especially age, social class and low income, sex and ethnic background) and students' attainments and development and, to a lesser extent, their progress during the junior years were found (Mortimore et al., 1986, Part A). These differences were already marked at age 7 at entry to junior school. Full account, therefore, had to be taken of these relationships before schools' effects on their students could be examined.

Children's attainment in any particular basic skill area (e.g., reading, maths, writing) at entry to junior school was found to be a good predictor of their attainment in that area several years later. This is as might be expected, since those who are good readers at entry tend to remain good readers as they grow older (though it was also found that some children made more or less progress than might be predicted, given their attainment at entry).

Even at entry to junior school at age 7, there were very marked differences between individual students in their attainment in basic skills (well over a two year gap in reading ages among the year group, for example). Moreover, some schools received intakes that contained students who had a much higher average attainment than other schools (see Table 6.1). These figures on differences in intake demonstrate why in studying school effectiveness it is essential to focus on the concept of 'value added' by looking at student progress individually (for further details of this argument see Mortimore et al., 1988a; McPherson, 1992). Unfortunately, the importance of the 'value added' concept has not been recognized in the legislation concerning the publication of National Curriculum assessment results for individual schools and classes. Without recognition of the impact of intake (especially prior attainment of pupils) the publication of 'raw' results will not provide a useful guide to the quality of schools. In fact, such results are likely to prove highly *misleading* because schools with advantaged intakes will appear to be successful, irrespective of their actual contribution to pupil progress.

In contrast, focusing on value added means that schools can be classified as effective whatever the absolute level of advantage/disadvantage of their pupil intake. Schools serving very *disadvantaged* populations can be highly effective. Similarly, schools serving very advantaged populations can fail to be effective and can foster underachievement amongst their pupils (Mortimore, 1991a).

School effects

Even after controlling for students' initial attainment at entry and for background factors, the JSP data show that the school made an important contribution to students' progress and development. In terms of the first question asked, it was found that schools *did* make a substantial difference to their students' progress and development. In fact, for many of the educational outcomes – especially progress in cognitive areas – the school was very much more important than background factors in accounting for variations between individuals. Analyses of students' oracy (speaking skills) and of the social outcomes (attendance, attitudes, behaviour etc.) confirmed the overriding importance of school.

The size of the effects of each of the sample schools on each of the measures of their pupils' educational outcomes was calculated. The differences were striking. For example, for reading the most effective school improved a student's attainment by an average of 15 points above that predicted by that child's attainment when he or she started junior school, taking into account her or his background. But in the least effective school, each child's attainment was, on average, 10 points lower than predicted. This compares with an overall average reading score for all students of 54 points, and a maximum possible of 100.

It was also important to establish whether some schools were *generally more effective* in promoting a broad range of educational outcomes than others. The results showed that a sizeable number of schools (fourteen in all) had positive effects on students' progress and development in most of the cognitive and most of the non-cognitive outcomes. These can be seen as the generally effective schools. In contrast, five schools were rather ineffective in most areas. Many schools were effective in a few, but not all, areas. However, there were very few schools that only had positive impacts on cognitive areas, but were unsuccessful at fostering non-cognitive development, or vice versa.

The findings demonstrate that schools often vary in their effectiveness in promoting different educational outcomes and this needs to be taken into account in any attempts to measure or improve the quality of schools. It is not appropriate, therefore, to divide institutions into simple categories of 'good' or 'bad'.

School effects on different groups

It was also possible to compare the effects of schools on the progress of different groups of students. Generally, for any particular outcome, schools that were effective in promoting the progress of one group of students were also effective for other groups, and those that were less effective for one group were also less effective for others. Thus, a school that was effective for boys was also effective for girls, one that was effective for a child from a non-manual worker's family was also effective for one from a working-class family. An effective school tended to 'jack up' the performance of all its students irrespective of their sex, social class origins or race. A recent re-analysis of the JSP data confirmed the original findings, although some evidence of differential effectiveness for pupils with different levels of prior attainment (an area not covered in the original study) was found; see Sammons, Nuttall and Cuttance, 1993. Moreover, the evidence indicates that although overall differences in attainment were not removed, on average a student from a working-class family attending one of the more effective schools ended up attaining more highly than one from a non-manual family attending one of the least effective schools.

Other more recent studies have suggested that there may be some differences within secondary schools in their effects on the examination results of particular groups (see Aitkin & Longford, 1986; Nuttall et al., 1989). The question of *differential effectiveness* is clearly of importance to those concerned with promoting equal opportunities and improving the quality of schooling for all pupils.

Understanding school effectiveness
The results of the JSP indicate that the particular school attended can make a substantial difference to the future educational prospects of individual students. Given this, the third question addressed by the Project is crucially important. What makes some schools more effective than others? In order to answer the third question it was necessary to establish what factors and processes were related to positive school effects. In other words, the aim was to identify the ways in which the more effective schools differed from those that were less effective.

A wide variety of process variables about the school as a whole, and ones specifically concerning individual classrooms, were investigated. These were divided into 'givens' and 'policies'. The 'givens' are aspects not directly under the control of the school (e.g., the size of its student roll, intake, stability of staff, pupil mobility). The policies, in contrast, are aspects that can be altered by the school (e.g., the headteacher's style of leadership, curriculum, the rewards and punishment system, organization, staff involvement and conditions). At the level of the individual classroom there are also a variety of 'givens' not under the direct control of the teacher (e.g., the size of the class, the composition of students in terms of balance of age, sex or ability, the resources available etc.). But many policy factors are under the class teacher's direct control (e.g., record-keeping, the system of rewards and punishments, the use of praise, the amount and type of communication with students, preparation and planning etc.).

Analyses indicated that much of the variation between schools in their effects on students' progress and development was accounted for by differences between schools in their policies and practices. Furthermore, a number of the significant variables were themselves associated. By a detailed examination of the ways in which classroom and school processes were interrelated, it was possible to gain a greater understanding of some of the important mechanisms by which effective education may be promoted.

Key factors for effective junior schooling
The JSP analyses identified a number of key factors that are important in accounting for differences in the effectiveness of schools. These factors are not purely statistical constructs. They were not obtained solely by means of quantitative analysis but from a combination of careful examination and discussion of the statistical findings, and the use of educational and research judgement. They

represent the interpretation of the research results by an interdisciplinary team of researchers and teachers.

Some schools were more advantaged in terms of their size, status, environment and stability of teaching staff. There was evidence that smaller schools tended to be more effective than larger ones. Class size is particularly relevant; smaller classes with less than 24 students (the average in the sample was 25, with a range of 16 to 34), had a positive association with student progress in maths, attainment in speaking skills, student behaviour, attitudes to school and self-concept, especially for progress in the earlier years (ages 7 to 9).

A good physical environment creates a positive location in which progress and development can be fostered. The stability of the school's teaching force is also important. Changes of headteacher and deputy headteacher, though inevitable in all schools at some stage, have an unsettling effect. Similarly, changes of class teacher during the school year had an adverse impact on students' progress and development. Gray (1990, p. 213) has also drawn attention to the importance of adequate resources and the absence of staffing difficulties:

> 'Adequate levels of resourcing, then, seem to be a necessary but not a sufficient condition for a school to be effective . . . in twenty years of reading research on the characteristics of effective schools I have only once come across a record of an "excellent" school where the physical environment left something to be desired.'

Nonetheless, although these favourable 'given' characteristics contribute to effectiveness, it appears that they do not, by themselves, ensure it. But they do provide a supporting framework within which the headteacher and teachers can work to promote student progress and development. The research suggests that it is the policies and processes within the control of the headteacher and teachers that are crucial. These are the factors that can be changed and improved.

Twelve key factors of effectiveness were identified (see Mortimore et al., 1988a, for details):

1 Purposeful leadership of the staff by the headteacher

Purposeful leadership occurred where the headteacher understood the needs of the school and was actively involved in the school's work, without exerting total control over the rest of the staff.

In effective schools, headteachers were involved in curriculum discussions and influenced the context of guidelines drawn up within the school, without taking total control. They also influenced the teaching strategies of teachers, but only selectively, where they judged it necessary. This leadership was demonstrated by an emphasis on monitoring pupils' progress through the keeping of individual records. Approaches varied: some schools kept written records, others passed on folders of pupils' work to their next teachers, some did both but a systematic policy of record keeping was important.

With regard to in-service training, those heads exhibiting purposeful leadership did not allow teachers total freedom to attend *any* course: attendance was allowed for a good reason. Nonetheless, most teachers in these schools had attended in-service courses.

2 The involvement of the deputy headteacher

The Junior School Project findings indicate that the deputy head can have a major role in the effectiveness of junior schools.

Where the deputy was frequently absent, or absent for a prolonged period (due to illness, attendance on long course, of other commitments), this was detrimental to pupils' progress and development. Moreover, a change of deputy head tended to have negative effects.

The responsibilities undertaken by deputy heads also seemed to be important. Where the head generally involved the deputy in policy decisions, it was beneficial to the pupils. This was particularly true in terms of allocating teachers to classes. Thus, it appeared that a certain among of delegation by the headteachers, and a sharing of responsibilities, promoted effectiveness.

3 The involvement of teachers

In successful schools, the teachers were involved in curriculum planning and played a major role in developing their own curriculum guidelines. As with the deputy head, teacher involvement in decisions concerning which classes they were to teach, was important. Similarly, consultation with teachers about decision on spending, was important. It appeared that schools in which teachers were consulted on issues affecting school policy, as well as those affecting them directly, were more likely to be successful.

4 Consistency among teachers

It has already been shown that continuity of staffing had positive effects. Not only, however, do pupils benefit from teacher continuity, but it also appears that some kind of stability, or consistency, in teacher approach is important.

For example, in schools where all teachers followed guidelines in the same way (whether closely or selectively), the impact on progress was positive. Where there was variation between teachers in their usage of guidelines, this has a negative effect.

5 Structured sessions

The Project findings indicate that pupils benefitted when their school day was structured in some way. In effective schools, pupils' work was organised by the teachers, who ensured that there was always plenty for them to do. Positive effects were also noted when pupils were *not* given unlimited responsibility for planning their own programme of work, or for choosing work activities.

In general, teachers who organised a framework within which pupils could work, and yet allowed them some freedom within this structure, were more successful.

6 Intellectually challenging teaching

Unsurprisingly, the quality of teaching was very important in promoting pupil progress and development. The findings clearly show that, in classes where pupils were stimulated and challenged, progress was greater.

The context of teachers' communications was vitally important. Positive effects occurred where teachers used more 'higher-order' questions and statements, that is, where their communications encouraged pupils to use their creative imagination and powers of problem-solving. In classes where the teaching situation was challenging and stimulating, and where teachers communicated interest and enthusiasm to the children, greater pupil progress occurred. It appeared, in fact, that teachers who more frequently directed pupils' work, without discussing it or explaining its purpose, had a negative impact. Frequent monitoring and maintenance of work, in terms of asking pupils about their progress, was no more successful. What was crucial was the *level* of the communications between teacher and pupils.

Creating a challenge for pupils suggests that the teacher believes they are capable of responding to it. It was evident that such teachers had *high* expectations of their pupils. This is further seen in the effectiveness of teachers who encouraged their pupils to take independent control over the work they were currently doing. Some teachers only infrequently gave instructions to pupils concerning their work, yet everyone in the class knew exactly what they were supposed to be doing, and continued working without close supervision. This strategy improved pupil progress and development.

7 Work-centred environment

In schools, where teachers spent more of their time discussing the *content* of work with pupils, and less time on routine matters and the maintenance of work activity, the impact was positive. There was some indication that time devoted to giving pupils feedback about their work was also beneficial.

The work-centred environment was characterised by a high level of pupil industry in the classroom. Pupils appeared to enjoy their work and were eager to commence new tasks. The noise level was also low, although this is not to say that there was silence in the classroom. Furthermore, pupil movement around the classroom, was not excessive, and was generally work-related.

8 Limited focus within sessions

It appears that learning was facilitated when teachers devoted their energies to one particular curriculum area within a session. At times, work could be undertaken in two areas and also produce positive effects. However, where many sessions were organised such that three or more curriculum areas were concurrent, pupils' progress was marred. It is likely, that this finding is related to other factors. For example, pupil industry was lower in classrooms where mixed activities occurred. Moreover, noise and pupil movement were greater, and teachers spent less time discussing work and more time on routine issues. More importantly, in mixed-activity sessions the opportunities for communication between teachers and pupils were reduced (as will be described later).

A focus upon one curriculum area did not imply that all the pupils were doing exactly the same work. There was variation, both in terms of choice of topic and level of difficulty. Positive effects tended to occur where the teacher geared the level of work to pupils' needs.

9 Maximum communication between teachers and students

It was evident that pupils gained from having more communication with the teacher. Thus, those teachers who spent higher proportions of their time *not* interacting with the children were less successful in promoting progress and development.

The time teachers spent on communications with the whole class was also important. Most teachers devoted the majority of their attention to speaking with individuals. Each child, therefore, could only expect to receive a fairly small number of individual contacts with their teachers. When teachers spoke to the whole class, they increased the overall number of contacts with children. In particular, this enabled a greater number of 'higher-order' communications to be received by *all* pupils. Therefore, a balance of teacher contacts between individuals and the whole class was more beneficial than a total emphasis on communicating with individuals (or groups) alone.

Furthermore, where children worked in a single curriculum area within sessions, (even if they were engaged on individual or groups tasks), it was easier for teachers to raise an intellectually challenging point with *all* pupils.

10 Record keeping

The value or record keeping has already been noted, in relation to the purposeful leadership of the headteacher. However, it was also an important aspect of teachers' planning and assessment. Where teachers reported that they kept written records of pupils' work progress, in addition to the Authority's Primary Yearly Record, the effect on the pupils was positive. The keeping of records concerning pupils' personal and social development was also found to be generally beneficial.

11 Parental involvement

The research found parental involvement to be a positive influence upon pupils' progress and development. This included help in classrooms and on educational visits, and attendance at meetings to discuss children's progress. The headteacher's accessibility to parents was also important, showing that schools with an informal, open-door policy were more effective. Parental involvement in pupils' educational development within the home was also beneficial. Parents who read to their children, heard them read, and provided them with access to books at home, had a positive effect upon their children's learning. One aspect of parental involvement was, however, not successful. Somewhat curiously, formal Parent-Teacher Associations (PTAs) were not found to be related to effective schooling. It was clear that some parents found the formal structure of such a body to be intimidating.

Nonetheless, overall, parental involvement was beneficial to schools with their pupils.

12   Positive climate

The Junior School Project provides confirmation that an effective school has a positive ethos. Overall, the atmosphere was more pleasant in the effective schools, for a variety of reasons.

Both around the school and within the classroom, less emphasis on punishment and critical control, and a greater emphasis on praise and rewarding pupils, had a positive impact. Where teachers actively encouraged self-control on the part of pupils, rather than emphasising the negative aspects of their behaviour, progress and development increased. What appeared to be important was firm but fair classroom management.

The first three factors concern the organization of the school and, in particular, relate to the quality of leadership of the headteacher and his or her management style. They point to the value of staff involvement in decisions. The fourth factor concerns the value of a 'whole-school' approach to the curriculum, with consistency among staff in the use of guidelines.

Factors 5 to 9 relate to the teacher's classroom behaviour, including the organization of pupils' work, the nature and level of communication with pupils, and the value of creating a work-centred environment and a limited focus within sessions. Recent work by Alexander (1992), conducted as part of the evaluation of the Leeds Primary Needs Programme, has provided confirmation of some of the JSP findings. Alexander (1992) has drawn attention to the advantages of maximizing communication with pupils, and to difficulties in managing a work-centred environment in classrooms where work in more than one curriculum area is organized in the same session.

The importance of the teacher's record-keeping was highlighted by the research (factor 10), as was the value of parental involvement in children's learning at home, in the school and as classroom helpers. The last factor drew attention to the benefits of a positive climate or ethos in the classroom and around the school. A positive climate appeared to be reflected in effective schools by happy, well-behaved students who were friendly towards each other and outsiders, and by the absence of graffiti.

From a detailed examination of the factors and processes that were related to schools' effects on their students, a picture evolves of what contributes towards effective education. The twelve factors identified by the research do not constitute a 'recipe' for effective schooling and should not be applied in a mechanistic way. It is important to remember that schools are not static institutions and that the JSP research took place prior to the changes surrounding the introduction of the National Curriculum. Nonetheless, the findings do provide a framework within which the various partners in the life of the school headteacher and staff, parents and students, and the community can operate and on which they can build towards improvement.

Links with other studies of school effectiveness

A detailed review of the links between the JSP findings and those of other studies of school effectiveness appears in Mortimore et al., 1986; Mortimore, Sammons and Ecob, 1988. This section draws particular attention to similarities with an earlier major study of secondary schooling.

The *Fifteen Thousand Hours* study examined the educational outcomes of students attending twelve inner London secondary schools. The research design was similar to that used in the later JSP. Four measures of student outcome were studied: behaviour in school; attendance; examination success; and delinquency. The results revealed that there were marked differences between schools in all these outcomes, even when account was taken of differences in the pupil intakes. A number of specific aspects of school process were found to be especially important. These included the academic emphasis, leadership of the head-teacher, teacher actions in lessons, the use of rewards and punishments, student conditions, student responsibility and participation, staff organization, and the skills of teachers.

The items found to contribute to an 'academic emphasis' in the *Fifteen Thousand Hours* study included the setting of homework by teachers (and checks by senior staff that this occurred), high teacher expectations for students, displays of students' work in classrooms and around the school, more teacher time devoted to teaching, group planning of courses by teachers and regular use of the library by students. The JSP also underlined the importance of high teacher expectations of students and the value of intellectually challenging teaching, a work-centred classroom environment with more teacher and student time on task, the involvement of teachers in developing curriculum guidelines and consistency in their use throughout the school.

There was agreement between the two studies on the importance of teacher actions in lessons. Both found that more teacher time spent communicating about work and less time on administration was beneficial. Good time-keeping (starting lessons on time and not finishing them early), planning and organization by the teacher were also found to be important. Another similarity was the positive impact of teacher time spent communicating with the whole class. The *Fifteen Thousand Hours* team concluded that if teachers spent too much time focusing on individuals, they tended to lose the attention of the class as a whole. The JSP also found that teachers who spent too high a proportion of time communicating with individuals spent more time on supervising and management comments and less time on work matters. It went further by pointing out that the use of a balance of whole class and individual communication maximized the *overall level of teacher-student communication* that occurred during lessons. Findings on the value of a greater emphasis on rewards and the use of praise in class for good work and behaviour by the *Fifteen Thousand Hours* team were also confirmed by the JSP. It can be concluded that students benefit from teacher actions that focus on *positive* rather than negative aspects of their work and behaviour.

Links between measures of the quality of the school's physical environment and student outcomes are also noteworthy. Both studies found that a pleasant, well-cared-for environment had a positive impact. In addition, the two studies point to the value of organizing extra-curricular activities for students (educational visits and trips, lunch-time and after-school clubs etc.).

The key role of the headteacher in providing purposeful leadership and in fostering school effectiveness was identified in the JSP. The importance of the quality of leadership given to the school by the headteacher, and his or her actions in developing the school's climate and goals, had also been demonstrated by *Fifteen Thousand Hours*. Both studies indicate that the school's climate or ethos has a significant role in creating effective education (see also Sammons, 1987a).

Conclusions

The findings of school effectiveness research consistently demonstrate that schools can make a difference to their students' educational outcomes, and that the difference can be substantial. The JSP finding of the existence of substantial differences between schools in their effects upon student progress has been supported by later studies of infant and secondary schools (Tizard et al., 1988; Smith & Tomlinson, 1989). They also provide a guide to what factors about schools and about classroom practice help to make that difference. In other words, it is possible to begin to explain why some schools are more effective than others.

An examination of the links between two major studies of school effectiveness reveals many areas of agreement. The findings of the JSP support many of those of the *Fifteen Thousand Hours* study, suggesting that the mechanisms and processes which lead to greater effectiveness are of general applicability. The consistency in the findings gives greater confidence in using them to develop programmes for school improvement and self-evaluation (see Stoll & Fink, 1989). It is stressed, however, that they should not be used as a way of holding individual schools to account.

In a recent contribution to the debate about appropriate frameworks for judging the quality of schooling, Gray (1990, p. 214) commented on current research on school effectiveness and factors that make a difference:

> 'In general terms it provides a relatively good introductory guide to the factors that make a difference. As a rule, schools which do the kinds of things the research suggests make a difference, tend to get better results (however these are measured or assessed). The problem is that these are tendencies not certainties. In betting terms the research would be right about seven out of ten times, especially if it could be supported by professional assessments.'

Because of this he also cautioned that school effectiveness research should not be treated as a blueprint for success (in line with the views of Mortimore, Sammons & Ecob, 1988, and Reynolds & Creemers, 1990).

The findings of school effectiveness research have sometimes been criticized for being just a matter of 'common-sense'. There is a grain of truth in this argument. Because school effectiveness research by its very nature sets out to identify the components of good practice that is already occurring in many schools, it is inevitable that some of the findings are unsurprising to practitioners. Renihan et al. (1986, p. 17) noted that 'The effective schools literature did not tell us anything startlingly revolutionary about organizing the school environment. Its major impact lay in highlighting factors vital to school success'. Similarly, Rutter et al. (1979, p. 204) concluded that 'Research into practical issues, such as schooling, rarely comes up with findings which are totally unexpected. On the other hand, it is helpful in showing which of the abundance of good ideas available are related to successful outcomes.'

Recently it has been suggested that the many changes schools are currently experiencing (particularly through LMS and the National Curriculum) mean that findings from school effectiveness research conducted in the 1970s and 1980s will be less relevant for improving practice and the quality of education during the nineties (Reynolds & Creemers, 1990). This conclusion is, however, open to question. The findings from studies such as the JSP and *Fifteen Thousand Hours* have demonstrated the importance of factors relating to overall school policy and individual teachers' classroom organization and practice rather than particular features of the curriculum.

Although it is likely that headteachers' roles may change following LMS, it is unlikely that their importance in fostering a common sense of purpose (a school mission), staff involvement and a positive school climate will diminish. As Gray (1990) commented, the importance of the head's leadership role is one of the clearest of the messages from school effectiveness research. Similarly, although there are changes to curriculum and assessment, factors such as a work-centred environment, intellectually challenging teaching, maximum communication with pupils and good record-keeping are likely to continue to be important in promoting pupils' progress and development.

School improvement has been described as

> 'a systematic, sustained effort aimed at changing learning conditions and other related internal conditions in one or more schools, with the ultimate aim of accomplishing educational goals more effectively' (Van Velzen, Miles & Ekholm, 1985, p. 48).

I believe that the methodology developed for school effectiveness research, particularly the focus on progress or 'value added' and the recognition of the importance of school intake, has an important contribution to make to the development of appropriate and valid performance indicators for schools. If such methodology is ignored there is a real danger that simplistic indicators will be used, which will result in unfair and misleading comparisons of the quality of institutions (McPherson, 1992). In particular, I believe that school effectiveness research clearly demonstrates that the policy of publishing schools 'raw' examination and National Curriculum assessment results will do nothing to improve the quality of schooling. On the contrary, it is likely to damage the reputations of institutions in disadvantaged communities in particular.

School effectiveness research methods offer a more appropriate basis for judging the quality of schools. The findings of such research should not be regarded as prescriptive, however. Rather they are a useful background for school self-evaluation and the development of school improvement programmes, which should be tailored to the needs of specific institutions.

Few programmes of school improvement have, as yet, been based directly on school effectiveness research (for an exception see Toews & Barker, 1985). However, an interesting example of the way the JSP and *Fifteen Thousand Hours* findings have been used to develop a practical school improvement programme in Ontario, Canada, has been described by Stoll & Fink (1989, 1991). Mortimore (1991b) has also put forward a number of postulates concerning ways of improving schools based on the results of school effectiveness research.

Many factors influence the extent to which schools can be successful in implementing change (see Fullan, 1988). The extent to which schools are able to set their own goals and priorities for improvement is important as are the extent of external changes and pressures affecting schools. The introduction of the National Curriculum and assessments, LMS and other educational reforms may well absorb much of the energy and commitment of schools and their staff, which are required for the successful development and implementation of improvement programmes. A whole school approach to improvement and the active involvement of staff in the development of an improvement plan and its implementation are essential if the goal of improving the quality of schooling is to be achieved.

7

Effective Teaching – Findings from the 'School Matters' Research

Introduction

School effectiveness research has commonly focused more on school level conditions affecting student performance than on teachers' practice in the classroom (Reynolds et al., 1994a). This feature of the field is at first sight somewhat surprising given the general acceptance that it is the quality of teaching and learning in the classroom which is likely to have the greatest, and certainly the most direct, impact upon student progress (Mortimore, 1993a, b, 1995; Creemers, 1994a, b; Luyten, 1995; Hill, Rowe & Holmes-Smith, 1995). The underlying reason for this state of affairs is surely related to the expense and time it is necessary to invest in order to obtain high quality information about classroom activities and interactions which can be related to measures of student attainment and, crucially, progress. Very few studies have been of sufficient size and scale to enable the collection of suitable classroom process measures concurrent with the measurement of student attainment at both the beginning and end of a school year. Only with such a design is it possible to attempt to investigate any associations between specific classroom practices and student outcomes. Of course, this is not to deny that many interesting studies of differences in organisation and classroom behaviour exist. The case study traditions of qualitative research are frequently illuminating. Nonetheless, for a study of what contributes to teacher effectiveness it is necessary to attempt to explore the relationships between teachers' classroom practice (including teacher-student interactions and relationships) and student outcomes, in a systematic fashion.

The question of teacher effectiveness is still a highly controversial one. The early work on teacher styles by Bennett (1976) was published at the time of public doubt about so called progressive methods in primary schools in the UK following an influential speech by the then Prime Minister, James Callaghan, calling for a back to basics approach – for an illuminating account of this period see Lawton, (1994). The Bennett (1976) research data, which had suggested that traditional or formal approaches were better than progressive ones, were subsequently reanalysed and the original findings challenged and revised (Aitkin, Anderson & Hinde, 1981; Aitkin, Bennett & Hesketh, 1981).

Later studies of primary classroom practice and pupil progress by Galton, Simon and Croll (1980, 1981) pointed to the value of using systematic classroom observation data and provided a more complex picture of the range in teachers' and students' classroom interactions. No studies which used multilevel methodology to study student progress and combined information about both school and classroom processes, however, had been conducted until the advent of the ILEA's Junior School Project which followed children's junior school careers from 1980 to 1984.

The last chapter offered a personal reflection on the Junior School Project research as published in School Matters *(Mortimore et al., 1986, 1988a, 1988b) in relation to two of the most topical concerns of UK education in the early 1990s – school quality and accountability.*

Inevitably, however, it could only give a flavour of detailed findings, particularly in relation to variations in teachers' classroom practice.

When published, the School Matters research design was unique in the wealth of information relating to the age group of children studied, including information about their teachers and schools tracked over four school years, and by its attempt to relate measures of school and class effects on children's progress and development, to information about school and classroom processes.

In particular, the study built on work such as that of Galton, Simon and Croll (1980) to include a substantial commitment to classroom observation using both systematic procedures (Boydell, 1974a, 1975) and more subjective rating scales (Powell & Scrimgeour, 1977). The research results pointed to the importance of using direct observation as well as teacher reports on their practice and organisation obtained through interviews and questionnaires. Inevitably teachers' perceptions of their classroom practice do not always coincide with external observations even in judgements about matters such as the relative extent to which the teacher believes they work with the whole class, groups or individual children. The research points to the value of collecting a range of different kinds of data to explore different aspects of teachers' classroom organisation and practice.

Classrooms are diverse places, varying in their size and the age and composition of their students as well as the aims, characteristics and experiences of teachers. Enormous variety in teaching approaches and student experiences is evident. Diversity in practice exists both between schools and between different classes in the same school (Sammons, 1998). Do these variations in conditions and practices and therefore in students' experiences of education, have an impact? The School Matters' research pointed to particular aspects of primary teachers' practice which suggested that these variations did matter. Studies of the patterns of associations of course cannot provide evidence of causal relationships. Nonetheless, the use of correlation and regression techniques enabled a number of statistically, and we believe educationally, significant interrelationships to be identified, which taken together and interpreted in the light of both researcher and teacher experience, provided a powerful commentary on what appeared to be more successful effective teacher practices. We recognised that 'the relationship between teaching and learning is neither straightforward nor obvious' (Mortimore, 1993, p. 59) and that learning can only be inferred from its results (Black, Hall & Martin, 1990). Given this, we strongly cautioned against the prescriptive use of the results. Nonetheless, four features of classroom practice briefly described in the last chapter, were, we believed, of particular interest to primary teachers.

In 1987 the School Matters team (led by Peter Mortimore and including myself, Louise Stoll, Russell Ecob and David Lewis) wrote a series of three articles for Forum, a journal aimed at practitioners and schools, describing the results of the School Matters research and one of these focused on these four features of classroom practice and their interrelationships. Several of these features came as something of a surprise, and not always a welcome one, to many professionals involved in the training of teachers, and to some concerned with inspection. At the time they appeared to challenge current orthodoxy in primary practice which emphasised certain approaches, particularly topic-based methods of covering the curriculum, and the encouragement of activities with groups of children working on different curriculum areas at the same time (mixed activities). Support for our findings which questioned this orthodoxy has, however, emerged in subsequent evaluations (for example, Alexander, 1992).

Our research also pointed to the complexity in teacher behaviour and the need to avoid simplistic descriptions of teaching style into opposing camps, such as informal or progressive versus formal or traditional, which had been the focus of heated but, in my view, not very fruitful debates about what constituted best classroom practice during the 1980s and which continue to receive wide emphasis today (Sammons, 1998). For example, the current endorsement of 'formal' whole class teaching approaches by some powerful voices fails to recognise the need to focus on the specific aspects of teacher behaviour which appear to be most important and has lead to fears by professionals that they will be disempowered and forced to adopt one narrow approach to teaching.

The Forum article was written with practitioners in mind and provides a more detailed elaboration of the School Matters findings concerning effective classroom practice of particular relevance to the primary sector. I believe it continues to have a useful message in stressing the value of evidence-based *approaches to the evaluation of different aspects of teaching behaviour. It also continues to highlight the complexity of different aspects of primary classroom practice and provides a timely caution against the advocates of simplistic approaches to teaching styles. Nonetheless, it does lend some support to the need for more focused, interactive approaches which have been noted by important recent reviews such as that of the UK's Literacy and Numeracy Task Forces (The Literary Task Force, 1997, DfEE, 1998) more than a decade after its publication.*

In this article, we intend to focus on four of the key factors concerning teachers' classroom practices which our research indicated affect pupils' progress and development during the junior years. The four key factors are 'structured sessions', 'intellectually challenging teaching', 'work-centred environment' and 'maximum communication between teachers and pupils'.

Structured sessions

One aspect of teachers' methods of organising pupils' work which we examined was the extent to which pupils were given responsibility for managing their own work. Our results indicate that most teachers favour giving pupils responsibility for managing individual pieces of work, but tend to organise the pupils' day by planning the sequence of pupils' work. In a minority of classes, however, pupils were given the responsibility for managing their own programme of work for extended periods – such as a whole day.

We found that the latter method was related negatively to pupils' progress in reading and mathematics and to a number of the non-cognitive outcomes (self-concept, attitude to school and attitude to writing). Our findings show that, where teachers provide a framework which gives order and facilitates the progression of work, pupils' learning and development benefits. Thus, sufficient teacher direction which offers a clear structure to the school day is important. Nonetheless, there is evidence that pupil responsibility and independence for managing particular pieces of work within sessions (rather than for longer periods of time) is beneficial.

The extent to which pupils were given responsibility for managing their own programme of work for long periods of time was related negatively to teachers' organisation, and to the level of pupil industry and involvement with their work. However, it was positively related to the amount of teacher time spent talking about routine (non-work) matters, rather than to the amount of teacher time spent communicating about work with pupils. These latter aspects are all related to the 'work-centred environment', which we discuss in more detail below.

From our results, it is evident that in classes where pupils were not given a sufficient structure to their day, the teacher's organisation of work was less effective and pupils were less engaged with their work. The teacher spent more time managing activities and keeping order (routine communication) and less time talking to pupils about their work. In addition, levels of noise and pupil movement were much higher (excessive noise and movement were also related to poorer pupil progress and development).[1]

The 'structured sessions' factor is thus associated with a more 'work-centred environment'. It is also related to other factors. There is evidence connecting it with a 'limited focus within sessions' and with some aspects of 'intellectually challenging teachers'. The link with the 'limited focus within sessions' factor is perhaps not surprising. The data demonstrate that the use, by the teacher, of a high proportion of activities (3 or more) in several different curriculum areas at the same time (mixed-activities), was related negatively to pupils' progress and development. Like 'structured sessions' it was also associated with more teacher time being spent on routine (non-work) communication, and higher levels of pupil noise and movement in the class. Where pupils have responsibility for managing their work over long periods it is, of course, much more likely that there will be activities in several different curriculum areas occurring in the classroom at any one time. Moreover, from our data there is also evidence that teaching sessions were rather less challenging and interesting in classrooms where pupils were given responsibility for managing their work over long periods.

Teachers who organise a classroom within which pupils can work with some – but not too much independence, appear to be better able to ensure that precious classroom time is not wasted, and that pupils' involvement with work remains high. In such classrooms it is less likely that important aspects of teaching and learning will be omitted. In our view this is what effective teachers have always done, and is what most teachers aim to do. The minority of teachers who try to give responsibility to pupils for managing their work over long periods, we would urge caution. In our assessment of pupils' self-perceptions in school, we found that over 40 per cent reported that they had difficulty in concentrating on their work all or most of the time. For such pupils, the encouragement of independence in managing particular pieces of work, rather than several pieces over an extended period of time, appears more likely to prove fruitful.

Intellectually challenging teaching

The content of teachers' communications is a vitally important aspect of intellectually challenging teaching. This use of what have been termed 'higher-order' questions and statements was found to contribute to effectiveness.[2] By the term 'higher order' questions and statements we mean the sorts of talk by the teacher

---

1 It is worth also reporting that we found that teachers could not easily be divided into distinctive groups operating particular 'styles'. The debate on 'styles' (e.g., 'informal/formal') has, in our view, distracted attention from the specific ways in which teachers' behaviours and approaches differ.
2 See Boydell (1974a, 1974b) for details of the scheme we adopted in our observations for the classification of different kinds of pupil-teacher contacts.

which encourages pupils to use their creative imagination and powers of problem-solving. Examples of questions of a high-order type—where there is not one right answer—might be *How many different ways can you think of to measure the length of this room?* or *What do you think would make a good end to this story?* Overall, and perhaps somewhat surprisingly, we found that teachers do not use higher-order questions and statements very often, on average only about two per cent of teachers' talk was observed to be of this kind.

This is much in accord with the findings of Galton and Simon (1980) and Galton, Simon and Croll (1980) in their study of teaching behaviour and pupil progress. Like them, however, we found there was considerable variation between individual teachers. Some make more use of this sort of communication than others (and a few teachers were never observed making use of these sorts of communication).

Our results indicate that greater use of 'highest-order' communications by the teacher has a positive impact. In classes where the teaching situation is challenging and stimulating, and where teachers communicate interest and enthusiasm to the children, greater pupil progress occurs. Our data reveal that there was a positive relationship between use of 'higher-order' communications, and the overall level of interest and challenge in teaching sessions. Analyses showed that 'higher-order' questions and statements were more frequently in evidence in sessions rated as being 'bright and interesting' (in terms of the Scots observation schedule developed by Powell and Scrimgeour, 1977). Furthermore, teachers who used class discussions as a regular teaching strategy also tended to make greater use of 'higher-order' communication.

We found that intellectually challenging teaching was also related to a number of aspects of the other key factors, including the 'work-centred environment'. Thus, pupils' industry and involvement with their work was greater when there was plenty for them to do and when the teacher's approach was interesting and challenging. This approach involved the more frequent use of 'higher-order' questions and statements. Teachers who devoted more of their time to discussing pupils' work also tended to offer more stimulating work for their classes.

Another important and interesting link can be seen in the fact that the percentage of teacher time spent on contacts with the class, as a whole, was related positively to teachers' use of 'higher-order' questions and statements and, to a lesser extent, with the incidence of questions as a whole. These findings support Galton and Simon (1980), who noted a link between the proportion of class contacts and the incidence of 'higher order' communications. The authors suggested that

> 'it may be the nature of the interactions which are shared by the whole class that differentiate between the successful and unsuccessful teachers' (p. 80).

Our data show that this is likely to be the case. When we discuss our factor 'maximum communication between teachers and pupils' the implications of using an appropriate balance of class and individual communications will be dealt with in more detail. However, it is worth noting here that, in communications with the whole class, teachers were more likely to raise 'higher-order' questions and statements than in comments with individuals. At first sight this may seem surprising: some teachers may see the greater opportunity to challenge pupils in individual work, rather than in group or class sessions.[3] Our data show that, in its extreme form, this view is mistaken. Intellectually challenging communications appear to be more likely to arise from group or class sessions than from interactions with an individual. These tend to be very brief (commonly under one minute's duration) and are often pre-occupied with classroom management issues. This is because most individual interactions observed were isolated and there was seldom time for a teacher to have an extended conversation with an individual pupil. Challenging comment can seldom be made without an appropriate building-up or focusing of ideas which we found was more common when teachers dealt with the class as a whole.

The implication for some class teachers of this finding is that they may need to reconsider aspects of their classroom practice and seek to use the opportunity of group and class sessions to promote, systematically, the sorts of 'higher-order' communications that, we have shown, challenge pupils.

Work-centred environment

A number of aspects of teacher behaviour and classroom management were found to be related negatively to pupils' progress and development. These included high levels of pupil noise and movement in the class especially noise and movement unrelated to work activityand a higher percentage of the teacher's time being spent on routine (non-work) communications. These three aspects were themselves interrelated. As might be expected, the incidence of high levels of pupil noise and movement were themselves associated. It was also found that, where pupil noise and movement were high, teachers spent more time managing the classroom using routine (non-work) communications, and less time actually talking about pupils' work.

In contrast, where teachers spent more time talking about the content of pupils' work, and giving them feedback about it, progress and development benefitted. Furthermore, where teachers spent more time talking about work matters and organised work so that there was always plenty for pupils to do, pupils' industry and involvement in work was observed to be greater and they appeared to be interested in what they were doing and eager to start new work. In such work-centred classrooms, levels of pupil noise and movement were lower.

The identification of consistencies in the relationships amongst these aspects of teacher behaviour and classroom practices reveal one of the important mechanisms of effectiveness, the creation of a work-centred environment. It should

---

3 Although in the vast majority of classes pupils almost always sat in groups, very little group work was observed to take place.

be noted, however, that our finding of the importance of the creation of a 'work-centred environment' does not imply that classrooms should be silent, with pupils permanently seated.

None of the classrooms we observed was silent or had immobile pupils. Rather, it was the *excessively high* levels of noise and movement which were associated with less emphasis on work both by pupils and by their teachers. In such classrooms, the teacher frequently was over-occupied with the management of activity, rather than communicating about pupils' work. Again, as noted in connection with 'structured sessions', pupils' responses to questions concerned with their self-perception of their own learning behaviour, showed that many experienced difficulties in concentrating on their work all or most of the time. A very noisy classroom with much pupil movement, must provide more distractions for pupils than one which is clearly focused on work activity.

Our data also reveal that levels of pupil noise and movement were higher and teachers spent more time managing work, where sessions were organised on a mixed-activity basis (those where a class of pupils were involved with work in three or more different curriculum areas at the same time). Moreover the amount of time teachers spent communicating with the pupils about work (including giving pupils feedback) was lower in classes where more sessions were devoted to mixed-activities than where work occurred mainly in one (or, on occasions, two, broad curriculum areas). Thus the creation of a 'work-centred environment' was positively associated with a factor we termed the 'limited focus with sessions'.

Our findings on the value of a work-centred environment are unlikely to surprise many teachers. It might appear self-evident that children (like adults) find it harder to work when surrounded by distractions, where the noise level is high, and where movement is excessive and disruptive. This is not to say that the most effective classes were silent or that pupils were kept seated on their chairs. Where talk was about work it was encouraged; where the learning task required movement, it was permitted; but the general atmosphere was, first and foremost, work-centred. Moreover, teachers themselves spent more time talking about the content of pupils' work and giving feedback. Effective classrooms were, on the whole, places where children were not only working, but enjoying their work.

Maximum communication between teachers and pupils

Our research highlighted the benefits of a high level of communication between teacher and pupils. Thus, overall the teachers who spent *less of their classroom time* communicating with pupils (and more time on other matters such as silently monitoring pupils, 'housekeeping', marking, in the stock room etc.) were less effective.

In addition, we found evidence that the average amount of her or his time that the teacher spent communicating with the class, rather than with groups (which in fact occurred very rarely, as noted earlier), or with individuals was associated with beneficial impacts on pupils' progress in reading, mathematics, writing and on behaviour, self-concept, attendance, and attitude to school.

Two points, however, need to be stressed. The first is that interacting with the class as a whole did not necessarily imply a 'whole class teaching' approach was adopted. In fact, our measures of the extent to which 'whole class teaching' was used were not related significantly to positive effects on pupils' progress and development. Thus, it was the proportion of the contacts directed to the class (e.g., introducing work to the class, telling and discussing a story, and other class-based discussions or feedback) which was important.

The second point to stress is that, from the analysis of the observational data, it was found that the majority of teachers' contacts were with individual pupils (over 60 per cent on average). This was found to be true even for teachers who claimed in interviews about their approach that they spent more of their time talking to *the class* rather than to individuals. However, some teachers frequently introduced topics, discussed work or raised points with the whole class. This was the case where pupils were working individually, or in groups on different tasks, or on the same activity. A sufficient balance of class to individual interactions appears to be more effective in promoting pupil progress. Effects tend to be positive where teachers spent at least a quarter or more of their contacts with pupils in communicating with the class as a whole.

It is interesting to note that our finding on the efficacy of class communications is in broad agreement with that reported earlier in the Oracle research (see Galton & Simon, 1980).

The links between the teacher's use of class communications and pupils' progress and development reflect, we think, the greater amount of attention received by members of the class through this mode of teaching. This is because when all, or nearly all, of a teacher's time is spent communicating with individual pupils, each child can receive only a relatively infrequent number of contacts with the teacher in any teaching day; even though the teacher is extremely busy. To illustrate this for our observational data, we found that, on average, pupils received only 11 individual contacts with their class teacher in any teaching day. Given that those with poor behaviour or particular learning difficulties tended to receive a higher number of contacts, other pupils generally received fewer than this average number of contacts. Moreover, it should be remembered that in larger classes than those found in the ILEA's project schools (the average size in our sample was 25 pupils) the number of individual contacts for pupils will tend to be even lower. The skill of the teacher, therefore, is to achieve a balance of class, group and individual communications where appropriate, which will maximise the *total* amount of communication with pupils in the class.

The use of a 'limited focus within sessions' (factor 8 of the 12 key factors identified in the Junior School Project) was related positively to the percentage of teacher time spent communicating with the class. This is not, perhaps, surprising. It is easier to discuss an issue with the whole class when all pupils are working within one broad curriculum area, than when pupils are engaged in work in three or more different curriculum areas. Our data also indicate, as reported earlier, that the use of 'higher-order' communication was associated with the amount of teacher time spent communicating with the whole class. Furthermore, where children work in a single curriculum area within sessions, it was more common for teachers to raise an intellectually challenging point with all pupils.

Maximum communication with pupils was also linked with the 'work-centred environment'. We found that the amount of time teachers spent on interactions with the class as a whole was correlated positively with the amount of time she or he spent talking about work, and with pupils' industry and movement.

The main implication of our finding on the importance of achieving maximum communication between teachers and pupils is that flexibility in approach is necessary. An over-reliance upon individual communication, paradoxically, with reduce inevitably the overall amount of teacher-pupil contact. Therefore, teachers, and those who train them, need to be aware of the benefits of using other forms of communication wherever appropriate. The majority of teachers' talk will still be with individual pupils but, by using class (not group) communications, especially for discussions of work, and for raising intellectually challenging points, the overall level and the quality of communications can be improved.

Conclusions

We think that some of the findings have a 'common sense' feel. For example, the value of 'intellectually challenging teaching' and of a 'work-centred environment' for pupils' progress and development will surprise no one. Other findings, however, may challenge some teachers', and their trainers' preconceptions for example, those on 'structured sessions' or 'maximum communication with pupils'. In particular, some readers may be surprised at the perhaps unexpected ways in which these different factors are interrelated and are linked with positive effects on pupils' progress and development.

8

Key Characteristics of Effective Schools

Introduction

The last chapter examined the particular findings of an individual research study focusing on one age group of children and conducted in a specific context – that of inner London primary schools during the early 1980s. An important question for policy makers as well as practitioners concerns the generalisability of school effectiveness research findings and the extent to which they can be used by those engaged in school improvement (see Reynolds et al., 1994a; Scheerens & Bosker, 1997). As noted earlier in this volume, far greater research attention has been devoted to the measurement and identification of differences in school effects on student outcomes than to the study of their organisational and process characteristics and the way they may influence such outcomes. Nonetheless, a fairly large body of evidence has accumulated over the last twenty years or so which allows these issues to be addressed. This chapter focuses, therefore, on what we know about the characteristics of effective schools (those which, given their intake, appear to add value to students' educational outcomes).

Interest in the possible contribution school effectiveness research can make to school improvement has grown markedly from 1993 in the UK, and has emerged as a particularly strong feature of Government education policy since the 1997 election of a Labour administration. As I discussed in the Introduction to this volume, there may be great potential for influence, but also dangers in the high profile school effectiveness now presents. The report chosen as the focus for the present chapter provides graphic example of this tension between the requirements of an academic piece of work and the uses to which research can be put once published.

In autumn 1994 a review of school effectiveness research was requested to inform the first revision of OFSTED's National Framework for the Inspection of Schools, which took place on a four yearly cycle. Dr Peter Matthews, then Head of Quality Assurance at OFSTED, wished to ensure that the revised Framework was compatible with the available evidence base. Given the generally unfavourable comments about the value of educational research common at the time in Government circles, this request for a review provided a very welcome opportunity for researchers to contribute to and highlight the positive contribution educational research could make to the development of policy. OFSTED had, for the first time in the history of UK national inspection, published the original Framework for the Inspection of Schools, an example of openness which was commended by several leading authorities in the school effectiveness research field (Gray & Wilcox, 1995).

The original research specification sought a review of 'key determinants' of school effectiveness. The specification perhaps took the lead from earlier US reviews commissioned to summarise 'What Works' in terms of school and teacher practices (US Department of Education, 1987). Nonetheless, the commissioners of the research readily accepted the view of colleagues (Peter Mortimore and Josh Hillman) and myself that the correlational nature of school effectiveness research evidence means that a deterministic approach was inappropriate. Our review, therefore, chose to focus on the key characteristics of effective schools, *as derived from the available evidence and did not attempt to draw causal inferences about how particular factors influenced students' outcomes. We were keen to produce a fairly accessible document which did not require a detailed knowledge of the academic literature from potential readers but which assessed and classified the main findings from a range of studies conducted in a variety of countries. The attempt to reach a wide audience – particularly of practitioners – was later strongly criticised by some academics, but we felt it was vital that the research summary seek to inform teachers and managers in schools, those most involved in the organisation and day-to-day process of teaching and learning in the classroom as well as the inspectorate. Our review attempted to demonstrate both the extent and also the limitations of the current research evidence and to caution against over-simplistic interpretations. The readers can judge for themselves the extent to which they feel we were successful in this endeavour.*

The Key Characteristics review received considerable media attention when published and as a result aroused much interest amongst practitioners. In some LEAs, for example, it featured in school improvement conferences and was used as a starting point for initiatives designed to encourage school-based review. The DfEE and OFSTED also incorporated a summary of the Key Characteristics findings in a document sent to all schools' Governing Bodies (DfEE, 1995).

Again, this was intended to provide guidance about important aspects of school which Governors could address in conducting evaluations of their own institution's performance. Some schools found it provided a focus for staff discussion in the process of preparing for an OFSTED inspection.

Perhaps an inevitable consequence of this high profile was a negative reaction from some educational researchers from other traditions, particularly that of action research (see Elliott, 1996; Hamilton, 1996; White, 1997). Proponents of the latter tradition argued that the educational process is all important, and denigrated the over-emphasis (as they perceived it) of school effectiveness research (SER) on the measurement of students' achievements as the basis for assessing effectiveness.

A flavour of the impassioned critiques of action researchers is given in the article published in Forum *by David Hamilton which is reproduced as a postscript to the Key Characteristics review, along with our response to his critique. Hamilton's (1996) commentary highlights a number of important theoretical and methodological issues which are of continued relevance today. Those interested in a more detailed discussion may wish to read Elliott's (1996) article on 'Alternative visions of schooling' which claims that the SER paradigm operates within*

> 'a mechanistic methodology, an instrumentalist view of educational processes and the belief that educational outcomes can and should be described independently of such processes' (p. 200).

Elliott also alleges that the SER model of schooling is underpinned by values which are 'anti-educational' and that its findings tell us nothing new about schools. He concludes that 'they are best viewed as ideological legitimisations of a socially coercive view of schooling' (Elliott, 1996, p. 200). A response to this, in our view, partisan evaluation (Sammons & Reynolds, 1997), disputed these allegations through reference to the historical context of SER, a discussion of the criteria of effectiveness and of 'good' schools and of the importance of linking outcomes and processes. The Elliott review, like that of Hamilton, has strong echoes of the often sterile debate about the relative merits of quantitative versus qualitative research styles. We believed it necessary to reiterate that

> 'SER seeks to go beyond reliance on cherished beliefs about "What Works" through the systematic and empirical (in the widest sense of the word) studies of the relationships between school and classroom processes and pupils' educational outcomes, the "touchstone" criteria for SER' (Sammons & Reynolds, 1997, pp. 131–132).

Thus our argument is that research evidence is necessary to unpack the detail of what on the surface may appear either 'obvious' from common sense or self-evident by deduction. Further discussion of these issues is provided in reflections by Reynolds (1997).

A major point of difference concerns the value placed by SER on measuring students' educational outcomes, particularly in the basic skills of literacy and numeracy or in terms of examination results. While we took care to stress in the Key Characteristics review that it is preferable to examine a broad range of outcomes (social and affective as well as cognitive) we remain convinced that the social empowerment argument is a vital one, particularly in relation to literacy, because the life chances of students from socio-economically disadvantaged backgrounds in particular will be enhanced by effective schools (those which foster progress in these crucial areas). Moreover, we believe that democracy is strengthened where academic achievement is distributed as widely as possible and students' post-school employment and further and higher education opportunities are maximised (Mortimore & Sammons, 1997). These arguments, we think, remain of continuing relevance given the concern evident in the UK but also in many other European countries about the problems of social exclusion in which education is seen to offer one important means of implementing policies intended to combat social disadvantage.

The Key Characteristics Review which forms the basis of this chapter provides an indication of the nature of the research evidence basis about the features of schools which are more successful in promoting student progress and which thus, it is argued, can help to promote equity and their subsequent life chances in the adult world.

In 1994 the Office for Standards in Education (OFSTED) commissioned the International School Effectiveness and Improvement Centre (ISEIC) to conduct a review of school effectiveness research summarising current knowledge about the factors identified in the literature as important in gaining a better understanding of effectiveness. The aim was to provide 'An analysis of the key determinants of school effectiveness in secondary and primary schools'.

Scheerens (1992) has identified five areas of research relevant to school effectiveness:

1   Research into equality of opportunity and the significance of the school in this (e.g., Coleman et al., 1966; Jencks et al., 1972).
2   Economic studies of education production functions (e.g., Hanushek, 1979, 1986).
3   The evaluation of compensatory programmes (e.g., Stebbins et al., 1977; and also reviews by Purkey & Smith, 1983 and Van der Grift, 1987).
4   Studies of effective schools and the evaluation of school improvement programmes (e.g., for studies of effective schools see: Brookover et al., 1979; Rutter et al., 1979; Mortimore et al., 1988a. For the evaluation of improvement programmes see the review by Miles, Farrar & Neufeld, 1983).
5   Studies of the effectiveness of teachers and teaching methods (see reviews by Walberg, 1984; Stallings, 1985; Doyle, 1985; Brophy & Good, 1986).

Although our primary focus is on the school effectiveness tradition, in conducting our review we have examined research in the related field of teacher effectiveness. Where appropriate, however, we also refer to work in the other three areas identified by Scheerens. It is important to take account of the relationships between school factors (such as policies, leadership and culture) and classroom processes, because in some institutions the former may provide a more supportive environment for teaching and learning than others (Purkey & Smith, 1983; Mortimore et al., 1988a; Fullan & Hargreaves, 1991; Scheerens, 1992; Reynolds et al., 1994a; Stoll & Fink, 1994). Where appropriate we refer to the results of previous reviews of literature in these fields (e.g., Purkey & Smith, 1983; Ralph & Fennessey, 1983; Rutter, 1983; Doyle, 1986; Walberg, 1986; Fraser et al., 1987; Rosenshine, 1987; Reid, Holly & Hopkins, 1987; Levine & Lezotte, 1990; North West Regional Educational Laboratory, 1990; Reynolds & Cuttance, 1992; Scheerens, 1992; Reynolds et al., 1994a; Tabberer, 1994). We draw attention to some of the limitations of existing school effectiveness research, particularly the weak theoretical basis (Scheerens, 1992; Reynolds & Cuttance, 1992; Creemers, 1994; Hopkins, 1994), and the fact that the number of empirical studies which focus directly on the characteristics of effective schools is exceeded by the number of reviews of the area.

We note the need for caution in interpreting findings concerning 'key determinants' of effectiveness based on evidence much of which, in the early research, is derived from studies of the characteristics of small numbers of outlier schools (selected as either highly effective or highly ineffective). The dangers of interpreting correlations as evidence of causal mechanisms are also highlighted. For example, reciprocal relationships may well be important, as may intermediate causal relationships. Thus, high expectations may enhance student achievement, which in turn promotes high expectations for succeeding age groups. Improved achievement may benefit behavioural outcomes which in turn foster later achievement. Conversely, lower expectations may become self-fulfilling, poor attendance and poor behaviour may lead to later academic under-achievement which exacerbates behavioural and attendance problems and so on. Despite these caveats, however, we conclude that such a review has value in synthesising current school effectiveness findings in an accessible format and providing an analysis of key factors likely to be of relevance to practitioners and policy-makers concerned with school improvement and enhancing quality in education.

Background

The major impetus for development of North American and British school effectiveness research is generally recognised to have been a reaction to the deterministic interpretation of findings by the US researchers Coleman et al. (1966) and Jencks et al. (1972) and, in particular, their pessimistic view of the potential influence of schools, teachers and education on students' achievement (Rutter et al., 1979; Mortimore et al., 1988a; Reynolds & Creemers, 1990; Firestone, 1991; Mortimore, 1993). These studies indicated that, although background factors are important, schools can have a significant impact also. More recently Creemers, Reynolds and Swint (1994) have pointed to the existence of different interpretations reflecting the intellectual ancestries of the school effectiveness research traditions in other national contexts. For example, in the Netherlands interest in school effectiveness grew out of research traditions concerning matters such as teaching, instruction, curriculum and school organisation, while in Australia the strong field of educational administration provided a stimulus.

The last 15 years has witnessed a rapid growth in the two related (albeit at times tenuously) areas of research and practice covering the fields of school effectiveness and improvement. In 1990, in a mission statement launching the first issue of a new journal devoted to these topics, Reynolds and Creemers (1990) argued that interest in the topics of school effectiveness and improvement had been

> 'fuelled by the central place that educational quality (and sometimes equity) issues have assumed in the policy concerns of most developed and many developing societies' (p. 1).

This review focuses primarily upon the results of school effectiveness research, but it is recognised that many school effectiveness researchers are profoundly concerned about the implications of their work for policy-makers, schools and their students. An interest in raising standards in the widest sense, improving the quality of education and opportunities available to students in all schools, and the implications of research results for practitioners is evident. It is, however, important to recognise that school effectiveness research results do not provide a blueprint or recipe for the creation of more effective schools (Reid, Hopkins & Holly, 1987; Sammons, 1987b; Mortimore et al., 1988a; Creemers, 1994; Sammons, 1994). School improvement efforts require a particular focus on the processes of change and understanding of the history and context of specific institutions (see Louis & Miles, 1991; Fullan, 1991; Ainscow & West, 1994; Stoll & Fink, 1994). Whilst it is recognised that,

> 'in many ways our knowledge of what makes a "good" school greatly exceeds our knowledge of how to apply that knowledge in programmes of school improvement to make schools "good"' (Reynolds & Creemers, 1990, p. 2).

there is growing acceptance that such research provides a valuable background and useful insights for those concerned with improvement (Reid, Hopkins & Holly, 1987; Mortimore, 1991a, b; Sammons, 1987b, 1994; Stoll & Fink, 1994). The findings should not, however, be applied mechanically and without reference to a school's particular context. Rather, they can be seen as a helpful starting point for school self-evaluation and review.

Aims and goals of effectiveness research

In reviewing early school effectiveness studies in the US context, Firestone (1991) highlighted the wide ranging impact of studies by Edmonds (1979) and

Goodlad et al. (1979). He noted that the effective schools movement was committed to the belief that children of the urban poor could succeed in school and that the school could help them succeed. Firestone (1991) recognised that

> 'Effectiveness is not a neutral term. Defining the effectiveness of a particular school always requires choices among competing values and that criteria of effectiveness will be the subject of political debate' (p. 2).

Early school effectiveness research incorporated explicit aims or goals concerned with equity and excellence. Three important features were:

- clientele (poor/ethnic minority children)
- subject matter (basic skills in reading and maths)
- equity (children of the urban poor should achieve at the same level as those of the middle class).

This early research therefore, had a limited and specific focus. As Ralph and Fennessey (1983) note, such research was often dominated by the perspectives of school improvers and providers of external support to schools. More recent research, especially in the UK context, has moved away from an explicit equity definition towards a focus on the achievements of all students and a concern with the concept of progress over time rather than cross-sectional 'snapshots' of achievement at a given point in time. This broadens the clientele to include all students, not just the disadvantaged, and a wider range of outcomes (academic and social). As in the US, however, the majority of UK studies have also been conducted in inner city schools. More recent research also recognises the crucial importance of school intake, and attempts to control, usually statistically, for intake differences between schools before any comparisons of effectiveness are made (Mortimore, 1991b; Mortimore, Sammons & Thomas, 1995).

Definitions of effectiveness
Although Reid, Hopkins and Holly (1987) concluded that

> 'while all reviews assume that effective schools can be differentiated from ineffective ones there is no consensus yet on just what constitutes an effective school' (p. 22),

there is now a much greater degree of agreement amongst school researchers concerning appropriate methodology for such studies, about the need to focus explicitly on student outcomes and, in particular, on the concept of the 'value added' by the school (McPherson, 1992). For example, Mortimore (1991a) has defined an effective school as one in which students progress further than might be expected from consideration of its intake. An effective school thus adds extra value to its students' outcomes in comparison with other schools serving similar intakes. By contrast, in an ineffective school students make less progress than expected given their characteristics at intake. Methodological developments have drawn attention to the need to consider issues of consistency and stability in effectiveness and the importance of caution in interpreting any estimates of individual school's effects. In particular, the need to take account of the confidence limits associated with such estimates is highlighted (Goldstein et al., 1993; Creemers, 1994; Sammons et al., 1994b; Mortimore, Sammons & Thomas, 1995).

Definitions of school effectiveness are thus dependent upon a variety of factors as Sammons (1994) has argued. These include:

- sample of schools examined (many studies have focused on inner city schools and this context may affect the general applicability of results);
- choice of outcome measures (studies which focus on only one or two outcomes may give only a partial picture of effectiveness, both in terms of effects and the correlates of effectiveness) a broad range reflecting the aims of schooling being desirable (for example the Mortimore et al., 1988a study examined several cognitive measures and a range of social/affective outcomes);
- adequate control for differences between schools in intakes to ensure that 'like is compared with like' (ideally, information about individual pupils, including baseline measures of prior attainment, personal, socio-economic and family characteristics are required, see Gray, Jesson & Sime, 1990; Willms, 1992; Goldstein et al., 1993; Thomas & Mortimore, 1994; Sammons et al., 1994b);
- methodology (value added approaches focusing on progress over time and adopting appropriate statistical techniques such as multilevel modelling to obtain efficient estimates of schools' effects and their attached confidence limits are needed, see Goldstein, 1987; Willms & Raudenbush, 1989; Gray et al., 1993; Goldstein et al., 1993); and
- timescale: longitudinal approaches following one or more age cohorts over a period of time rather than cross sectional 'snapshots' are necessary for the study of schools' effects on their students, to allow issues of stability and consistency in schools' effects from year to year to be addressed (see Gray et al., 1993; Sammons, Mortimore & Thomas, 1993a).

Evidence of effectiveness
The central focus of school effectiveness research concerns the idea that

> 'schools matter, that schools do have major effects upon children's development and that, to put it simply, schools do make a difference' (Reynolds & Creemers, 1990, p. 1).

Although Preece (1989) looked at research pitfalls in school effectiveness studies and made a number of criticisms of selected studies, Tabberer (1994) concludes that 'Despite Preece's critique there is little argument now that schools can and do have an effect'.

During the last two decades a considerable body of research evidence has accumulated which shows that, although the ability and family backgrounds of students are major determinants of achievement levels, schools in similar social circumstances can achieve very different levels of educational progress (e.g., Reynolds, 1976, 1982; Edmonds, 1979; Brookover et al., 1979; Madaus et al., 1979; Rutter et al., 1979; Gray, 1981; Mortimore et al., 1988b; Tizard et al., 1988; Smith & Tomlinson, 1989; Willms & Raudenbush, 1989; Nuttall et al., 1989; Gray, Jesson & Sime, 1990; Daly, 1991; FitzGibbon, 1991; Jesson & Gray, 1991; Stringfield et al., 1992; Goldstein et al., 1993; Sammons et al., 1994a, 1994b, 1994c; Thomas & Mortimore, 1994; Thomas, Sammons & Mortimore, 1994). Such studies, conducted in a variety of different contexts, on different age groups, and in different countries confirm the existence of both statistically and educationally significant differences between schools in students' achievements.

Most school effectiveness studies have focused on academic achievement in terms of basic skills in reading and mathematics, or examination results (Goodlad, 1984). However, a few have also provided evidence of important differences in social/affective outcomes such as attendance, attitudes, and behaviour (Reynolds, 1976; Rutter et al., 1979; Mortimore et al., 1988a).

There is some indication from recent British research (Sammons et al., 1994a; Goldstein & Sammons, 1995) following up the *School Matters* cohort that primary school effects may be larger than those identified in the secondary sector, and that primary schools can have a significant long term impact on later attainment at GCSE (in other words evidence of a continuing primary school effect). In this connection, Teddlie and Virgilio's (1988) research in the USA, which indicates that the variance in teacher behaviour at the elementary grade levels is greater than that at the secondary level, may be relevant.

Measuring effectiveness
Methodological advances, particularly the development of multilevel techniques (e.g., Goldstein, 1987) have led to improvements in the estimation of school effects (Scheerens, 1992; Creemers, 1994). These have enabled researchers to take better account of differences between schools in the characteristics of their pupil intakes and facilitated exploration of issues such as consistency and stability in schools' effects upon different kinds of outcome and over time (see reviews by Gray et al., 1993; Sammons, Mortimore & Thomas, 1993a; Reynolds et al., 1994; Thomas & Mortimore, 1994; Mortimore, Sammons & Thomas, 1995). The need to examine subject differences, whether at A-level (FitzGibbon, 1991, 1992; Tymms, 1992) or at GCSE (Sammons et al., 1994c), as well as overall levels of attainment in terms of total A-level or GCSE points scores, is becoming an important focus of recent studies. These highlight the importance of multilevel analyses which examine departmental as well as school effects (see also Dutch work by Luyten, 1994, and Witziers, 1994).

In addition, multilevel techniques also allow investigation of the concept of differential effectiveness, whether some schools are more or less effective for particular student groups (boys or girls, low or high ability students, those from specific ethnic groups).

Issues such as stability and consistency in effects over time and across multiple outcomes, departmental differences and differential effectiveness for particular student groups clearly have important implications for interpreting the effectiveness of individual schools (Nuttall et al., 1989; Sammons, Mortimore & Thomas, 1993b). Thus Tabberer's (1994) discussion of the possibilities of differential effectiveness argued that if it exists to a notable extent, then single feature measures of school effectiveness such as are considered for league tables are brought further into question.

The importance of taking note of the confidence limits attached to estimates (residuals) which give a measure of the relative value added to or subtracted from their students' achievements by individual schools, also has implications for the use of league tables. It is not appropriate to produce detailed rankings of value added estimates because the confidence limits overlap (Goldstein et al., 1993; Sammons, Mortimore & Thomas, 1993b; Sammons et al. 1994b, 1994c; Thomas & Mortimore, 1994). Rather, the methodology allows the identification of schools where results are significantly different from those predicted on the basis of intake over one or more years.

Size and importance of school effects

The increasing sophistication of school effectiveness research has provided strong evidence that individual student background characteristics account for a much larger proportion of the total variance in students' academic outcomes than does the particular school attended (Coleman et al., 1966; Jencks et al., 1972). This is especially true of the impact of prior attainment. However, gender, socio-economic, ethnicity and language characteristics (which, of course are also strongly correlated with prior attainment, see Sammons, Nuttall & Cuttance, 1993) also have a small but continuing influence. Creemers (1994) states that

> 'About 12 per cent to 18 per cent of the variance in student outcomes can be explained by school and classroom factors when we take account the background of the students' (p. 13).

Other authors have produced slightly more modest estimates (between 8 per cent and 10 per cent, Daly, 1991). Expressed as percentages, school and classroom effects do not appear exceptionally large, but in terms of differences between schools in students' outcomes they can be highly significant both educationally and statistically. For example, Thomas & Mortimore (1994) report differences between schools' value added scores of between 7 Grade E results and 7 Grade C results (over 14 points) at GCSE.

Whilst there are strong arguments against producing detailed rankings or league tables of schools results even using value added techniques (Goldstein et al., 1993), the size of the differences between schools identified as statistically significantly more or less effective is not trivial and can be striking (Mortimore et al., 1988b; Gray, Jesson & Sime, 1990; Sammons, Mortimore & Thomas, 1993b; Sammons et al., 1994b). Furthermore, Mortimore et al., (1988a, b) have shown that in terms of pupil progress (the value added) school effects are much more important than background factors such as age, gender, and social class (being roughly four times more important for reading progress, and ten times for mathematics progress). In terms of equity differences, this study also showed that, although no school removed social class differences in attainment, the absolute achievement in basic skills of working class pupils in the most effective schools was higher than those of middle class pupils in the least effective schools after three years of junior education. Again, such findings point to the educational significance of differences between schools in their effectiveness in adding value to student outcomes, and highlight the importance of using longitudinal rather than cross-sectional approaches.

There is also some evidence from American, British and Dutch studies that schools' effects may vary for different kinds of outcomes, being larger for subjects such as maths or science primarily taught at school, than for reading or English which are more susceptible to home influences (Scheerens, 1992). Fuller and Clarke's (1994) recent review of school effects in developing countries reaches similar conclusions.

Unfortunately, less attention has been paid to social than to the academic affective outcomes of education. Further research on these is needed focusing on questions of consistency, stability and differential effectiveness (Sammons, Mortimore & Thomas, 1993a).

Context and transferability

There is increasing recognition that, although much can be learned from international and comparative studies of school and teacher effectiveness conducted in different countries, the results of such studies are unlikely to be directly transferable to other contexts (see the discussion by Wimpleberg et al., 1989). For example, early results from the on-going International School Effectiveness Research Programme (ISERP) investigating primary mathematics achievement, provide indications of differences between five countries in the impact of pupil background factors and the effects of certain aspects of teacher behaviour (Creemers, Reynolds & Swint, 1994). Although, the sample size is severely limited, this research also suggests that the proportion of variance in achievement attributable to schools and classes may vary in different countries.

Creemers (1994) reports findings which point to the contingent nature of school effectiveness research, and the importance of distinctions such as primary/secondary, and high versus low socio-economic status (SES) of student intakes. Riddell, Brown and Duffield (1994) likewise draw attention to factors such as policy context (national and local) and SES context in case studies of Scottish secondary schools. Reynolds's (1994) international review of school effectiveness research has also highlighted differences in traditions and findings, and the importance of awareness of the contextual dimension of national educational context, which is often subject to rapid change. Fuller and Clarke (1994) likewise draw attention to the importance of context in attempts to analyse school effects in developing countries.

Given the likely importance of contextual factors, particularly national context, the present review has given a particular emphasis to the results of British school effectiveness research because this is likely to be of greatest relevance to schools in the UK. Other research has also been examined and summarised and, where appropriate, attention is drawn to any differences in the emphasis given to specific findings.

It now widely recognised that there is no simple combination of factors which can produce an effective school (Willms, 1992; Reynolds & Cuttance, 1992). Indeed, there is very little research especially in Britain, which is explicit about 'turning round' so-called 'ineffective' schools' as Gray and Wilcox (1994) note. These authors go on to argue that

> 'in the search for the correlates of effectiveness, the correlates of ineffectiveness have been assumed to be the same. It is by no means clear, however, that they are. How an "ineffective" school improves may well differ from the ways in which more effective schools maintain their effectiveness' (p. 2).

Sammons et al., (1994c) have drawn attention to the need for further case studies of ineffective as well as of more effective schools to enhance our understanding of the processes of effectiveness. Recipes for success and 'quick fixes' are not supported by the research base. In contrast to the ambitiously entitled United States department review it is not intended to present deterministic conclusions about *'What Works'* in education. In many ways every school is unique because each has its own characteristics shaped by such factors as location, pupil intake, size, resources and, crucially, the quality of its staff (Reid, Hopkins & Holly, 1987). To this list we can add its particular history, as well as Governing Body, LEA and national influences. As Chubb (1988) argues, school performance is unlikely to be significantly improved by any set of measures that does not recognise that schools are institutions, complex organisations composed of interdependent parts, governed by well established rules and norms of behaviour, and adapted for stability.

Nonetheless, given these reservations, a number of reviewers, ourselves included, have identified certain common features concerning the processes and characteristics of more effective schools (e.g., Purkey & Smith, 1983; Reid, Hopkins & Holly, 1987; United States Department of Education, 1987; Gray, 1990; NREL, 1990; Firestone, 1991; Mortimore, 1991a, b, 1993). As Firestone (1991) observed,

> 'There is a core of consistency to be found across a variety of studies conducted here and abroad with a wide range of different methodological strengths and weaknesses. Moreover, there is considerable support for the key findings in related research on organizational behaviour in a variety of work settings and countries.' (p. 9)

Key characteristics of effective schools

In this section we provide a description of some of the key factors (or correlates) of effectiveness identified by our review. These factors should not be regarded as independent of each other, and we draw attention to various links between them which may help to provide a better understanding of possible mechanisms of effectiveness. Whilst our list is not intended to be exhaustive, it provides a summary of relevant research evidence which we hope will provide a useful background for those concerned with promoting school effectiveness and improvement, and the processes of school self-evaluation and review.

Professional leadership

Almost every single study of school effectiveness has shown both primary and secondary leadership to be a key factor. Gray (1990) has argued that

> 'the importance of the headteacher's leadership is one of the clearest of the messages from school effectiveness research' (p. 214).

Table 8.1 Eleven Factors for Effective Schools.

1	Professional leadership	Firm and purposeful
		A participative approach
		The leading professional
2	Shared vision and goals	Unity of purpose
		Consistency of practice
		Collegiality and collaboration
3	A learning environment	An orderly atmosphere
		An attractive working environment
4	Concentration on teaching and learning	Maximisation of learning time
		Academic emphasis
		Focus on achievement
5	Purposeful teaching	Efficient organisation
		Clarity of purpose
		Structured lessons
		Adaptive practice
6	High expectations	High expectations all round
		Communicating expectations
		Providing intellectual challenge
7	Positive reinforcement	Clear and fair discipline
		Feedback
8	Monitoring progress	Monitoring pupil performance
		Evaluating school performance
9	Pupil rights and responsibilities	Raising pupil self-esteem
		Positions of responsibility
		Control of work
10	Home-school partnership	Parental involvement in their children's learning
11	A learning organisation	School-based staff development

He draws attention to the fact that no evidence of effective schools with weak leadership has emerged in reviews of effectiveness research. Reviews by Purkey and Smith (1983) and the United States Department of Education (1987) conclude that leadership is necessary to initiate and maintain school improvement.

However, the importance of the headteacher's leadership role (rather than that of other staff members such as heads of department) may be sensitive to context, particularly patterns of school organization (see Hallinger & Leithwood, 1994). Thus the headteacher's leadership is a marked feature of British (e.g., Rutter et al., 1979; Mortimore et al., 1988a; Caul, 1994; Sammons et al., 1994c) and American research (e.g., Edmonds, 1979; Brookover et al., 1979; Stringfield & Teddlie, 1987) but specific aspects (assertive principal leadership and quality monitoring) have not been found important in the Netherlands (Scheerens, 1992). Hallinger and Leithwood (1994) have argued for further comparative research in this domain.

Leadership is not simply about the quality of individual leaders although this is, of course, important. It is also about the role that leaders play, their style of management, their relationship to the vision, values and goals of the school, and their approach to change.

Looking at the research literature as a whole, it would appear that different styles of leadership can be associated with effective schools, and a very wide range of aspects of the role of leaders in schools have been highlighted. As Bossert et al., 1982 concluded

> 'no simple style of management seems appropriate for all schools... principals must find the style and structures most suited to their own local situation' (p. 38).

However, a study of the literature reveals that three characteristics have frequently been found to be associated with successful leadership: strength of purpose, involving other staff in decision-making, and professional authority in the processes of teaching and learning.

Firm and purposeful
Effective leadership is usually firm and purposeful. Although case studies have shown isolated examples of schools where the central leadership role is played by another individual, most have shown the headteacher (or principal in American studies) to be the key agent bringing about change in many of the factors affecting school effectiveness (Gray, 1990; United States Department of Education, 1987).

The research literature shows that outstanding leaders tend to be proactive. For example, effectiveness is enhanced by the vigorous selection and replacement of teachers according to Levine and Lezotte, 1990, although research in Louisiana (Stringfield & Teddlie, 1987) emphasised that this mainly takes place in the early years of a principal's term or of an improvement drive. Once a staff has been constituted that is capable of working together towards effectiveness, staff stability tends to be resumed in effective secondary schools. Interim results reported by Sammons et al. (1994c) also suggest that in effective schools, heads place a great emphasis on recruitment and also point to the importance of consensus and unity of purpose in the school's senior management team.

Another aspect of firm leadership is brokerage, the ability to mediate or 'buffer' the school from unhelpful change agents, to challenge and even violate externally-set guidelines (Levine & Lezotte, 1990; Hopkins, Ainscow & West, 1994). The increasing autonomy of schools in recent years has reduced the need for this type of activity, but it has increased the scope for another factor in effective leadership which some studies have shown to be important, namely successful efforts to obtain additional resources, for example through grants, or contributions from local business and the community (Venezky & Winfield, 1979; Murphy, 1989; NREL, 1990; Levine & Lezotte, 1990).

A number of studies have pointed to the key role of leadership in initiating and maintaining the school improvement process (Trisman, Waller & Wilder, 1976; Berman & McLaughlin, 1977; Brookover & Lezotte, 1979; Venezky & Winfield, 1979; Lightfoot, 1983; McLaughlin, 1990; Louis & Miles, 1992; Stoll & Fink, 1994; Sammons et al., 1994c). Improving many of the school effectiveness factors or making fundamental changes may require support from outside agencies, such as local education authorities, universities or consultants (Purkey & Smith, 1983; Weindling, 1989), and successful leaders will establish and sustain regular contact with these networks (Louis & Miles, 1990). However, the message from school improvement programmes, synthesised most exhaustively by Fullan (1991), is that effective change comes from within a school.

Whilst some case studies have pointed to the long hours worked by effective principals (Venezky & Winfield, 1979; Levine & Stark, 1981), the impact of this factor is difficult to determine: it is *only* effective when accompanied by other factors. It can fluctuate widely over short periods of time; and it is almost impossible to separate its direct impact on improvement from its role as a means of building a shared vision and as a signal of ethos to other staff.

A participative approach
A second feature of effective headteachers is the sharing of leadership responsibilities with other members of the senior management team and the involvement more generally of teachers in decision-making. Mortimore et al. (1988a), in their study of primary schools mentioned, in particular, the involvement of the deputy head in policy decisions, the involvement of teachers in management and curriculum planning, and consultation with teachers about spending and other policy decisions, as all being correlates of school effectiveness. This is tied to another important characteristic of a school: the extent to which its culture is a collaborative one.

In larger primary schools and secondary schools, there may be an even greater need for delegation of some of the responsibilities of leadership. Smith and Tomlinson's (1989) study of secondary schools stressed the importance of leadership and management by heads of departments, a finding which has been borne out by recent research showing substantial differentials in departmental effectiveness within schools (Sammons et al., 1994c). In case studies of schools in Northern Ireland, Caul (1994) drew attention both to the need for clear leadership and delegated authority. His study noted the importance of good middle managers in the school at head of department level. Research in the Netherlands has also pointed to the importance of the departmental level in secondary schools (Luyten, 1994; Witziers, 1994).

Summing up these first two features, effective leadership requires clarity, avoidance of both autocratic and over-democratic ways of working, careful judgement of when to make an autonomous decision and when to involve others, and recognition of the efficacy of the leadership role at different levels of the school. Such leadership is also important for the development and maintenance of a common school mission and a climate of shared goals (see the discussion under factor 2 Shared vision and goals below).

The leading professional
An effective headteacher is in most cases not simply the most senior administrator or manager, but is in some sense a leading professional. This implies involvement in and knowledge about what goes on in the classroom, including the curriculum, teaching strategies and the monitoring of pupil progress (Rutter et al., 1979; Mortimore et al., 1988a). In practice this requires the provision of a variety of forms of support to teachers, including both encouragement and practical assistance (Levine & Stark, 1981; Murphy, 1989). It also involves the head projecting a 'high' profile through actions such as frequent movement through the school, visits to the classroom and informal conversation with staff (Sizemore, Brossard & Harrigan, 1983; Mortimore et al., 1988a; Pollack, Watson & Crishpeels, 1987; Teddlie, Kirby & Stringfield, 1989). It also requires assessing the ways teachers function, described by Scheerens (1992) as 'one of the pillars of educational leadership'. Of course, this type of approach in itself can have little bearing on effectiveness. It is when it is in conjunction with other factors mentioned, such as emphasis on teaching and learning and regular monitoring throughout the school, that it can have such a powerful impact. Indeed every one of the eleven key factors that we have identified have implications for effective leaders. This is borne out in Murphy's (1989) distillation of the literature on instructional leadership. The impact headteachers have on student achievement levels and progress is likely to operate indirectly rather than directly by influencing school and staff culture, attitudes and behaviour which, in turn, affect classroom practices and the quality of teaching and learning.

Shared vision and goals
Research has shown that schools are more effective when staff build consensus on the aims and values of the school, and where they put this into practice through consistent and collaborative ways of working and of decision-making. For example, Lee, Bryk and Smith's (1993) review of literature concerning the organization of effective secondary schools points to the importance of a sense of community

'Such elements of community as cooperative work, effective communication, and shared goals have been identified as crucial for all types of successful organizations, not only schools.' (p. 227).

Others have reached similar conclusions concerning primary schools (e.g., Cohen, 1983; Mortimore et al., 1988a). Whilst the extent to which this is possible is partly in the hands of the headteacher (see leadership), it also relates to broader features of schools which are not necessarily determined by particular individuals.

Unity of purpose
Most studies of effective organisations emphasise the importance of shared vision in uplifting aspirations and fostering a common purpose. This is particularly important in schools which are challenged to work towards a number of difficult and often conflicting goals, often under enormous external pressure (Purkey & Smith 1983; Levine & Lezotte 1990). Both school effectiveness research and evaluations of school improvement programmes show that consensus on the values and goals of the school is associated with improved educational outcomes (Trisman, Waller & Wilder, 1976; Rutter et al., 1979; Venezky & Winfield, 1979; Lightfoot, 1983; MacKenzie, 1983; Lipsitz, 1984; California Assembly, 1984; United States Department of Education, 1987; Stoll & Fink, 1994). Rutter et al., (1979) stressed that the atmosphere of a school was greatly influenced by the degree to which it functions as a coherent whole and that a school-wide set of values was conducive to both good morale and effective teaching. Similarly, Edmonds (1979) emphasised the importance of school-wide policies and agreement amongst teachers in their aims. Unity of purpose, particularly when it is in combination with a positive attitude towards learning and towards the pupils, is a powerful mechanism for effective schooling (California State Department of Education, 1980). Cohen (1983) has also highlighted the need for clear, public and agreed instructional goals.

In their discussion of Catholic schools' relatively greater effectiveness in promoting students' academic and social outcomes (e.g., low drop out) in the US context, Lee, Bryk & Smith (1993) drew attention to the importance of strong institutional norms and shared beliefs producing an 'educational philosophy that is well aligned with social equity aims' (pp. 230–231). In Northern Ireland, Caul (1994) has also concluded that more effective schools share common goals including a commitment to quality in all aspects of school life and clear sets of organisational priorities.

Consistency of practice
Related to the notion of consensus amongst staff is the extent to which teachers follow a consistent approach to their work and adhere to common and agreed approaches to matters such as assessment, and the enforcement of rules and policies regarding rewards and sanctions. (See also the discussions concerning factor 7 Positive Reinforcement and factor 8 Monitoring Progress.) Of

course, consistency across the school will be much more amenable in a context underpinned by unity of purpose as noted above. Work by Cohen (1983) concludes that the need for curriculum and instructional programmes to be interrelated, especially in elementary (primary) schools, implies that in more effective schools, prevailing norms which grant considerable autonomy to individual teachers carry less weight than do the shared goals of professional staff.

Mortimore et al. (1988a) found that in schools where teachers adopt a consistent approach to the use of school curriculum guidelines there was a positive impact on the progress of pupils. Glenn (1981) had similar findings. Rutter et al. (1979) focused in particular on consistent approaches to discipline, and demonstrated that pupils are more likely to maintain principles and guidelines of behaviour when they understand the standards of discipline to be based on general expectations set by the school rather than the whim of the individual teacher. The authors also pointed to the importance of teachers acting as positive role models for the pupil, in their relationships with pupils and other staff and in their attitude to the school. In his study of Welsh secondary schools Reynolds (1976) likewise drew attention to the importance of avoiding a rigid and coercive approach to discipline.

Collegiality and collaboration
Collegiality and collaboration are important conditions for unity of purpose (Rutter et al., 1979, Lightfoot, 1983, Purkey & Smith, 1983; Lipsitz, 1984; United States Department of Education, 1987). As was seen in the section on leadership, effective schools tend to have a strong input from staff into the way that the school is run. For example, Rutter et al. (1979) found that pupil success was greater in schools with a decision-making process in which all teachers felt that their views were represented and seriously considered. In the primary sector Mortimore et al. (1988a) also drew attention to the importance of teacher involvement in decision-making and the development of school guidelines creating a sense of 'ownership'. However, such involvement represents only one aspect of collegiality. To some extent, the contribution to achievement comes through a strong sense of community among staff and pupils, fostered through reciprocal relationships of support and respect (Rutter et al., 1979; Wynne, 1980; Lightfoot, 1983; Finn, 1984; Lipsitz, 1984; Wilson & Corcoran, 1988). It also comes through staff sharing ideas, observing each other and giving feedback, learning from each other, and working together to improve the teaching programme (NREL, 1990).

A learning environment
The ethos of a school is partly determined by the vision, the values and the goals of the staff, and the way that they work together, as discussed above. It is also determined by the climate in which the pupils work: the learning environment. The particular features of this appear to be an orderly atmosphere and an attractive working environment.

An orderly atmosphere
Successful schools are more likely to be calm rather than chaotic places. Many studies have stressed the importance of maintaining a task-oriented, orderly climate in schools (Weber, 1971; Stalling & Hentzell, 1978; Brookover et al., 1979; Edmonds, 1979, 1981; Rutter et al., 1979; Coleman, Hoffer & Kilgore, 1982; Lightfoot, 1983). Mortimore et al. (1988a) also pointed to the encouragement of self-control amongst pupils as a source of a positive ethos in the classroom, and the disadvantages of high levels of pupil noise and movement for pupil concentration. What the research in general shows is not that schools become more effective as they become more orderly, although this may well be the case, but rather that an orderly environment is a prerequisite for effective learning to take place. Creemers (1994) also reports on Dutch research by Schweitzer (1984) which concluded that an orderly atmosphere aimed at the stimulation of learning was related to students' academic achievement. The most effective way of encouraging order and purpose amongst pupils is through reinforcement of good practice of learning and behaviour (see also factor 7 Positive Reinforcement).

An attractive working environment
School effectiveness research suggests that the physical environment of a school can also have an effect on both the attitudes and achievement of pupils. Rutter et al. (1979) found that keeping a school in a good state of repair and maintenance was associated with higher standards of academic attainment and behaviour, and other studies have shown similar links (Pablant & Baxter, 1975; Chan, 1979). Rutter (1983) suggested two explanations for this: attractive and stimulating working conditions tend to improve morale; and neglected buildings tend to encourage vandalism. At the primary level, Mortimore et al. (1988a) have also pointed to the importance of creating a pleasant physical environment, including the display of children's work.

Concentration on teaching and learning
The primary purposes of schools concern teaching and learning. These would appear to be obvious activities in an effective school but research suggests that schools differ greatly in the extent to which they concentrate on their primary purpose. Cohen (1983) noted that school effectiveness is clearly dependent upon effective classroom teaching. Similar conclusions about the importance of teaching and learning at the classroom level are evident in reviews by Scheerens (1992), Mortimore (1993) and Creemers (1994). A number of studies have shown correlations between the focus on teaching and learning and school and teacher effectiveness. In some cases this focus has been defined by quantifying teachers' and pupils' use of time, and in others it has been defined in terms of different measures of the school's concentration on the actual process of learning and on achievement. It is clearly vital for schools and teachers to focus on the quality as well as the quantity of teaching and learning which takes place.

Maximisation of learning time
Some studies have examined the use of time in schools, and a number of measures of learning time have been shown to have positive correlations with pupil outcomes and behaviour. The measures include:

- proportion of the day devoted to academic subjects (Coleman, Hoffer & Kilgore, 1981), or to particular academic subjects (Bennett, 1978);
- proportion of time in lessons devoted to learning (Brookover et al., 1979; Brookover & Lezotte, 1979; Rutter et al., 1979; Sizemore, 1987), or to interaction with pupils (Mortimore et al., 1988a, Alexander, 1992);
- proportion of teachers' time spent discussing the content of work with pupils as opposed to routine matters and the maintenance of work activity (Galton & Simon, 1980; Mortimore et al., 1988a; Alexander, 1992);
- teachers' concern with cognitive objectives rather than personal relationships and affective objectives (Evertson, Emmer & Brophy, 1980);
- punctuality of lessons (Rutter, 1979; deJong, 1988);
- freedom from disruption coming from outside the classroom (California State Department, 1980; Hersh et al., 1981).

Collectively, they point to the need for teachers to manage the transition of activities actively and efficiently. Each of these factors has been seen to have a positive relationship with school effectiveness. Researchers who have combined these variables into a single measure of instruction or academic learning time (Rosenshine & Berliner, 1978; Good, 1984; Carroll, 1989) or those who have reviewed this literature as a whole (United States Department of Education, 1987; NREL, 1990; Levine & Lezotte, 1990) have also demonstrated a clear impact of the maximisation of learning time on effectiveness. Of course, measures of time provide only a crude indication of focus on learning. As Carroll (1989) cautioned 'time as such is not what counts, but what happens during that time' (p. 27), nonetheless academic learning time and time on task remain powerful predictors of achievement.

In a recent review of British literature on teaching and learning processes, Sammons et al. (1994d) drew attention to findings concerning single subject teaching and the management of teaching and learning time

> 'teachers can have great difficulties in successfully managing children's learning in sessions where work on several different curriculum areas is ongoing. In particular, lower levels of work-related teacher-pupil communication and more routine administrative interactions and lower levels of pupil engagement in work activity have been reported in primary school research studies' (p. 52).

Academic emphasis
A number of studies, including some mentioned above, have shown effective schools to be characterised by other aspects of academic emphasis: as judged by teachers and pupils (McDill & Rigsby, 1973); through high levels of pupil indus-

try in the classroom (Weber, 1971; Mortimore et al., 1988a); and through regular setting and marking of homework (Ainsworth & Batten, 1974), with checks by senior staff that this had occurred (Rutter et al., 1979). Reviews (Walberg, 1985; United States Department of Education, 1987) have pointed to the importance of both quantity and quality [appropriateness] of homework set as well as the need for good teacher feedback.

Numerous studies of primary schools have also found that unusually effective schools tend to emphasise 'mastery of academic content' as an important aspect of their teaching programmes (Levine & Lezotte, 1990). In Northern Ireland, Caul's (1994) work has drawn attention to the importance of universal entry to GCSE, and an emphasis on academic standards in effective secondary schools. Work by Smith and Tomlinson (1989) has also pointed to examination entry policies as a key feature in secondary school effectiveness. Sammons et al. (1994c) reported that academic emphasis (including regular setting and monitoring of homework) and high GCSE entry rates were features of more highly academically effective secondary schools.

An important factor influencing academic emphasis concerns teachers' subject knowledge. For example, Bennett, Summers & Askew (1994) have clearly demonstrated that, at the primary level teachers' knowledge of subject content is often limited particularly in areas such as science. Adequate knowledge was seen as a necessary prerequisite, although not in itself a sufficient condition, for effective teaching and learning. In case studies contrasting highly effective and highly ineffective secondary schools, Sammons et al. (1994c) report that the ineffective schools had experienced high staff turnover and severe staff shortages in specialist subjects which were seen to have acted as barriers to effectiveness.

Curriculum coverage is also important. For example, Bennett (1992) has demonstrated wide variations in curriculum coverage both for pupils within the same class and in different schools. Likewise, Tizard *et al.*'s (1988) work on infant schools pointed to a wide range between schools and classes in what children of the same age were taught which could not be accounted for by intake differences. These researchers emphasised the importance of curriculum coverage:

'it is clear that attainment and progress depend crucially on whether children are given particular learning experiences' (p. 172).

Focus on achievement
Some researchers have examined the extent to which a school concentrates on the achievement of pupils as a measure of academic emphasis. For example, case studies of American primary schools and reviews have shown emphasis on the acquisition of basic skills or 'achievement orientation' to have a positive influence on school effectiveness (Brookover & Lezotte, 1979; Brookover et al., 1979; Venezky & Winfield, 1979; Glenn, 1981; Edmonds, 1979, 1981; Schweitzer, 1984). The problem with highlighting this type of factor is that outcome measures tend to be at least partly based on tests in these skills for primary

schools, or examination achievement for secondary schools, making factors associated with focus on achievement partially self-fulfilling prophesies. This is particularly true in relation to class-level data, but less of a problem when examining the effect of a *shared* acceptance of a commitment to a focus on achievement throughout a school.

So while a focus on teaching and learning is at the heart of an effective school, researchers have approached it from a number of different angles. One interesting attempt to consolidate this work is that of Scheerens (1992) who, drawing on a vast range of international school effectiveness literature, judged effective learning time to be one of only three factors for which there is multiple empirical research confirmation. He considered four aspects to be relevant:

- institutionalised time spent on learning (length of school day/week/year),
- amount of homework,
- effective learning time within institutional constraints, and
- learning time for different subjects.

Whilst this typology may not entirely capture the essence of 'focus on teaching and learning', it provides a useful framework for pinning down measurable factors that indicate important practical manifestations of this focus.

Purposeful teaching

It is clear from the research literature that the quality of teaching is at the heart of effective schooling. Of course, this is partly determined by the quality of the teachers in the school, and as we have seen, recruiting and replacing teachers is an important role in effective leadership. However, high quality teachers do not always perform to their full potential, and teaching styles and strategies are important factors related to pupil progress. Whereas learning is a covert process and cannot be directly observed, teaching is an overt activity which is easier to describe and evaluate (Mortimore, 1993), although Levine and Lezotte (1990) have pointed to a number of problems in drawing general conclusions on effective teaching practices. Examining the findings on teaching practices in effective schools research, the outstanding factor that emerges is what we call purposeful teaching. This has a number of elements: efficient organisation, clarity of purpose, structured lessons and adaptive practice.

Efficient organisation

Several studies have shown the importance of teachers being well-organised and absolutely clear about their objectives. For example, Evertson, Emmer and Brophy (1980) found positive effects on achievement when teachers felt 'efficacy and an internal locus of control', and where they organised their classrooms and planned proactively on a daily basis.

Rutter et al. (1979) drew attention to the beneficial effects of preparing the lesson in advance, and Rutter (1983) later pointed out that the more time that teachers spend organising a lesson after it has begun, the more likely it is that they will lose the attention of the class, with the attendant double risk of loss of opportunity to learn and disruptive behaviour. Various studies and reviews have also stressed the importance of appropriate pacing of lessons to make sure that their original objectives are achieved (Powell, 1980; Brophy & Good, 1986; Levine & Lezotte, 1990).

Clarity of purpose
Syntheses of effective schools research highlight the importance of pupils always being aware of the purpose of the content of lessons (Brophy & Good, 1986; United States Department of Education, 1987; NREL, 1990). In summary, the research shows that effective learning occurs where teachers clearly explain the objectives of the lesson at the outset, and refer to these throughout the lesson to maintain focus. These objectives should be related to previous study and to matters of personal relevance to the pupils. The information of the lesson should be structured such that it begins with an overview and transitions are signalled. The main ideas of the lesson should be reviewed at the end.

Structured lessons
A review by Rosenshine & Stevens (1981) highlighted the importance of structured teaching and purposefulness in promoting pupil progress. The NREL review (1990) drew particular attention to effective questioning techniques where questions are structured so as to focus pupils' attention on the key elements of the lessons. Stalling (1975) pointed to improvements in pupil outcomes through systematic teaching methods with open-ended questions, pupil answers, followed by teacher feedback. Supporting earlier findings by Galton and Simon (1980), Mortimore et al. (1988a) likewise noted positive effects on progress through teachers spending more time asking questions and on work-related communication in their study of junior education. They also found positive outcomes to be associated with efficient organisation of classroom work with plenty for pupils to do, a limited focus to sessions, and a well-defined framework within which a degree of pupil independence and responsibility for managing their own work could be encouraged. Clearly, for older age groups greater stress on independence and responsibility is appropriate.

A summary of research on effective teachers by Joyce and Showers (1988) concludes that the more effective teachers:

- teach the classroom as a whole
- present information or skills clearly and animatedly
- keep the teaching sessions task-oriented
- are non-evaluative and keep instruction relaxed
- have high expectations for achievement (give more homework, pace lessons faster, create alertness)
- relate comfortably to the students, with the consequence that they have fewer behaviour problems.

Scheerens (1992) in his analysis of the international body of effective schools research highlights 'structured teaching' as one of three factors which have been convincingly demonstrated to promote effectiveness. His definition of structured teaching is slightly different from other researchers but it is worth looking at some of the examples of what he means by it:

- making clear what has to be learnt;
- splitting teaching material into manageable units for the pupils and offering these in a well-considered sequence;
- much exercise material in which pupils make use of 'hunches' and prompts;
- regularly testing for progress with immediate feedback of the results.

Scheerens admits that this exemplification of structured teaching is more applicable to primary schools, in particular in subjects that involve 'reproducible knowledge'. However, he suggests that a modified and less prescriptive form of structured teaching can have a positive effect for the learning of higher cognitive processes and in secondary schools, and he cites a number of studies to confirm this (Doyle, 1985; Brophy & Good, 1986). However, Gray (1993) is not convinced that this factor is necessarily appropriate beyond the earlier years of schooling, and suggests the need for caution, given that so much of the early school effectiveness research is focused on disadvantaged schools thus giving particular weight to the teaching of basic skills.

Adaptive practice
Although school effectiveness research shows a number of factors to be consistently correlated with better outcomes, it also shows that application of mandated curriculum materials and teaching procedures does not often bring out gains in achievement. Pupil progress is enhanced when teachers are sensitive to differences in the learning styles of pupils and, where feasible, identify and use appropriate strategies (NREL, 1990). In many cases this requires flexibility on the part of the teachers in modifying and adapting their teaching styles (Armor et al., 1976; Sizemore, Brossard & Harrigan, 1983).

High expectations
Positive expectations of pupil achievement, particularly amongst teachers but also pupils and parents, is one of the most important characteristics of effective schools (United States Department of Education, 1987). However, care is needed in interpreting the relationship between expectations and achievement, since the causal process can run in the reverse direction, with high achievement enhancing optimism amongst teachers. However, the weight of the evidence suggests that if teachers set high standards for their pupils, let them know that they are expected to meet them, and provide intellectually challenging lessons to correspond to these expectations, then the impact on achievement can be considerable. In particular, low expectations of certain kinds of student have been identified as an important factor in the under-achievement of students in disadvantaged urban schools (OFSTED, 1993).

High expectations all round
A large number of studies and review articles in several countries have shown a strong relationship between high expectations and effective learning (Trisman, Waller & Wilder, 1976; Brookover et al., 1979; Edmonds, 1979, 1981; Rutter et al., 1979; California State Department of Education, 1980; Schweitzer, 1984; Stringfield, Teddlier & Suarez, 1986; United States Department of Education, 1987; Tizard et al., 1988; Mortimore et al., 1988a; Scheerens, 1992; Stoll & Fink, 1992; Caul, 1994; Sammons et al., 1994c). High expectations have also been described as a crucial characteristic of virtually all unusually effective schools described in case studies (Levine & Lezotte, 1990). The important point as far as teachers are concerned is that low expectations go hand in hand with a sense of lack of control over pupils' difficulties and a passive approach to teaching. High expectations correspond to a more active role for teachers in helping pupils to learn (Mortimore, 1994) and a strong sense of efficacy (Armor et al., 1976).

As with most of the factors identified in this report, high expectations alone can do little to raise effectiveness. They are most likely to be operationalised in a context where there is a strong emphasis on academic achievement, where pupils' progress is frequently monitored, and where there is an orderly environment, conducive to learning. In addition, high expectations are more effective when they are part of a general culture which places demands on everyone in the school, so that, for example, the headteacher has high expectations for the performance and commitment of all of the teachers (Murphy, 1989).

Communicating expectations
Expectations do not act directly on pupil performance, but through the attitude of the teacher being communicated to pupils and the consequent effect on their self-esteem (Bandura, 1992). The expectations may be influenced by factors other than the perceived ability or actual attainments of children. For example, Mortimore et al. (1988a) found that teachers had lower expectations for younger pupils in the class and for those from lower social classes, even when account was taken of children's attainment in areas such as reading and mathematics. But even if teachers do not believe success is possible, conveying conviction that achievement can be raised may have a powerful effect. Teachers may need to monitor either or both their beliefs and behaviour to make sure that this takes place (NREL, 1990). It should also be noted that raising expectations is an incremental process and demonstrated success plays a critical role (Wilson & Corcoran, 1988). Reinforcing this success through praise (see Positive Reinforcement) is a key opportunity for communicating high expectations.

Providing intellectual challenge
There seems little doubt that a common cause of under-achievement in pupils is a failure to challenge them. The implications of this are that when schools have high expectations of their pupils they attempt, wherever possible, to provide

intellectually challenging lessons for all pupils in all classes. This approach has been shown by several studies to be associated with greater effectiveness.

A British piece of research had some important findings which go some way to explaining the processes through which expectations have an effect. Tizard et al. (1988) in a study of infant schools in inner London found that teachers' expectations of both individual pupils and of classes as a whole had a strong influence on the content of lessons, which to a large extent explained differences in curriculum between classes with similar intakes. These expectations were not just influenced by academic considerations but also by the extent to which a child or a class was felt to be 'a pleasure to teach'. The result was that different levels of expectations of pupils were translated into differing requirements for their work and their performance.

Mortimore et al. (1988a) in their study of the junior years of primary schools found that in classes where the pupils were stimulated and challenged, progress was greatest. They particularly mentioned the importance of teachers using more higher-order questions and statements and encouraging pupils to use their creative imagination and powers of problem-solving. Levine and Stark (1981) also stressed the importance of the development of higher-order cognitive skills in effective primary schools, mentioning in particular reading comprehension and problem-solving in mathematics. Levine and Lezotte (1990) and NREL (1990) pointed to a number of other studies with similar findings.

Positive reinforcement
Reinforcement, whether in terms of patterns of discipline or feedback to pupils, is an important element of effective schooling (Brookover et al., 1979; Rutter et al., 1979). Indeed, Walberg's (1984) major review of studies of teaching methods concluded that reinforcement was the most powerful factor of all. As will be seen, school effectiveness research has tended to show that not all forms of reinforcement have a positive impact. Rewards, other positive incentives and clear rules are more likely than punishment to be associated with better outcomes.

Clear and fair discipline
Good discipline is an important condition for an orderly climate (see ethos), but is best derived from 'belonging and participating' rather than 'rules and external control' (Wayson et al., 1988). For example, too frequent use of punishment can create a tense and negative atmosphere with counterproductive effects on attendance and behaviour (Rutter, 1983). Indeed, a number of studies have found that formal punishments are either ineffective or have adverse effects (Clegg & Megson, 1968; Reynolds & Murgatroyd, 1977; Heal, 1978; Rutter et al., 1979; Mortimore et al., 1988a). These and other studies show that effective discipline involves keeping good order, consistently enforcing fair, clear and well-understood rules and infrequent use of actual punishment (National Institute of Education, 1978; Rutter et al., 1979; Coleman, Hoffer & Kilgore, 1981).

Feedback
Feedback to pupils can be immediate (in the form of praise or reprimand) or to some extent delayed (in the form of rewards, incentives and prizes). Two large reviews of effective schools research showed that school-wide or public recognition of academic success and of other aspects of positive behaviour contribute to effectiveness (NREL, 1990; Purkey & Smith, 1994). The British study of secondary schools by Rutter et al. (1979) showed that direct and positive feedback such as praise and appreciation had a positive association with pupil behaviour, but that prizes for work had little relationship with any outcome measure. The researchers posited three explanations for the greater effect of praise: it affects a greater number of pupils; the lack of delay allows more definite links to incentives; and is more likely to increase the *intrinsic* rewards of that which is being reinforced.

Mortimore et al. (1988a) had similar findings for primary schools showing that praise and indeed neutral feedback were more effective than a reliance on control through criticism. It should be noted that the NREL synthesis of the literature (1990) pointed out that the research shows that praise and other reinforcements should be provided for correct answers and progress in relation to past performance, but that use should be sparing and must not be unmerited or random. A number of studies have also shown that rewards and praise need not necessarily be related solely to academic outcomes, but can apply to other aspects of school life such as attendance and citizenship (Rutter et al., 1979; Hallinger & Murphy, 1986; Levine & Lezotte, 1990). Brophy & Good's (1986) review of teacher behaviour and student achievement provides a set of guidelines for effective praise. Amongst other aspects these stress the need for praise to be specific, contingent, spontaneous and varied and to use students' own prior accomplishments as a context for describing present accomplishments and to attribute success to effort and ability.

Monitoring progress
Well-established mechanisms for monitoring the performance and progress of pupils, classes, the school as a whole, and improvement programmes, are important features of many effective schools. These procedures may be formal or informal, but either way they contribute to a focus on teaching and learning and often play a part in raising expectations and in positive reinforcement. There appear to be particular benefits from active headteacher engagement in the monitoring of pupil achievement and progress.

Monitoring pupil performance
Frequent and systematic monitoring of the progress of pupils and classes by itself has little impact on achievement, but has been shown to be an important ingredient of the work of an effective school (see Weber, 1971; Venezky & Winfield, 1979; Edmonds, 1979, 1981; Sizemore, 1985). First, it is a mechanism for determining the extent to which the goals of the school are being realised. Second, it focuses the attention of staff, pupils and parents on these goals. Third,

it informs planning, teaching methods and assessment. Fourth, it gives a clear message to pupils that teachers are interested in their progress. This last point relates to teachers giving feedback to pupils, which we discuss under the factor Positive Reinforcement.

Levine and Lezotte (1990) recognised monitoring of student progress as a factor often cited in effective schools research but argued that there has been little agreement about defining the term or providing guidance for practice. They also pointed to a number of studies that have shown that some schools waste time or misdirect teaching through too frequent monitoring procedures. In their list of effective school correlates they used the phrase 'appropriate monitoring' in view of the need for more work on the form and frequency of its use.

A large British study of primary schools (Mortimore et al., 1988a) concentrated on a well-established form of monitoring pupil performance. These researchers examined record-keeping by teachers as a form of continual monitoring of the strengths and weaknesses of pupils, combining the results of objective assessments with teachers' judgement of their pupils. In many effective schools these records relate not only to academic abilities but also to personal and social development. The researchers found record-keeping to be an important characteristic of effective schools.

Evaluating school performance
Effective schools research also shows that monitoring pupil performance and progress at the school-level is an important factor. In discussing leadership we already mentioned the importance of the headteacher having active involvement and detailed knowledge of the workings of the school, for example through visiting classrooms. On a more formal basis, Murphy's (1989) review of studies of effective leaders showed that they practice a range of monitoring procedures, feed back their interpretation of these to teachers and integrate these procedures with evaluation and goal-setting.

Scheerens (1992), in a review of school effectiveness research, argued that proper evaluation is an essential prerequisite to effectiveness-enhancing measures at all levels. Evaluating school improvement programmes is particularly important. For example, Lezotte (1989) emphasised the importance of the use of measures of pupil achievement as the basis for programme evaluation, indeed, this was one of his five factors for school effectiveness.

It can be concluded that the feedback and incorporation of monitoring and evaluation information routinely into decision-making procedures in the school ensures that information is used actively. Such information also needs to be related to staff development (see also factor 11 – the Learning Organisation).

Pupil rights and responsibilities
A common finding of effective schools research is that there can be quite substantial gains in effectiveness when the self-esteem of pupils is raised, when they have an active role in the life of the school, and when they are given a share of responsibility for their own learning.

Raising pupil self-esteem
Levels of self-esteem are significantly affected by treatment by others and are a major factor determining achievement (Helmreich, 1972; Bandura, 1992). In the case of pupil self-esteem, the attitudes of teachers are expressed in a number of ways: the way that they communicate with pupils; the extent to which pupils are accorded respect, and feel they are understood; and the efforts teachers make to respond to the personal needs of individual pupils. Trisman, Waller & Wilder (1976) found student-teacher rapport to have a beneficial influence on outcomes, and a number of other studies have shown positive teacher-pupil relations to be a dimension linked with success (Rutter et al., 1979; Coleman, Hoffer & Kilgore, 1982; Lightfoot, 1983; Lipsitz, 1984). Mortimore et al., (1988a) found positive effects where teachers communicated enthusiasm to pupils, and where they showed interest in children as individuals.

Teacher-pupil relationships can be enhanced out of the classroom. British studies of secondary schools have found that when there were shared out-of-school activities between teachers and pupils (Rutter et al., 1979; Smith & Tomlinson, 1989) and where pupils felt able to consult their teachers about personal problems (Rutter et al., 1979), there were positive effects on outcomes.

Positions of responsibility
British studies have also shown positive effects on both pupil behaviour and examination success through giving a high proportion of children positions of responsibility in the school system, thus conveying trust in pupils' abilities and setting standards of mature behaviour (Ainsworth & Batten, 1974; Reynolds, 1976; Reynolds & Murgatroyd, 1977; Rutter et al., 1979).

Control of work
Some studies have shown that when pupils respond well when they are given greater control over what happens to them at school, enhancing a number outcomes, even at the primary level (National Institute of Education, 1978; Brookover et al., 1979). A British study of primary schools showed that there are positive effects when pupils are encouraged to manage their work independently of the teacher over short periods of time, such as a lesson or an afternoon (Mortimore et al., 1988a), but that when this occurred for extended periods of time it was disadvantageous.

Home-school partnership
Effective schools research generally shows that supportive relations and co-operation between home and schools have positive effects. Coleman, Collinge & Seifer (1993) have drawn particular attention to the benefits of schools fostering parents' involvement in their children's learning. The question of whether higher levels of parental involvement have an impact is a difficult one, since it can mean a multitude of things in different contexts and there are likely to be marked differences between primary and secondary schools in the nature of parental involvement. As yet, there has been no research into the relationship

between the level of accountability of schools to parents in the UK (increased under the provisions of the Education Reform Act 1988) and their effectiveness.

Parental involvement in their children's learning
The particular ways in which schools encourage good home-school relations and foster parents' involvement with their children's learning will be affected by pupil's age and marked differences are likely to be identified between primary and secondary schools.

Mortimore et al.'s (1988a) junior school study found positive benefits where parents helped in the classroom and with school trips, where there were regular progress meetings, where there was a parents' room and where the headteacher had an 'open door' policy. Interestingly, they found a negative effect for Parent-Teacher Associations, and suggested that this more formalised type of parental involvement was not sufficient in itself to engender involvement and in some cases, could present barriers to those not within the 'clique'. Tizard, Schofield and Hewison (1982) showed that parental involvement in reading had more effect than an extra teacher in the classroom. Epstein (1987), Weinberger et al. (1990) and Topping (1992) have also drawn attention to the value of parental involvement in reading projects in primary schools.

Armor et al. (1976) showed that parental presence in the school buildings, and participation in committees, events and other activities all had positive effects on achievement. On the other hand, Brookover and Lezotte (1979) found no support for a relationship between parental involvement and effectiveness.

More recent work on school improvement by Coleman, Collinge and Seifert (1993); Coleman, Collinge and Tabin (1994), and Coleman (1994) has drawn attention to the importance of positive and supportive teacher, student and parent attitudes for the development of pupil responsibility for learning.

Parental involvement is often highly correlated with socio-economic factors, and concern that highlighting it as an important factor might unfairly pass responsibility for effectiveness to parents partly explains why some researchers have avoided defining or measuring it. However, the studies above did control for socio-economic intake. Interestingly, at least one study has shown that parental involvement can be *more influential* in schools enroling more poor or working-class pupils (Hallinger & Murphy, 1986).

Interim results by Sammons et al. (1994c) indicate that there was a tendency for staff in less effective secondary schools to attribute lack of parental interest as a major factor contributing to under-achievement, whereas in more effective secondary schools serving similar intakes there were more favourably perceptions of parental interest and more active relations with parents.

The actual mechanisms by which parental involvement influences school effectiveness are not entirely clear. It might be speculated that where parents and teachers have similar objectives and expectations for children, the combined support for the learning process can be a powerful force for improvement (Jowett, Baginsky & MacDonald 1991; Mortimore, 1993; Coleman, 1994). Parents who are involved may expand pupils' active learning time (e.g., by working with children themselves especially for younger children, or by supervising homework) and, in the case of difficulties arising at school, perhaps in attendance or behaviour, being more likely to support the school's requirements and standards. As MacBeath (1994) has argued successful schools are likely to be those 'which not only "involve" but support and make demands on parents' (p. 5). He further argues for a more active role for parents in school self-evaluation and development planning. Coleman, Collinge & Tabin (1994) draw particular attention to the interconnectedness of the affective and cognitive domains in the triad of relationships between teacher, parent and student. They argue

> 'it is the relationship between the individual teacher and the parent(s) that is critical in enlisting the home as ally, or rendering it the enemy of the educative (or not) activities of the classroom' (p. 30).

A learning organisation
Effective schools are learning organisations, with teachers and senior managers continuing to be learners, keeping up to date with their subjects and with advances in understanding about effective practice. We use the term 'learning organisation' in a second sense which is that this learning has most effect when it takes place at the school itself or is school-wide, rather than specific to individual teachers. The need for schools to become 'learning organisations' is increasingly important given the pace of societal and educational change (Hopkins, Ainscow & West, 1994). Southworth (1994) provides a helpful review of the features of a learning school which stresses the need for learning at five interrelated levels – children's, teacher, staff, organisational and leadership learning.

School-based staff development
Almost every single research study which has looked at the impact of staff development on school effectiveness has pointed to the need for it to be school-based. For example, Mortimore et al. (1988a) found that in-service training courses only had a positive effect on outcomes when they were attended for a good reason. Stedman (1987) stressed the importance of training being tailored to the specific needs of staff and being 'an integral part of a collaborative educational environment'. Coleman & LaRocque's (1990) research in Canada also points to the positive impact which support from administrative bodies at a local level (School Boards, which are equivalent to LEAs) can provide.

Levine and Lezotte (1990) and Fullan (1991) cite a number of studies that show that one-off presentations by outside experts can be counterproductive. Their review of unusually effective schools had similar conclusions to other reviews and studies. Staff development in effective schools is generally at the school site, is focused on providing assistance to improve classroom teaching and the instructional programme, and is ongoing and incremental (Armor et al., 1976; Venezky & Winfield, 1979; California State Department, 1980; Glenn, 1981; Purkey & Smith, 1983; Hallinger & Murphy, 1985; NREL, 1990).

Studies have also stressed the value of embedding staff development within collegial and collaborative planning, and ensuring that ideas from development activities are routinely shared (Purkey & Smith, 1983; NREL, 1990; Stoll & Fink, 1994).

Conclusions

The majority of effectiveness studies have focused exclusively on students' cognitive outcomes in areas such as reading, mathematics or public examination results. Only a relatively few (mainly British) studies have paid attention to social/affective outcomes (e.g., Reynolds, 1976; Rutter et al., 1979; Mortimore et al. 1988a; Teddlie & Stringfield, 1993). Because of this focus the results of our review, inevitably, tell us more about the correlates of academic effectiveness. As Reynolds (1994) has observed, we have less evidence about school and classroom processes that are important in determining schools' success in promoting social or affective outcomes such as behaviour, attendance, attitudes and self-esteem. Barber (1993) has drawn particular attention to the major problem of low levels of pupil motivation in British secondary schools, and combatting this is likely to be especially important for raising standards in deprived urban areas. Further research on the ways effective schools influence social and affective outcomes including student motivation and commitment to school would be desirable. Having said this, we feel that enhancing academic outcomes and fostering pupils' learning and progress remain crucial tests of effective schooling. For this reason, identifying the correlates of effectiveness, especially academic effectiveness has an important part to play in making informed judgements about schools.

The eleven interrelated and, in many ways, mutually dependent factors identified in this review appear to be generic. In other words, evidence for their importance is derived from both secondary and primary school studies. Initially, an attempt was made to produce separate analyses for the two sectors. However, the degree of overlap identified in findings would, in our view, make the presentation of separate summaries repetitious.

Despite the broad agreement in findings for both sectors, however, it should be noted that the emphasis or means of expression will often differ. For example, the ways in which a school pays attention to the factors 'Pupil Rights and Responsibilities' and 'Positive Reinforcement' will clearly be strongly influenced by pupils' age. Appropriate forms of praise and reward and the manner and extent to which pupils are encouraged to take responsibility for their own learning and to become involved in the school's life will vary for different age groups. Nonetheless, the need for appropriate feedback and positive reinforcement and a concern with pupil rights and responsibilities is important at all stages in education. Ways of focusing on teaching and learning and teaching techniques will also differ for different age groups, but careful and appropriate planning and organisation, clarity of objectives, high quality teaching and maximisation of learning time remain crucial for effective teaching at all stages. Likewise, ways of fostering parental involvement in their children's learning, and with the school will also vary markedly between the primary and secondary sectors.

The centrality of teaching and learning

Scheerens (1992) has drawn attention to the centrality of teaching and learning and of classroom processes in determining schools' academic effectiveness in particular. The eleven factors identified in this review focus on aspects to do with whole school processes (leadership, decision-making, management, goals, expectations and so on) and those to do with, and directly related to, classroom organisation and teaching. Ultimately, the quality of teaching (expressed most clearly by Factors 4 and 5) and expectations (Factor 6) have the most significant role to play in fostering pupils' learning and progress and, therefore, in influencing their educational outcomes. Given this, school processes, including professional leadership, remain highly influential because they provide the overall framework within which teachers and classrooms operate. They are important for the development of consistent goals and ensuring that pupils' educational experiences are linked as they progress through the school. In some schools (those that are more effective) the overall framework is far more supportive for classroom practitioners and pupil learning than in others.

The results of our review do not support the view that any one particular teaching style is more effective than others. Mortimore et al.'s (1988a) analysis of observational and other data about primary school teachers indicated that teacher behaviour was too complex and varied for the application of simple descriptions of teaching style or approach and that 'teachers could not validly be divided into a number of categories on the basis of differences in teaching style' (p. 81). Re-analysis of the Bennett (1976) data by Aitkin, Bennett and Hesketh (1981) and Aitkin, Anderson and Hinde (1981) also points to problems in the use of divisions such as 'formal' or 'informal', 'traditional' versus 'progressive' and the separation of teachers into groups operating distinctive styles. Joyce and Showers's (1988) analysis of ways staff development can foster student achievement concludes that a number of educational practices

> 'ranging across ways of managing students and learning environments, teaching strategies or models of teaching . . . can affect student learning' (p. 56).

Recent reviews highlight the importance of effective management, clarity of objectives, good planning, appropriate and efficient organisation of pupils' time and activities, and emphasis on work communication and intellectually challenging teaching (Gipps, 1992; Sammons et al., 1994d) and suggest that flexibility, the ability to adapt teaching approaches for different purposes and groups is more important than notions of one single 'style' being better than others. Indeed, in our view debates about the virtues of one particular teaching style over another are too simplistic and have become sterile. Efficient organisation, fitness for purpose, flexibility of approach and intellectual challenge are of greater relevance.

Commonsense

The findings of school effectiveness research have sometimes been criticised for being just a matter of 'common sense'. Sammons (1994) notes

> 'There is a grain of truth in this argument. Because school effectiveness research by its very nature sets out to identify the components of good practice . . . it is inevitable that some of the findings are unsurprising to practitioners' (p. 46).

Rutter et al. (1979) likewise pointed out that

> 'research into practical issues, such as schooling rarely comes up with findings that are totally unexpected. On the other hand it is helpful in showing which of the abundance of good ideas available are related to successful outcomes' (p. 204).

In a discussion about appropriate frameworks for judging the quality of schooling Gray (1990) commented

> 'As a rule, schools which do the kinds of things the research suggests make a difference, tend to get better results (however these are measured or assessed). The problem is that these are tendencies not certainties. In betting terms the research would be right about seven out of ten times, especially if it could be supported by professional assessments' (p. 214).

In connection with Gray's comments on the importance of professional assessments, it is interesting to note the links between the findings of this review of school effectiveness research and some of the conclusions reached in studies by inspectors. For example, the influential HMI report *'Ten Good Schools'* (DES, 1977) explicitly drew attention to common features in a sample of secondary schools judged to be 'good'. This report suggested that

> '"success" does not stem merely from the existence of certain structures of organisation, teaching patterns or curriculum planning, but is dependent on the spirit and understanding that pervades the life and work of a school, faithfully reflecting its basic objectives' (p. 7).

In particular, the creation of a 'well-ordered environment', levels of expectation which are at once realistic and demanding, whether in academic performance or in social behaviour and the need for functions and responsibilities to be clearly defined and accepted were highlighted. Other aspects emphasised include the professional skills of the headteacher, the importance of team work, and systems for monitoring progress and pastoral care of students. In connection with the quality of teaching aspects such as variety of approach, regular and constructive correction of work, and consistent encouragement were seen as the hallmarks of successful teaching. School climate, leadership, and links with the local community were also noted.

Comparisons with *Ten Good Schools* are useful because this report pre-dates much of the school effectiveness research we have reviewed and, therefore, is less likely to have been influenced by the dissemination of research findings than more recent inspection documents which often refer to the effectiveness research explicitly. The professional judgements evident in this report draw attention to many of the aspects covered by the eleven key factors which have emerged from our review of school effectiveness research.

Resources

Most studies of school effectiveness have not found the level of resources allocated to schools to be a major determinant of effectiveness. However, this does not imply that resources are unimportant. Mortimore et al. (1988a) cautioned that the schools in its sample

> 'were all relatively well resourced (under the arrangements of the former ILEA). Because all schools were well funded, we did not find resourcing to be a key factor. Had our sample been drawn from a range of LEAs with both high and low spending traditions, it is unlikely this would have been the case' (p. 264).

The importance of a good physical environment of staffing stability and absence of staff shortages were also noted.

Influential US research by Hanushek (1986, 1989) involving meta analyses of many studies concluded that there was little relationship between levels of resources and the accomplishments of students in schools, but many of the studies included suffered from significant limitations. A recent re-analysis of Hanushek's synthesis of the literature (using the same set of studies with their limitations) has questioned this view. Hedges, Laine and Greenwald's (1994) study indicates that the impact of resource allocations (especially per pupil expenditure) has been under-estimated. These authors reject Hanushek's conclusion that resources are unrelated to outcomes, noting that

> 'the question of whether more resources are needed to produce real improvements in our nation's schools can no longer be ignored' (p. 13).

Whilst a new appreciation of evidence concerning the positive impact of resources is timely, our review suggests that the aspects of school and classroom processes summarised under the headings of the 11 key factors exert more powerful and direct influences. Our review confirms Gray's (1990) observation that

> 'adequate levels of resourcing, then, seem to be necessary but not a sufficient condition for a school to be effective. . . . in twenty years of reading research on the characteristics of effective schools I have only once come across a record of an "excellent" school where the physical environment left something to be desired' (p. 213).

Educational markets and other changes

It is important to recognise that the evidence accumulated concerning the correlates of effectiveness during the last twenty years does not allow any firm conclusions to be drawn about the impact of recent legislative changes in the UK which were intended to improve quality and raise standards by extending diversity and choice and stimulate the development of educational markets (DFE, 1992). For example, the increased powers and role of governors in school management and the changing role of the headteacher under LMS have not, as yet, featured in school effectiveness studies. Similarly, whilst parental involvement has been found to be important, it is not possible as yet to establish the impact of increased choice and availability of greater information (e.g., the publication of league tables) intended to increase accountability on school performance. Further research addressing such changes is require before their impact on schools can be evaluated (Sammons & Hillman, 1994). Other changes in the UK context which are also likely to prove important in future school effectiveness research include the impact of development planning (MacGilchrist, 1995) and the impact of the National Curriculum and national assessment. Many other education systems in different parts of the world have or are in the process of introducing similar kinds of changes to those evident in the UK and studies which explicitly examine the consequences of such changes in context for school effectiveness are urgently needed.

Appendix 8.1

'Peddling feel-good fictions' David Hamilton[1]

Effective schooling has become an international industry. Its activities embrace four processes: research, development, marketing and sales. Research entails the construction of new prototypes: development entails the commodification of these prototypes: marketing entails the promotion of these commodities: and sales entails the effort to ensure that market returns exceed financial investment. The school effectiveness industry, therefore, stands at the intersection of educational research and a much broader political agenda – social engineering.

There is another perspective on school effectiveness research. Its efforts cloak school practices in a progressive, social-Darwinist, eugenic rationale. It is progressive because it seeks more efficient and effective ways of steering social progress. It is social-Darwinist because it accepts survival of the fittest. And it is eugenic because it endorses the desirable and, consequently, depreciates the exceptional.

But something else lurks beneath this liberal veneer. School effectiveness research underwrites, I suggest, a pathological view of public education in the late twentieth century. There is, it appears, a plague on all our schools. Teachers have been infected: school organisation has been contaminated; and classroom practices have become degenerative and dysfunctional.

In short, schools have become sick institutions. They are a threat to the health of the economic order. Their decline must be countered with potent remedies. Emergency and invasive treatment are called for. Schools need shock therapy administered by outside agencies. Terminal cases merit organ transplants (viz new heads or governing bodies). And, above all, every school requires targeted INSET therapy. Senior management teams deserve booster steroids to strengthen their macho leadership, while their rank and file colleagues should received regular appraisal-administered HRT (human resource technology) to attenuate their classroom excesses.

From this last perspective, then, school effectiveness research hankers for prototypes, in the form of magic bullets or smart missiles, that are the high-tech analogues of the lobotomies and hysterectomies of the nineteenth century. It is no accident that Professor David Reynolds (Newcastle upon Tyne) who co-authored a 'mission statement' on school effectiveness and school improvement in 1990, was moved five years later to caution against quackery: 'we need to avoid peddling simplistic school effectiveness snake oil as a cure-all' (*The Times Educational Supplement*, 16 June 1995).

For these reasons, school effectiveness research is technically and morally problematic. Its research findings are associated prescriptions cannot be taken

---

1 Originally published in *FORUM*, 38, (2): 54–56 Summer 1996.

on trust. They are no more than sets of assumptions, claims and propositions. They are arguments to be scrutinised, not prescriptions to be swallowed.

Key Characteristics of Effective Schools illustrates these problems. It is a 'review of school effectiveness research', commissioned in 1994 by the Office for Standards in Education (OFSTED). The reviewers, based at the International School Effectiveness and Improvement Centre of the University of London Institute of Education, saw their task as two-fold. First, to summarise 'current knowledge' about school effectiveness: and secondly, to respond to OFSTED's request for 'an analysis of the key determinants of school effectiveness'.

This task redefinition is noteworthy. The extension of OFSTED's remit – the attention to 'current knowledge' as well as 'key determinants' – suggests that the reviewers were reluctant to focus unilaterally on causality. There was, they imply, a 'need for caution' in interpreting 'findings concerning key determinants'.

The redefinition also suggests that the sponsors and researchers did not share the same view of causality. OFSTED appears to espouse a straightforward, linear model of causality. In linear systems, a straightforward cause leads to a straightforward effect. In non-linear systems the outcome is so sensitive to initial conditions that a minuscule change in the situation at the beginning of the process results in a large difference at the end.

OFSTED assumes that, in cases of straightforward causality, outcomes can be linked directly to inputs. OFSTED believes, in effect, that it is possible to predict the final resting place of a clutch of billiard balls on the basis of the prior cue stroke.

The reviewers, however, shared a more elaborate view of causality. They recognise that schooling cannot be reduced to the dynamics of the billiard table. If several balls are simultaneously impelled by separate cues, the play remains straightforward; but it is much more difficult to distinguish the key determinants. Yet, if it is assumed that schools and classrooms are complex, non-linear, adaptive systems, their behaviour ceases even to be statistically straightforward.

The reviewers carefully acknowledge such problems of prediction. Yet, having voiced a series of caveats, they proceed to dilute or disregard them. The notion of key determinants is abandoned, to be immediately replaced by 'key factors'.

Semantic sleight of hand continues. The key factors are packaged in an 'accessible [i.e., tabular] format'. The preamble to this table denoted them as 'correlates of effectiveness' whereas the table itself is headed 'eleven factors *for* effective schools' (emphasis added). Social engineering assumptions are smuggled back into the analysis. The factors, that is, provide a better understanding of possible 'mechanisms' of effectiveness.

Once the factors have been identified, however, their aggregation presents further problems. The tacit OFSTED assumption seems to be that causal factors are independent, universal and additive. The OFSTED reviewers, in return, fully acknowledge that these conditions rarely apply in the multivariate world of education. Yet, as before, they appear disinclined to confront OFSTED's innocent assumptions. First, they aggregate results from different studies conducted at different times in different countries. And secondly, they aggregate factors into a summary table. The aspiration to simplify, in the interests of communication (or packaging and marketing), becomes self-defeating.

The reviewers run into difficulties because they conflate clarification (achieving 'better understanding') with simplification (the extraction of 'key determinants'). They are careful to identify recurrent problems in school effectiveness research. They report, for instance, that previous reviews had commented that 'there is no consensus yet on just what constitutes an effective school'. And they quote another author to the effect that 'defining the effectiveness of a particular school always requires choice among competing values' and that 'criteria of effectiveness will be the subject of political debate'. Overall, the reviewers seem to accept that current school effectiveness debates are as liable to disagreement as any other area of human endeavour. But they make no effort to insert this caveat in their analysis. Clarification is about the honouring of complexity, not its obfuscation.

The conflation of simplification and clarification is also evident elsewhere in the reviewers' arguments. Effective schools, they suggest, are characterised by 'shared vision and goals' (Key Factor Two) which, in turn, are contingent upon notions of 'a sense of ownership', 'strong input from staff' and 'reciprocal relationships of support and respect' among pupils and staff.

Elsewhere, however, the review projects a different model of collegiality. Key Factor One is 'professional leadership', a characteristics that, among other things, should 'usually' be 'firm and purposeful'. Under this criterion as a sub-heading, the reviewers go on to quote a US study which suggested that, 'in the early years of . . . an improvement drive', effectiveness is also enhanced by 'vigorous selection and replacement of teachers'. Thus, it seems, school effectiveness depends on two kinds of reciprocity: 'strong' input *from* staff, and 'purposeful' output *of* staff. Such reciprocity is clearly asymmetrical. Its elaboration and retention services a rhetorical purpose in the OFSTED review – as a feel-good fiction.

To conclude: *Key Characteristics of Effective Schools* relates to an ill-defined policy field where, the authors admit, reviews outnumber empirical studies. The search for better understanding, it seems, is repeatedly swamped by the desire for policy prescriptions. Such imbalance arises because, as the reviewers also acknowledge, school effectiveness research suffers from a 'weak theoretical base'. The associated demands of social engineering and human resource management outstrip the capacity of the research community to deliver the necessary technical wisdom.

In these circumstances, research is pulled by the market place rather than steered by axioms and principles. It becomes product-oriented. It is expected to supply prototypes configured, in this case, as a package of 'key characteristics'. Sponsored by powerful quasi-governmental agencies, this package is placed – and generously hyped – on the global cash-and-carry market for educational products. Bundled with a franchising deal and/or a complementary package of technical support, it is then disseminated around the world (e.g., east of Berlin, south of Rome and north of Euston).

I reject both the suppositions and conclusions of such research. I regard it as an ethnocentric pseudo-science that serves merely to mystify anxious administrators and marginalise classroom practitioners. Its UK manifestations are shaped not so much by inclusive educational values that link democracy, sustainable growth, equal opportunities and social justice but, rather, by a divisive political discipline redolent of performance-based league tables and performance-related funding.

The enduring lessons of the school effectiveness literature are to be found in its caveats, not its cure-alls. The OFSTED review should have given greater attention to the value suppositions as well as to the empirical outcomes of such research; to its diversities as well as its central tendencies; and to its exceptions as well as to its 'common features'. By such means, the more enduring aspiration of the reviewers – a 'better understanding' of schooling – might result.

Appendix 8.2

Key characteristics of effective schools: a response to peddling feel-good fictions
Pam Sammons, Peter Mortimore & Josh Hillman[1]

Introduction

The last issue of *Forum* included David Hamilton's reflections on our recent *Key Characteristics of Effective Schools* research Review published jointly in April 1995 by the Institute of Education and OFSTED. We welcome the opportunity to comment on Hamilton's article which we consider fails to provide an accurate account of the nature, purpose and conclusions of our Review. The tone of the critique, with references to 'social Darwinist eugenic rationale' and accusations of 'ethnocentric pseudoscience', is somewhat intemperate but we have endeavoured to respond to the issues raised in a constructive fashion.

Background

School effectiveness research commenced 30 years ago largely in response to the pessimistic interpretation of findings by researchers in the US (Coleman et al., 1966; Jencks et al., 1972) about the possible influence of schooling on students' achievement. In the UK seminal studies were conducted in the late seventies and mid-eighties (Rutter et al., 1979; Reynolds, 1982; Gray, McPherson & Raffe, 1983; Mortimore et al., 1988a; Smith & Tomlinson, 1989). The research base was thus established well before the Government's market driven educational reforms were introduced.

Hamilton claims that school effectiveness research is 'ethnocentric' and unconcerned with democracy, equal opportunities or social justice, which suggests that it ignores the powerful impact of socio-economic factors, gender and race. This is untrue as even a cursory reading of much published work shows. In fact, we and other researchers in the field have highlighted the nature of such influences.[2] Furthermore school effectiveness research has led to the development of a methodology for separating and identifying the impact of school from the influences of student background factors such as age, low income, social class, gender and race, and their prior achievement levels at entry to school. These studies demonstrate the vital importance of taking account of differences between schools in their intakes so that any comparisons made are done on a 'like with like' basis thus highlighting the need for the concept of *'value added'*.

---

1 Originally published in FORUM, 38, (3): 88–90 Autumn 1996.
2 See, for example, the three articles published in this Journal (Mortimore *et al.*, 1987a, b & c) or Gray, Jesson and Sime, 1990.

Such value added approaches have provided a powerful critique of the simplistic use of raw league tables to measure school performance. We have consistently demonstrated that such tables cannot provide accurate information about the contribution of the school and are especially misleading in relation to the performance of inner city schools (e.g., Sammons et al., 1993a, 1993b, 1994b; Mortimore, Sammons & Thomas, 1994 and Goldstein & Thomas, 1996).

Acknowledging the powerful impact of intake factors, however, does not mean that schools can exert no influence on pupils' educational outcomes. Our work has consistently revealed the existence of both educationally and statistically significant school effects at both secondary and primary levels. In a detailed study of inner London comprehensives, for example, the difference between the most and least effective schools was over 12 GCSE points – equivalent to 6 Grade Bs instead of 6 Grade Ds – for a student of average prior attainment (Sammons et al., 1995a). At the primary level the differences can be even more striking (Mortimore et al., 1988a; Sammons, West & Hind, 1997). Indeed, although no schools overcame the social class difference in attainment between working and middle class pupils, our *School Matters* study revealed that, because they made greater progress over three years, working class pupils in the most effective schools attained more highly than middle class pupils in the least effective ones. In terms of further education and life chances such differences are highly significant, especially for disadvantaged groups.

The key characteristics review

Our Review was commissioned by OFSTED to inform its revision of the *Framework for the Inspection of Schools*. It was conducted independently and at no point were we requested to make any alterations to the text. Involving an analysis of over 160 publications, it was intended to summarise current knowledge and was based on studies conducted in a variety of contexts and countries. Unlike Hamilton we feel it is a strength rather than a weakness to adopt an international perspective, a failure to do so would indeed merit the charge '*ethnocentric*'! With respect to Hamilton's *billiard ball analogy* although we do not advocate a linear model of causality, neither do we accept his alternative proposition that schools and teaching are too complex for analysis to reveal any patterns or consistencies. Our combined research experience leads us to conclude that the study of such patterns is important[3], and our Review provides strong evidence of the existence of 'common features concerning the processes and characteristics of more effective schools'.

We stressed throughout that the Review should not be seen as prescriptive and certainly cannot be viewed as a simplistic recipe for effectiveness. It is

---

3 See Mortimore, 1995; Sammons, Thomas and Mortimore, 1997; National Commission, 1996

regrettable that Hamilton fails to report that the summary table to which he takes exception is introduced by the following paragraph:

> 'These factors should not be regarded as independent of each other, and we draw attention to various links between them which may help to provide a better understanding of possible mechanisms of effectiveness. Whilst our list is not intended to be exhaustive, it provides a summary of relevant research evidence which we hope will provide a useful background for those concerned with promoting school effectiveness and improvement and the processes of school self-evaluation and review'.

It is true that in writing the Review we attempted to provide information in a format which would be *accessible* to non-researchers but we see this as a positive rather than a negative feature and reject the claim that we conflate clarification with simplification. Of course there are dangers of over-simplification in summarising research findings but we believe strongly that research should be made available to practitioners and policy-makers. Such accessibility does not have to be simplistic however. For example, with regard to the centrality of teaching and learning we argue that 'the results of our review do not support the view that any one particular teaching style is more effective than others' and went on to conclude

> 'Indeed in our view debates about the virtues of one particular teaching style over another are too simplistic and have become sterile. Efficient organisation, fitness for purpose, flexibility of approach and intellectual challenge are of greater relevance.'

Democracy and research
Hamilton claims that UK manifestations of school effectiveness research

> 'are shaped not so much by inclusive educational values that link democracy, sustainable growth, equal opportunities and social justice, but rather, by a divisive political discipline redolent of performance-based league tables and performance-related funding!'

We reject this view. We hope our Review demonstrates that the field has, and continues to have, a strong focus on equity and, as we have noted, that it provides forceful evidence against the simplistic use of league tables. In fact we think that school effectiveness methods will provide particularly valuable tools for evaluating the impact of recent policy changes concerning educational markets, school status and admissions policies and the (as yet untested) claims that such changes in themselves will raise standards.

We also contest Hamilton's claim that such research 'underwrites . . . a pathological view of public education in the late twentieth century'. In reality, studies have focused much more on the identification of effective schools and effective practices for raising student achievement than on failure a trend followed in our

Review. A more telling criticism would be that we have tended to ignore the less effective spectrum of schools and practices in favour of the more effective! Only recently have studies examined so called 'failing' schools (Reynolds & Packer, 1992; Gray & Wilcox, 1995; Barber & Dann, 1995; Myers & Goldstein, 1996; Stoll, Myers & Reynolds, 1996). As Gray and Wilcox have argued '... the correlates of ineffectiveness have been assumed to be the same. It is by no means clear, however, that they are' and further work is needed in this area.

We are aware that reviewing research to inform policy-makers, practitioners and lay audiences may be regarded as controversial in a climate in which education is often treated as a political football. Nevertheless, as argued recently in the *British Educational Research Association's Research Intelligence* editorial, we believe the virtue of research needs to be vigorously asserted.

> '... we can mobilise rational argument, empirical evidence, critical debate and creative insights. These are of the essence of democracy ... the social responsibility of researchers ... should be to try to disseminate findings not only to fellow researchers, practitioners and policy-makers but also to the general public ... difficulties in simplifying complex findings and fears of misrepresentation by the press are insufficient grounds for trying to hide in simulated ivory towers' (February 1996).

We think that Hamilton's comments about 'mystifying' administrators and 'marginalising' practitioners are misplaced and there are greater dangers in viewing research as suitable only for an academic elite.

Our claims for *Key Characteristics* remain modest: we hope it provides a useful summary for those interested in the results of three decades of school effectiveness research. So far the reactions we have received from practitioners to the Review have been overwhelmingly positive. Of course, the findings must not be seen as a panacea and we strongly caution against prescriptive interpretations. However, we hope they will stimulate debate and encourage heads and teachers in the process of evaluating their institutions. We are committed to playing our part in improving understanding of the processes of schooling and we believe that the school effectiveness tradition can make a valuable contribution to this aim.

9

Theory and School Effectiveness Research

Introduction

As a relatively new field of educational enquiry, school effectiveness research is commonly acknowledged to possess a stronger empirical than theoretical basis. Indeed, as illustrated in the Hamilton (1996) critique which formed a postscript to the last chapter, some critics of the field have chosen the weak theoretical underpinnings as a focus for questioning the validity of the field as a proper focus for educational research (Pring, 1995). Although such criticisms have some justification in relation to early school effectiveness studies, the theoretical basis of the field has been much strengthened during the last decade and the charge of a theorism can no longer be substantiated. Important advances have been made by the Dutch research community in particular (see Scheerens, 1992; Creemers & Scheerens, 1994; Creemers, 1994a; Scheerens & Bosker, 1997). In the US work by Slavin (1987, 1996) on school and classroom organisation and instructional effectiveness, Stringfield (1994a,b) and Lee, Bryk and Smith (1993) has proved influential, while in the UK Reynolds et al., 1994a,b; Hopkins, Ainscow and West, 1994; Gray and Wilcox, 1995 and Hargreaves, 1995 have all made important contributions, particularly in seeking to integrate the school effectiveness and improvement traditions.

The Key Characteristics Review which formed the focus of the last chapter, while by no means comprehensive, serves to illustrate the empirical basis of the field in terms of what is known about the features of more effective schools, and indicates that such schools tend to possess particular characteristics, irrespective of phase. Moreover, the review suggests that these characteristics can be identified to a greater or lesser degree in research conducted in a variety of different countries and contexts.

The review, however, was intended to answer a particular question 'What is known about the key characteristics of effective schools?' rather than the more complicated and challenging questions of 'How such characteristics influence student outcomes' or 'Why some schools are more effective than others?' These are questions which seek to improve our understanding of the ways in which school processes are related to students' learning, and thus to their educational outcomes. The development of theoretical understanding, though important in its own right, is also argued to provide a much sounder base for the successful development and implementation of school improvement initiatives and as such has much to offer the reflective practitioner as well as policy makers in education (Scheerens, 1995).

This chapter examines the theoretical underpinning of the school effectiveness field from the perspective of a recent study of secondary schools. The Differential School Effectiveness Project focused on departmental differences in secondary school academic effectiveness and was conducted with colleagues (Peter Mortimore & Sally Thomas) at the Institute of Education in London during the mid-1990s. The chapter seeks to integrate the findings from this study with the existing theoretical perspectives in SER, and presents a model of academic effectiveness in secondary schools. Our book, Forging Links (Sammons, Thomas & Mortimore, 1997) describes the main findings of the study while further details of the research are provided in a variety of articles (see Sammons, Thomas & Mortimore, 1998; Sammons et al., 1998b,c; Thomas et al., 1997a, b). A brief account of the research design is provided below as a background to the model of academic effectiveness developed from the study.

The Differential School Effectiveness Project arose as a result of a review of SER discussed in a seminar series by UK researchers in the field in 1993 (later published by Sammons, Mortimore & Thomas, 1996). This review pointed to the need to 'unpack' the notion of consistency in school effectiveness and focus in particular on three sub-themes: consistency in promoting different educational outcomes, stability over time, and differential effects (for particular student groups). The research also sought to account for variations (both between schools and internal within school differences) in effectiveness by examining differences in school processes and their relation to students' academic outcomes, as measured in national public examination results at age 16 (the end of compulsory schooling in the UK). The research project chose to explore the topic of school and departmental differences in secondary school academic effectiveness because at the time there was no recent published research which focused on the explanation of variations in effectiveness, nor on both the school and departmental levels simultaneously. In particular, we wished to use the research to inform the development of public policy in the light of the greater emphasis *given to the schools' raw GCSE examination results following the first publication of school 'league tables' in 1992 (DFE, 1992).*

The research had three major aims:

- *To extend current knowledge about the size, extent and stability over time between secondary schools in their overall effectiveness in promoting students' GCSE attainments.*
- *To explore the extent of internal variations in school effectiveness*
  [a]   at the departmental level, and
  [b]   for different groups of students.
- *To investigate in detail the reasons underlying any differences in effectiveness in reaction to school and departmental processes.*

It was intended to make a contribution to the continuing debate about how schools' performance may best be judged by exploring internal variations in effectiveness as well as measuring trends over time. Our study was designed to investigate the applicability of the concept of overall school effectiveness to the secondary sector and whether schools could be divided validly into broadly effective or ineffective groups. In other words, it was an attempt to test the validity of UK policy developments in the 1990s which had increasingly labelled schools with adjectives such as 'good', 'bad', 'successful' and 'failing'.

It was recognised that more is known about the measurement of school effectiveness than about its underlying causes. We hoped that the Forging Links *study would improve understanding of the processes which influence effectiveness and feed into the theoretical development of the field, by the formulation of a new model of secondary school effectiveness. The research was designed to go beyond correlational studies of associations between process indicators and measures of school effectiveness, and to test out the continued applicability of the results of earlier seminal studies of secondary schools such as* Fifteen Thousand Hours *(Rutter et al., 1979).*

Equally important was our concern to increase the value and accessibility of SER for practitioners and policy makers. At the time of conducting the research and writing Forging Links *the quality of education and school improvement was high on the political agenda in the run up to a general election. We believed it vital that policy and practice be informed by the results of carefully conducted empirical studies which investigate the impact of different school and classroom policies and practices on students' educational outcomes.*

The research project was designed to investigate the examination outcomes of three successive cohorts involving over 17,000 candidates and 90 secondary schools in inner London. Overall GCSE performance and results in different subjects were analysed using multilevel methodology (Goldstein, 1995). In addition to detailed quantitative analyses of students' GCSE results, qualitative approaches were adopted to specific schools and subject departments. The empirical methods attempted to identify patterns in data and examine relationships between variables through the study of large numbers of students, departments and schools using statistical concepts of probability to assist in generalisation. *By contrast, the qualitative case studies sought to illuminate understanding of practice by a focus on individual institutions in their specific context. We believed that a combination of these two approaches would be more fruitful than reliance on only one.*

In designing the Differential School Effectiveness Project we recognised that, as with all tests and assessments, GCSE examination results can only provide indicators of past performance. Students' learning experiences over the five preceding years of secondary schooling (from age 11 to 16) will influence their GCSE performance, so too will their prior attainments and background characteristics. A retrospective approach was adopted in the case studies, with a strong emphasis on comparing the present situation in the school or department with that five years previously. Inevitably, this had a number of limitations including the continued availability of staff and, problems of recall and post-hoc rationalisation. Nonetheless, it had some advantages in allowing an explicit focus on the extent of change over a five year period to be added to the project and providing an indication of the variation in school and departmental histories. In addition to qualitative case studies of outlier schools and departments which were found to be highly effective or, by contrast, very ineffective in value added terms, the research used multilevel methods to test the relationship between measures of specific school and departmental process characteristics and schools' academic effectiveness. These multilevel analyses were used to establish whether such measures made a statistically significant contribution to the explanation of school and departmental variation in GCSE performance over three years (1990–92) after controlling for differences between schools in their intakes.

The project research design thus had the strength of combining quantitative multilevel and qualitative case study approaches in order to improve our understanding of the concept of academic effectiveness and enhance the development of school effectiveness theory. The chapter from Forging Links *which is incorporated in this volume illustrates our attempt to link the findings of the Differential School Effectiveness research with existing SER theoretical frameworks. It elaborates a model of secondary school academic effectiveness which we believe recognises the importance of the 'nested' organisation of secondary schools and helps to unpack some of the important linkages in the input-process-output equation which has characterised school effectiveness studies over the last quarter of a century.*

The school effectiveness research tradition has been criticised on a variety of grounds – philosophical and moral as well as technical and empirical (e.g., Sirotnik, 1985; Preece, 1989; Ball, 1994; Hamilton, 1996; White, 1997). Indeed Pring (1995) has argued that the field should not be seen as part of educational research at all because, in his view, its context is not 'educational', although this is not an argument we accept! It is beyond the scope of this chapter to discuss all these criticisms (see Mortimore, Sammons & Ecob, 1988; Reynolds, 1995; Sammons, Mortimore & Hillman, 1996; Mortimore & Sammons, 1997;

Sammons & Reynolds, 1997 for responses to some of these critiques). However, one of the major weaknesses of school effectiveness research is still seen to be its weak theoretical base (Creemers, 1991; Mortimore, 1991a; Scheerens, 1992) and it is the criticism of atheorism which we seek to address in this chapter. Thus Creemers (1994a) observed 'until recently the state of the art of school effectiveness research was excellently described by the fact that the research reviews outnumbered the total of empirical investigations' (p. 9). Fortunately, the number of empirical studies of school effectiveness has grown rapidly during the 1990s and has focused on new and important research questions. For example, there is increasing interest in the topics of departmental effectiveness, consistency and stability as our present study (and recent work by Ainley, 1994; Luyten, 1994; Witziers, 1994; Harris, Jamieson & Russ, 1995) illustrates. A major study in Australia has also drawn increasing attention to the extent of within school variation at the classroom level, and this by implication the importance of teacher effects (Hill & Rowe, 1994, 1996; Rowe & Hill, 1996). Nonetheless, there remains a need to focus on the integration and synthesis of the 'increasingly varied, disparate and complex bodies of knowledge which have become known collectively as school effectiveness research' (Reynolds et al., 1994b, p. 1).

Some work, however, has been done on the drawing together of the field and the important contribution made by the Dutch school effectiveness tradition to the theoretical development should not be underestimated (e.g., Scheerens & Creemers, 1989; Scheerens, 1990; Creemers, 1991; Scheerens, 1992; Bosker & Scheerens, 1994; Creemers, 1994a, b; Scheerens & Bosker, 1997). Despite this, however, relatively few studies have attempted to investigate the strength of the links between the results of specific empirical studies and theories of educational effectiveness so that more can be learnt about their generalisability/context specificity (Bosker & Scheerens, 1994; Creemers & Reezigt, forthcoming, provide innovative examples of such investigations). The growth of school effectiveness research in many parts of the world is encouraging researchers to conduct collaborative international comparative studies (Creemers et al., 1996) to investigate the impact of social and cultural context. Moreover, reviews of school effectiveness research (NWREL, 1995; Sammons, Hillman & Mortimore, 1995; Scheerens & Bosker, 1997) have attempted to establish the extent to which findings about the characteristics of effective schools are consistent across sectors – primary, secondary, post-school – and different contexts (socio-economic as well as national). For example, the possibilities of learning from 'high performing schools' with very different educational systems – such as in the Pacific rim – has recently been highlighted through international and national comparisons of mathematics and science performance (Reynolds & Farrell, 1996).

In this chapter we seek to examine the findings from the Differential School Effectiveness project in relation to existing theories and models of educational effectiveness and in doing so hope to increase understanding of the ways in which schools' influence their students' educational outcomes. But first we address the question of why we need better theories? Some practitioners and policy makers see theoretical developments as an essentially academic exercise of little or no relevance to their concerns. Given consistent findings which can be applied, the question 'But does it work in theory?' can seem both esoteric and irrelevant. However, the alternative view can also be justified as Scheerens (1995) argues:

> 'there is nothing more practical than a good theory and there is nothing more relevant for school improvement than a well-designed evaluation of a programme ... built upon the knowledge-base of school effectiveness' (p. 9).

The examination of current theory in the light of research results can be especially beneficial for those concerned with school improvement because it may help to disentangle the causal chains and mechanisms which influence student achievement. This is important for those seeking to apply school effectiveness results for the purposes of improvement since it provides a guide as to the most influential factors and the way action in one area can affect other areas, and thus can be used to select specific foci for the initial phases of an improvement initiative.

Theories and models of school effectiveness

Our definition of school effectiveness is a student-achievement centred one, as we argued in Chapter 2. For us an effective school is thus one in which students progress more than might be expected on the basis of their intake characteristics (Mortimore, 1991a). We are not talking in general terms about 'good' schools, but those which are effective in their promoting students' educational outcomes (Silver, 1994, and Gray, 1995, provide a more detailed discussion of this issue). It is important, therefore, that we take a longitudinal perspective, controlling for students' attainment at entry to school, and so investigate progress over subsequent years to get a measure of the *value added* by the school. We do not accept that IQ type assessments of students' ability provide the best measures of intake control, particularly if they are measured some time after students join a school. Baseline assessments of students' skills in areas relevant to the curriculum, such as reading or mathematics, provide a better basis for value added studies of school effectiveness. Given this focus on student achievement Reynolds (1995) has commented that the '"touchstone criteria" to be applied to all educational matters concern whether children learn more or less because of the policy or practice. Fads, fallacies and policy and practice fantasies hopefully pass us by because we try to form our views of the educational world on a scientific basis' (p. 59). In other words, we are concerned with finding out about the school and classroom processes which help to make some schools and departments more effective than others in advancing their students' educational achievements.

In connection with the growing interest in theory development, Scheerens (1995) has argued that 'we do not just want to know what *works* in education but also *why* certain things seem to work. Not only is there an interest in the causal determinants of educational achievement but also in possible underlying explanatory mechanism' (p. 6). Similar arguments concerning the practical value of theory for practitioners have been proposed by Bush (1995) in connection with theories of educational management.

It is important to recognise that there are few generally agreed theories of educational effectiveness, although attempts have been made to borrow theories from existing disciplines such as economics, psychology and sociology to help in its interpretation. For example, organisational, contingency, catastrophe, systems and public choice theories have all been drawn on in the development of ones about educational effectiveness (Scheerens, 1992; Creemers & Scheerens, 1994; Scheerens & Bosker, 1997). School effectiveness research can be distinguished from the wider educational effectiveness field by its focus on schools and explicit attempts to specify *how* school and classroom processes influence students' educational outcomes. School effectiveness research has thus resulted in the development of causal models of educational attainment which attempt to demonstrate the nature and direction of links between particular school processes and student outcomes.

The framework of Input-Process-Output has been commonly adopted, and in recent years the importance of Context has also been widely recognised (Teddlie, 1994a). In the Differential School Effectiveness project, we have focused on the concept of *academic effectiveness* at the secondary level, using various measures of students' public examination results at GCSE as outcome measures and controlling for prior attainment and other relevant intake characteristics.[1] Although these examination results are not the only important goals of education, there are strong arguments for emphasising academic goals, due to the 'high stakes' nature of UK public examinations as determinants of young people's future educational and employment life chances (Mortimore & Sammons, 1997). Because we only have limited – though important – measures of students' academic outcomes at the end of compulsory schooling, we will restrict our discussions to the development of a model of 'secondary school academic effectiveness'. We use the term 'model' to indicate the partial nature of our propositions. As Creemers and Scheerens (1994) have noted, 'models and theories can be thought of as positions on a continuum. Thus theories can be seen as "improved" models, where improvement means that central propositions gain in precision and generalisability and relationships become more formalised' (p. 135). We do not think that the current 'state of the art' of school effectiveness research as yet allows more than the elementary outlines of a comprehensive educational effectiveness theory to be sketched.

The basic structure of models of school effectiveness has been outlined by Creemers and Scheerens (1994) and is shown in Figure 9.1. Such models

---

1 West and Hopkins (1996) have argued for a broader definition of effectiveness encompassing a range of students and teacher outcomes. We accept the value of such a broad perspective but suggest that this relates to more general concepts such as what constitutes a 'good' or 'successful' school.

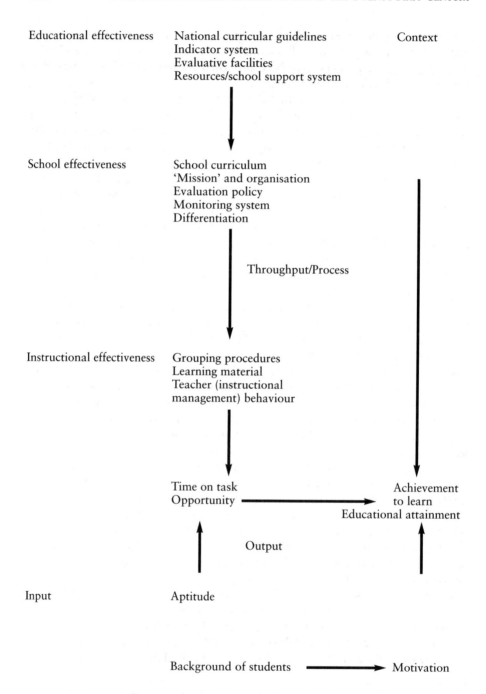

Fig. 9.1   Comprehensive framework for educational effectiveness [after Creemers & Scheerens, 1994].

explicitly attempt to describe the multilevel structure and linkages between levels of the Context-Input-Process-Output chain.

The levels involved comprise the individual student, the classroom, the school and the school environment (the latter covers matters such as the national or local context which would include in the UK the influence of the NC and NA, publication of league tables of examination results, the OFSTED inspection cycle, all of which can be seen as accountability mechanisms. Creemers and Scheerens (1994) have argued that 'theories of learning and instruction, such as the Carroll model, are at the core of multilevel educational effectiveness models' (p. 137). The model they propose defines school-level factors as facilitating conditions for classroom-level factors. Due to this restriction the only school-level factors that are viewed as relevant are those 'conditional for and directly related to quality of instruction or to time allowed/opportunity to learn'. In addition, Creemers (1992) has argued that school-level factors should either promote cohesion between teachers (stimulate similar effective teacher behaviour in all classrooms) or control what is going on in classrooms. He also stresses the importance of continuity in terms of school rules and policies over several years.

In a more recent educational effectiveness model developed from theories about how students learn, Creemers (1994a) stresses the impact of three key concepts – quality, time for learning and opportunity – which are seen to be relevant to each level. Stringfield (1994b) also provides an example of a model of primary (elementary) school effects influenced by learning theories and stressing Slavin's (1987) QAIT (quality, appropriateness, incentive structures and time for instruction) research. Stringfield also emphasises the notion of 'high reliability organisations' (HROs) in which failure is not tolerated and also draws attention to research evidence which indicates 'that the routes to becoming highly effective and highly ineffective are not mirror images of one another' (Stringfield, 1994b, p. 183).

Scheerens (1995) has pointed out an additional complexity to model building, namely that 'combinations of effectiveness enhancing conditions may work better given specific contextual contingencies' (p. 8) i.e., some factors may have beneficial effects only in combination or may only produce positive effects in particular contexts. In commenting on the theoretical base of school effectiveness research, Mortimore (1995a) has stressed that little is known about contingency effects and unintended consequences of different improvement approaches. However, the importance of the SES context of the student body has been highlighted by work in the US (Teddlie et al., 1989) and Scotland (Brown, Riddell & Duffield, 1996). Of course, interpreting this result is not easy; it may be due to lower teacher expectations for some social classes, or lower expectations amongst the peer group or students' parents. Quite probably it represents a combination of such influences. Contextual influences have important policy and practical implications which we discuss elsewhere (Chapter 9 of Forging Links).

School versus classroom effects
Early school effectiveness research concentrated on the identification of more and less effective schools (often by means of outlier studies) and school processes related to effectiveness but paid little attention to the question of teacher effectiveness (Stringfield, 1994a). There has been an increasing recognition of the importance of classroom/teacher effects in recent years and attempts to utilise three level models to separate effects at these levels because so much of the educational process – the teaching of students – takes place in the classroom (Rowe & Hill, 1994, 1996; Luyten, 1995; Creemers & Reezigt, forthcoming (a)). However, it is important to recognise that students commonly attend only one school at a given time, usually for a period of several years. They will experience a succession of teachers and classes during this time, especially at the secondary level, where, as in our own study, the departmental level may be particularly influential. We believe that studies which examine the effects on students of being a member of a particular secondary school over a period of years remain of considerable practical and theoretical relevance. This is because any student's educational experience involves being taught by a number of subject specialists in any year. Although students and their parents may have (usually quite heavily constrained) opportunities to choose schools in the UK system, this is not the case for individual classes or teachers. Moreover, students usually spend several years in one institution and therefore the question of whether over several years the particular school attended has an impact on their later educational outcomes is of prime importance. The period covered by our study from age 11 to 16 years represents nearly a third of their lifetime for secondary students.

Research which adopts a longitudinal approach, for example, using baseline measures of student achievement at entry, and which follows-up academic outcomes several years later, is more likely to identify school effects than studies (particularly in primary schools) conducted over only one year where class/teacher effects are likely to be much stronger. As Teddlie (1994b) has argued, there is a need to integrate school and teacher effectiveness research traditions 'since true change in elementary and secondary schools must occur at both the school and class levels simultaneously and since neither level (school or classroom) can be adequately studied without considering the other'' (p. 113).

In our Differential School Effectiveness project, we assessed student progress from age 11 to age 16, the end of compulsory schooling. This approach provides an indication of the cumulative effect of attending a particular institution over five years. Also we used three successive cohorts of 16-year-olds so that the impact of schools could be distinguished from annual fluctuations. In this way we could establish the extent of stability in patterns of schools' academic performance at GCSE, as we described in Chapters 3 and 4. In the light of this, it is perhaps not surprising that many of the factors found to be related to academic effectiveness concerned school processes related to leadership, organisation and policy. (See part 2 of Forging Links)

The departmental level

On the basis of our empirical research (and other research examining the impact of the department at the secondary level such as that by Ainley, 1994; Luyten, 1994; Harris, Jamieson and Russ, 1995; Witziers, 1994) we conclude that models of secondary school effectiveness need to analyse the impact of the department explicitly. As we have argued earlier in Forging Links, judgements about effectiveness are complex. Effectiveness should be seen as both outcome (i.e., subject) and time specific. Our results show that, in the vast majority of cases, secondary schools cannot be readily classified as either 'effective' or 'ineffective', although the study of unusually effective outliers may be illuminating. In addition, our investigation of differential effectiveness draws attention to the need to focus on effectiveness for specific student groups, thus adding an important equity dimension to any model. The concept of school effectiveness needs to be qualified at the secondary level to the term school and departmental effectiveness (Sammons et al., 1995a,b; Thomas et al., 1997a,b).

In addition, in line with Bosker and Scheerens' (1994) conclusion, models of school effectiveness need to consider the influence of potential 'feedback loops'. Thus student intake factors may influence the nature of school and classroom experiences as well as having a direct impact on students' educational outcomes. For example, the nature of students' characteristics or achievements at intake can affect teacher expectations. There is evidence that teachers tend to rate middle class students more highly in terms of ability, even when control is made for their current attainments (Mortimore et al., 1988a), and this phenomenon is likely to influence their teaching approaches for such students. In addition, teachers' expectations of subsequent age groups may be influenced by the particularly good or poor results of earlier cohorts. It is in this sense that contextual or compositional effects may operate. For example, a number of studies have shown that, in addition to the impact of a student's own socio-economic status (SES), the composition of the student body in his/her school is also related to measures of academic achievement. Our current study indicates that a measure of low family income (eligibility for free school meals) is related to English GCSE performance at both the level of the individual student and at the *school level* where a contextual or compositional effect, related to the level of disadvantage in the intake as a whole, is identified. Put simply, students attending schools which serve poor areas have an additional educational disadvantage. Again this has important policy implications which we discuss in the next chapter.

Cross-level influences

If we accept that schools are best studied as organisations which are made up of nested layers – students within classrooms, departments within schools – the question of how we can conceptualise the ways in which one level influences another becomes especially important. The most pervasive view on cross-level influences in nested (i.e., multilevel) models of school effectiveness is that higher-level conditions, aspects concerning school leadership, policy and organisation, for example, in some way facilitate conditions at lower levels (the quality of teaching and learning in classrooms). These, in turn, have a direct impact on students' academic outcomes (see Bosker & Scheerens, 1994).

While some have argued that it is unlikely that aspects of overall school functioning have a direct impact on students' educational outcomes, we conclude that this is not impossible. For example, a clear school policy on homework may have a direct impact on students' learning and therefore on their educational outcomes by increasing both the time on task and opportunity to learn, including curriculum coverage. We found qualitative evidence suggesting the value of a consistently applied school policy which stressed the importance of regularly set and marked homework in our case studies of academically more and less effective schools. Likewise, aspects of organisation and policy at the departmental level, whilst most likely to operate indirectly, by affecting the quality of teaching and learning in the classroom, may in some circumstances have a direct effect. For example, departmental policies on GCSE examination entry can be viewed as having a direct impact on academic outcomes. As an illustration, our research found clear differences between the more and less academically effective departments in terms of the emphasis given to GCSE entry. This can directly affect students' opportunities for academic achievement (as we described in Part 2 of Forging Links).

Bosker and Scheerens (1994) examine a number of alternative interpretations of cross-level facilitation. They highlight six possibilities concerning the nature of the impact of school-level processes:

- *Contextual effects* – in this case it is suggested that in a school with a majority of 'effective' teachers and feedback amongst staff, the performance of the *less effective minority will be improved*;
- *Mirrors* – in this case the congruence between evidence on effective schools and effective classrooms is highlighted. Congruence of factors (e.g., orderly climate, high expectations, achievement pressure etc.) *helps to create a consistent school culture* which provides a general supportive background;
- *Overt measures* – in this case specific measures are taken to create effectiveness enhancing conditions at lower levels (e.g., the classroom). Examples given include the positive impact of instructional leadership, increasing allocated learning time, recruiting 'effective' teachers, selecting teaching materials, keeping records of student progress etc.; and
- *Incentives* to promote effectiveness enhancing conditions at lower levels. This view would cover rewards for 'effective' teachers from senior managers and monetary grants from their districts for schools if they achieve certain standards.

The application of 'market forces' via open enrolment and publication of league tables can be seen as a crude focus of incentive-based approach in the UK context.

- *Material facilities* for conditions at lower levels. In this case the example given is a computerised school-monitoring system implemented at the

school-level which gives teachers better information on student progress; and
- *Buffers* to protect effectiveness enhancing conditions at the classroom level. This view implies minimal expectations of the direct influence of school management on what the authors call the 'education production process' and covers administrative functions such as student involvement, dealing with government regulations, external pressure etc.

Of course these different interpretations are not necessarily mutually exclusive and it seems likely that higher level conditions can operate in more than one way simultaneously. Given this, it may be hard to distinguish between them in practice.

Understanding academic effectiveness

In Chapter 5(of Forging Links), we described the results of our detailed case studies of six schools and 30 academically more and less effective departments. The conclusions we drew concerning the characteristics of more and less effective schools and departments in the results of our larger questionnaire survey of project schools are summarised in Chapter 7. The quantitative analysis of the questionnaire responses of HTs and HoDs support and extend the findings of the qualitative case studies and enhance our confidence in the conclusions. They highlighted nine factors which we think help to increase our understanding of the mechanisms of academic effectiveness at the secondary level. These are:

- high expectations;
- strong academic emphasis;
- shared vision/goals;
- clear leadership by both HTs and HoDs;
- an effective SMT;
- consistency in approach;
- quality of teaching;
- a student-centred approach; and
- parental involvement and support.

We were particularly interested to examine whether the main conclusions of the much earlier *Fifteen Thousand Hours* research on 12 inner London secondary schools remain valid, given the lapse in time and many significant changes to the educational system over the last 20 years. As we discussed in Chapter 5, the results of the two studies do not diverge greatly. In terms of school processes, Rutter et al. (1979) highlighted the concept of school climate in interpreting their results. Leadership was identified as important, although the impact of middle managers (the HoDs) is a feature of our results and the coherent functioning and unity of the SMT is also significant. The positive impact of shared vision and goals and consistency in approach received emphasis in both studies, but the benefits of home/school partnership are a more notable feature of our results.

Harris, Jamieson and Russ's (1995) qualitative case studies of accelerating departments in a west country city were conducted much more recently but in a different socio-economic context. This research focused explicitly on the departmental level and highlighted 11 aspects. There are a number of important similarities between our results and theirs, which we describe in Chapter 5. Our research, however, draws more attention to the importance of several school-level factors – namely high expectations, strong academic emphasis, shared vision and goals, clear leadership by the HT and an effective SMT. This is perhaps because both more and less effective departments and schools were covered in the case study phase, and for the questionnaire survey we received responses from 55 schools and 79 departments covering a wide range of academic effectiveness. While the department level is undoubtedly very influential for students' GCSE performance in specific subjects, our results indicate that in some schools the benefits of a more supportive context (including a whole school emphasis on the central importance of student learning and achievement, high expectations and consistency in approach, particularly in relation to policy and practices concerning student behaviour) fostered the academic effectiveness of all departments. This conclusion supports Bosker and Scheerens' (1994) suggestion relating to mirrors creating a consistent culture where there is congruence of factors at the school and, in this case, the departmental level.

Interdependence of factors

In terms of empirical results, there is now a fair degree of agreement as to the factors which are important in determining the academic effectiveness of secondary schools at GCSE. However, as noted earlier, we need to explore the relationships between our empirical findings and models of academic effectiveness further to improve our understanding of the way schools and departments influence student outcomes. Reezigt, Guldemond & Creemers (forthcoming) have argued that educational effectiveness research has yielded a lot of classroom and school factors which are correlated with achievement to some extent. However, although there is broad agreement about the lists of factors which are relevant, they comment that such lists,

> 'suggest that the effectiveness factors, whether they are classroom or school factors all have their own independent effects on student achievement. Moreover, the lists suggest that all factors are of equal importance and that their effects are of the same size' (p. 3).

They go on to argue that:

> 'it is possible that not all effectiveness factors have their own independent effects. Their effectiveness may be due to their interrelatedness, at least partly. Also factors at the classroom level may influence achievement in another way than factors at the school level. Finally, factors of different levels may interact' (p. 4).

Some authors have suggested that school effectiveness factors are not independent and have investigated the links between factors at different levels in order to improve understanding of the mechanisms of effectiveness (see Mortimore et al., 1988a). We certainly concur with Creemers' and his colleagues' comments concerning the possibility, and we suggest the strong probability, that the various factors interact. The results of our secondary school case studies, which investigate more extreme (outlier) examples of academically highly effective as well as highly ineffective schools and departments, also support this contention.

Reezigt and colleagues (forthcoming) present an interesting attempt to test empirically an educational effectiveness model of primary schools (Creemers, 1994a) by defining factors at different levels (school and classroom). This model focuses on the key concepts of quality, time and opportunity which are held to affect student learning, and also stresses the concept of consistency. Disappointingly, the results of the multilevel analysis of the model provide only a very fragmentary picture of classroom and school effects on student achievement. Factors did not always have similar effects on different subjects (language and mathematics). Reezigt, Guldemond and Creemers (forthcoming) note that, due to measurement problems in the information available about some factors, further research is needed before their school effectiveness model can be confirmed or refuted by empirical evidence.

An expanded model

In this section we seek to relate the findings from our research to the kinds of educational effectiveness models such as those developed by Creemers and Scheerens (1989) and Scheerens (1990), while incorporating an explicit departmental level into our secondary school model. Ainley (1994) has drawn attention to the importance of the subject department in secondary school effectiveness research. He observes: 'The nature of secondary schools raised additional issues because of their greater organisational complexity and because the outcomes of learning may involve a much wider range of areas of learning' (p. 14) and concludes that 'in terms of research it is important to incorporate the department as a central component of high school organisation' (p. 15).

Current school effectiveness models have drawn heavily on the traditions of primary school research (e.g., Scheerens, 1990; Creemers, 1994a; Stringfield, 1994b) and seem to be particularly relevant to younger children's achievement in the basic skills. In our view such models need extension to incorporate the departmental level in order to describe adequately the educational process at the secondary level.

The model we outline (Figure 8.2) suggests that congruence between factors operating at different levels (school, department and classroom) is an important feature of academically effective schools. In particular, academic emphasis and high expectations are mirrored at the school, department and class levels, while consistency, shared vision and goals and a student-focused approach are mirrored at two levels (school and departmental). Both senior and middle management can be seen to influence academic effectiveness through the leadership of

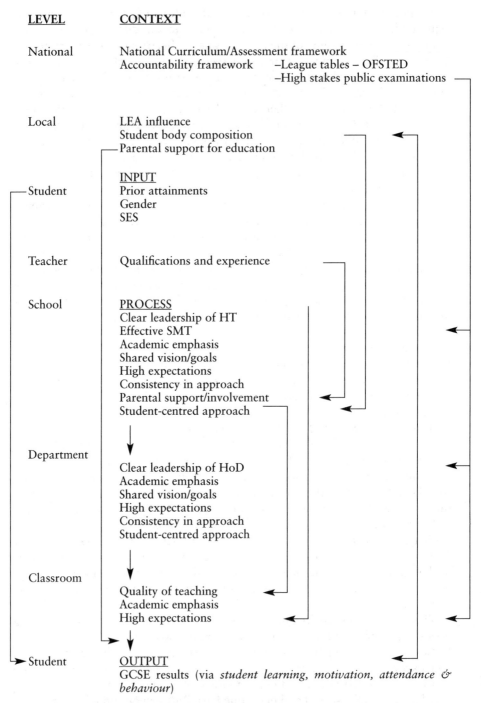

Fig. 9.2 A model of secondary school academic effectiveness.

the HT and the functioning of the SMT at the school level and leadership by HoDs at the department level.

In contrast to other factors, that of parental support/involvement should be viewed as a feature only partially under the control of the school. Thus some schools are more advantaged in terms of the value placed on education and support for the school amongst the parents of students in their intakes. Nonetheless, our case studies indicate that even schools serving very socio-economically disadvantaged intakes differed markedly in staff views of the extent of parental interest and support. Some reported much higher levels of interest, support and involvement than others and had worked hard to build better relationships and encourage such involvement.

Student attitudes/motivation, behaviour and attendance were noted as important outcomes in their own right by many practitioners in our case study schools and questionnaire survey. They can also be seen as intermediate outcomes which facilitate (or hamper) academic achievement. A safe, orderly school environment and a clear and consistently applied school policy on behaviour appeared to be necessary, though not in themselves sufficient conditions for academic effectiveness. Classroom processes, particularly the quality of teaching, can be seen to exert a direct impact on students' learning and motivation which, in turn, affects academic outcomes. Behaviour and attendance, however, may be influenced by both school and classroom processes. Behaviour, motivation and attendance can also influence student learning directly. For example, at the student-level, studies have shown significant correlations between attendance, behaviour and attainment (Mortimore et al., 1988a; Sammons, 1996). Our own data also revealed significant relationships between secondary schools' impact on academic outcomes and the extent of staff satisfaction with past levels of student motivation, attendance and behaviour, (reported in Part 2 of Forging Links).

Teacher qualifications and experience can be viewed as INPUTS to the educational process which can directly influence the quality of teaching at the classroom level. We found significant correlations between English HoDs' assessment of factors such as staff knowledge of the content of subject and GCSE syllabus, their experience of teaching the subject and their qualifications and effectiveness in promoting overall GCSE performance and English performance, although associations were weaker in mathematics. Our study also pointed to the adverse impact of high levels of staff absence in the school or specific departments, and of shortages of qualified teachers. Staff shortages and high levels of absence inevitably affect the quality of teaching in individual classrooms and in departments teaching specified subjects (e.g., mathematics, science or languages in which the supply of qualified teachers has been problematic for many schools in inner city areas). However, high levels of staff absence and difficulties in recruitment/retention of good teachers may also be a symptom as well as a cause of academic ineffectiveness, being influenced by staff morale and by school and departmental leadership. Historically, staff shortages and absence levels have been higher in disadvantaged inner city contexts and this factor has important policy implications which we consider in Chapters 9 and 10 (of Forging Links).

Importance of different factors

In interpreting the results of our Differential School Effectiveness project, we integrated findings from the qualitative case studies, and from simple analyses of patterns of correlations between process variables and value added measures of school or departmental performance at GCSE. In addition, we used multilevel analyses to test the *explanatory power* of process variables in accounting directly for school level variance in academic effectiveness using three measures of student outcomes (total GCSE performance score as an overall measure of achievement, English, and mathematics GCSE scores) after control for differences in student intake (Chapter 7). In this section we revisit these multilevel analyses in an attempt to increase our understanding of the relative importance of different factors as determinants of student outcomes. Headteachers' and English and mathematics HoDs' questionnaire responses were tested in relation to total GCSE score. In English only HTs and HoDs responses were analysed, and for mathematics only HTs and mathematics' HoDs responses, as we reported in Chapter 7. By using this approach our analyses were able to throw light on the perspectives of different players (HTs and subject leaders). In addition, because of differences in the number of respondents from the three groups in different schools and the need to exclude non-respondents to specific questions, our analyses were based on sub-samples of the 94 project schools. Given this, the extent of correspondence in results based on different categories of respondent and rather different sub-samples enhances our confidence in the main findings. However, where results were different it is not possible to say whether this is due to slightly different sample of schools included in the analysis or to 'real' differences in the impact of particular process factors for different GCSE outcome measures.

School culture

Our findings point to the importance of *school and departmental ethos or culture* in determining the academic effectiveness of secondary schools. Most of the items found to be significant in accounting for variations in academic effectiveness in our multilevel analysis fall into three broad categories. The three aspects or dimensions of culture concern are: order (behaviour policy and practice); task achievement (academic emphasis); and relationships (a student-focused approach). In academically more effective schools the school and departmental cultures mirror and reinforce one another.

Order – behaviour, policy and practice

The empirical results reveal the positive impact of behaviour policy and practice. The benefits of a clear and consistently applied whole-school approach to student behaviour and discipline and high staff expectations for student behaviour were evident in the explanation of better than predicted total GCSE and English performance. In contrast, significant behaviour problems and an inconsistent approach were related to poorer than predicted mathematics results.

In order to achieve positive student behaviour, it is evident that staff at all levels need to share a common visions/goals and accept the need for consistency in approach. The outcome is a safe, orderly working environment both in the classrooms and around the schools, maximising time and energy for teaching and learning.

It should be noted that efforts to create this environment may be a particular requirement for academic effectiveness at the secondary level for schools which serve socio-economically disadvantaged communities in inner city contexts, where social fragmentation and disorder may be a greater feature of students' lives outside school. Such an environment can be viewed as a necessary pre-condition for effective teaching and learning. Its importance is likely to be most evident in studies which analyse student progress over several years where cumulative effects may be anticipated. The positive effects on student learning are likely to operate through patterns of better attendance, behaviour and motivation providing greater learning time and opportunities.

Academic emphasis
While a safe and orderly working environment can be viewed as a necessary – but not in itself sufficient – condition for effectiveness, an academic emphasis must be seen as absolutely essential. Such an emphasis involves agreement amongst all staff, teachers, middle and senior managers on the importance of teaching and learning in the school and acceptance that examination uptake and results are important in judging both school and departmental effectiveness. A weak academic emphasis in the school and in the English department were found to be significant in accounting for poorer than predicted English and total GCSE scores in particular. Our value added analyses demonstrated a somewhat closer relationship between schools' overall academic effectiveness in terms of total GCSE performance score and English than mathematics scores. Nonetheless, for mathematics, the emphasis placed on a high level of achievement in examinations as a factor for judging school effectiveness and staff and students' shared belief that the school is primarily a place for teaching and learning were important. From a 'common sense' view, it is not very surprising that academic emphasis is important for academic effectiveness! Yet the importance ascribed to public examination results, as reported by HTs and HoDs, did vary markedly amongst our sample of schools, especially in the past. For example, where the English HoDs thought that the creation of confident, articulate people was a key factor in judging school effectiveness rather than academic outcomes, English results were significantly poorer than predicted. However, where uptake at GCSE/A-level was regarded by the HoD as a key factor which ought to be taken into account in judging departmental performance, the opposite occurred. Too little emphasis on homework – identified as a key factor holding the department back from greater effectiveness by the English HoD – was also found to be significant in the explanation of poorer than predicted English results.

As with behaviour and the creation of safe, orderly working environment, achieving a strong academic emphasis also implies the existence of shared vision/goals, high expectations and consistency in approach, and supportive conditions at different levels. In other words, the school's academic emphasis is mirrored at the departmental and classroom level.

Student-focused approach
This third strand of school culture or ethos also exerts a powerful impact on academic effectiveness. Items related to this aspect focus on the student's experience of schooling. 'Students feel valued as people' (identified as a key factor contributing to their school's effectiveness) was significant in accounting for better than predicted total GCSE performance. 'A caring pastoral environment' and 'student satisfaction' identified as key factors which ought to be take into account in judging school effectiveness were relevant for English. In addition, 'creating a positive climate for learning', 'promoting student responsibility' and 'promoting students' ability to learn independently' all found to be significant for English GCSE results can be seen to contribute to a student-focused approach.

A student-focused approach is indicative of a positive affective environment, an emphasis on the quality of staff-student relationships and enjoyment in the process of learning. It is related to higher levels of student motivation. Our results demonstrate that a student-focused approach is important for academic outcomes. Task achievement (enabled by an academic emphasis and an orderly environment) needs to be supplemented by an acknowledgement of the importance of individuality, the quality of the learning experience and attendance and behaviour.

Teaching
At first sight it is rather surprising that none of the items specifically related to the quality of teaching were found to be significant in explaining differences in academic effectiveness at GCSE when tested in combination with other items in the multilevel analysis. It should be remembered that, when tested individually, these showed strong relationships with schools' value added results and all correlations were in the directions expected from the existing school effective knowledge base (Sammons, Hillman & Mortimore, 1995). However, our multilevel analyses did provide clear evidence of the adverse impact of shortages of teaching staff and high levels of staff absence in relation to students' subsequent GCSE performance. There may be a number of explanations for our findings. The first is related to our focus on students' academic progress over five years from secondary transfer at 11 to GCSE performance at 16, using three consecutive age cohorts in the study. It is likely that school level factors may be more influential than classroom factors in accounting for overall measures of academic achievement. This is because students would have been taught by many teachers in a variety of subjects and classes during their five years at secondary school. In addition, items related to quality of teaching were themselves negatively associated with staff shortages, high levels of staff absence and a weak academic emphasis, as might be expected. Of course, our results are not meant to imply that quality of teaching is unimportant – we identified it as one of nine

key factors because clear differences were evident in both the case studies and questionnaire findings. Also two general items – 'insufficient high quality teaching in some classes' and 'dissatisfaction with the performance of the teaching staff' – as assessed by HoDs were relevant for the explanation of poorer value added results in mathematics and overall GCSE scores. Specific aspects of quality of teaching, however, are likely to be more variable over time and between classes than measures of the items found to contribute to the culture or ethos of the school. The research message from our attempts to explain statistically variations in academic effectiveness at GCSE and the importance of different aspects of school processes is that, although quality of teaching is highly relevant, the three aspects of school culture we highlight appear, in combination, to be more powerful. Where these are positive at each level – school, department *and* classroom – it is likely that the quality of teaching will also be favourable.

Leadership

Our multilevel analyses drew attention to the impact of leadership at different levels within the school. For example, items related to the 'quality of the HT's leadership', 'a strong cohesive SMT' were related to greater value added in total GCSE scores when tested with other process items. In contrast, 'considerable teacher involvement in decision making' was not. Items concerning the leadership of subject leaders were also significant in the analyses of subject (English and mathematics) performance. Leadership, of course, has an important part to play in the creation and maintenance of positive school and departmental cultures.

External influences

Two other items found to be relevant in the explanation of academic effectiveness relate to possible external influences and to the impact of national or local context. One was the negative impact of a falling student roll. Of course, this may be viewed as a symptom as much as a cause of ineffectiveness. Parents may choose not to send their children to a secondary school perceived to have 'problems' in terms of academic results or student behaviour. Thus there is likely to be a negative feedback loop between low examination results in absolute terms and the composition of student intakes. Indeed, the government's explicit intention in *Choice and Diversity* (DFE, 1992) was stated as the 'withering away' of poor schools via the application of market forces. However, no reference was made to the consequences of this policy for students currently attending such schools. Our results suggest that, controlling for intake and school processes, a falling roll tends to have a detrimental impact on the education of existing students. This may be due to the financial consequences affecting the school's ability to retain teachers (especially older, experienced and therefore more expensive staff), or to recruit good teachers, which, in turn, can lower staff morale. The second item concerns the apparently positive impact of 'pressure of external changes' (NC, LMS, etc.). This item was included as a possible barrier to greater effectiveness but in the questionnaire survey was found to be important in accounting for better value added results in overall GCSE performance. Again, the interpretation of this is not straightforward. As we noted in our case studies, the respondents in less effective schools generally stressed internal problems as barriers to effectiveness whereas in the more effective schools internal problems were fairly minor with external pressures identified as problem areas. Nonetheless, given that intake and process items were included in our analyses, the result may indicate that schools highlighting external changes as a barrier have reacted more strongly to the perceived need to enhance their academic emphasis. This interpretation remains very tentative but would suggest that the national context has an impact on standards and is in line with Gray, Goldstein & Jesson's (1996) report on the jump in raw GCSE results which occurred over and above a broadly rising trend associated with the first annual publication of secondary school league tables (1992). This can be interpreted as indicating that as a direct consequence of the use of league tables for accountability purposes, schools entered more students and gave greater attention to GCSE performance.

Discussion

In an extensive American research review on the topic of effective secondary schools, Lee, Bryk and Smith (1993) have pointed out that the faculty is an important feature of school organisation. Nonetheless, these authors found evidence that schools with a common sense of purpose and a strong communal organisation (involving collegial relationships among staff and positive adult-student relationships) are effective in promoting a range of student academic and social outcomes, reflecting student engagement and commitment. They stress the importance of student and staffs' experience of the school as a social organisation and the quality of human relationships experienced within it. Lee, Bryk and Smith (1993) also concluded, however, that attention to social relations reflecting a common aim and perspective is not sufficient: 'it is clear to us that 'good' or 'effective' schools must couple concern for social relations with an appreciation for the structural and functional aspects that instrumentally affect instruction and academic learning' (p. 228). In the very different context of Hong Kong, Ming and Cheong's (1995) primary research has also drawn attention to the benefits of a caring and supportive climate and a cohesive student-centred philosophy of teaching for the entire school.

A somewhat similar thesis to that of Lee, Bryk and Smith (1993) is argued by Hargreaves (1995). He also draws attention to the relevance of the concept of school culture to both school effectiveness and school improvement research. He argues that the school as a social institution has two domains which are potentially in tension. The *instrumental* is concerned with task achievement and social control, and requires teachers and students to work together in orderly ways; whereas the *expressive* – social cohesion – involves the maintenance of good relations both amongst staff and between staff and students. Hargreaves suggests that there is an optimal level of both domains which avoids four extremes which he typifies as A 'traditional', B 'welfarist', C 'hothouse' and D 'anomic'.

Hargreaves acknowledges that few actual school cultures fall into these extremes. However, although Weberian 'ideal types' do not exist in reality, he suggests that they are helpful in interpreting real institutional cultures. Over time, he argues, a real school moves its position and, because schools are 'loosely coupled' institutions different parts of the school could be located in a different segment from the rest of the school. In his model the effective school's culture

> 'is around the centre (E), striving to hold its optimal position in the social control and social cohesion domains. Expectations of work and conduct are high – the principal's expectations of staff and teachers' of students. Yet these standards are not perceived to be unreasonable: everyone is supported in striving for them and rewarded for reaching them. For both teachers and students, such a school is a demanding but enjoyable place to be.' (p. 28)

The Hargreaves typology avoids the common implication that there is a linear continuum from the least to the most effective school. Moreover, it recognises the potential of internal variation in school effectiveness (different departments or classes may vary in their positions across the two dimensions), and suggests that there may be 'different ways of being ineffective excessive formalism, welfarisms and survivalist' (p. 29). It also implies the notion of change, because over time a school (or sub-units within the school) may vary in its position. The evidence from our study gives some support for these arguments. Certainly our case studies of academically ineffective schools and departments provided evidence of ineffectiveness related to the both 'welfarist' and the 'anomic' school cultures. Given the inner city and, by national standards, socio-economically disadvantaged nature of our sample it is perhaps not surprising that we found no evidence suggestive of the 'traditional' and 'hot house' school cultures. If our sample had included selective schools, this might not have been the case.

Our research, particularly the case studies of mixed schools in which highly academically effective and ineffective departments coexist, also supports the view that many schools are 'loosely coupled' institutions and that some departments may successfully create a positive culture which enhances students' academic results, whereas others for a variety of reasons do not. Our case studies indicate the importance of individual department and school histories and draw attention to the need to treat effectiveness as a time as well as an outcome specific concept (Sammons, 1996).

In contrast to Hargreaves arguments, however, we found less evidence for the view that there may be many ways to be effective, although our results are likely to be most applicable to schools in inner city contexts. Whilst effective schools and departments were not identical, consistent results both from our qualitative case studies and the empirical testing of questionnaire data concerning the correlates of effectiveness were found. Moreover, we were able to go beyond interpreting patterns of correlations and through our multilevel analysis demonstrate that it is possible to account for much of the variation in academic effectiveness. Such analyses help to provide firmer evidence of causal connections, as we demonstrated in Chapter 7.

In the context of schooling, there seems to be a variety of paths to ineffectiveness. As Reynolds and Packer (1992), Gray and Wilcox (1995) and Stoll and Myers (1997) have argued, we need to know more about the functioning of ineffective institutions, although we believe that our research provides some important pointers about aspects which hinder secondary school academic effectiveness. Nonetheless, there appears to be greater consensus about the requirements for academic effectiveness. Our findings are in accord with Reynolds' (1995) argument that 'the ineffective school may also have inside itself "multiple schools' formed around cliques and friendship groups... there will be none of the organisation, social, cultural and symbolic "tightness' of the effective school' (p. 61). Such tightness appears to be a particular requirement for academic effectiveness in the context of the inner city.

Conclusions

In this chapter, we have attempted to tease out the most important factors in the explanation of academic effectiveness, linking our empirical results with models of school effectiveness. Our findings strongly suggest that effective secondary schools are *not* simply schools with effective teachers. Although the quality of teaching is undoubtedly very important, it is not by itself seen to be sufficient for general academic effectiveness at the secondary level. Our proposed model of secondary school academic effectiveness highlights the importance of the school and additional departmental levels. This is likely to be, at least in part, due to the longitudinal approach we have adopted looking from entry to performance five years later. Students may experience in excess of 30 different subject teachers during their secondary school careers. Our analyses also cover three consecutive age groups. This repeated measures design inevitably provides a much better indication of the impact of the school than studies which look at student progress over one or two years. Again the greater stability of overall measures of academic effectiveness (like total GCSE performance score) over time also supports the contention that the school attended can have a consistent impact on students' overall academic performance, for good or ill. We also think that current models of school effectiveness are most applicable to the primary sector where the importance of the classroom level is likely to be much greater.

In interpreting our model it is important to stress its multilevel nature. Our results suggest that cross-level relationships are fundamental (school to department, department to classroom). Teaching and learning takes place in the classroom and most of the school and departments' influence is thus likely to be indirect, operating through the culture or ethos. Where culture and ethos at different levels –

school, department and classroom – are mirrored, their impact becomes more powerful, due to consistency in students' educational experiences over several years. Nonetheless, overt measures may both contribute to the culture and also have a direct impact on student learning – for example – school-wide policies consistently applied in areas such as behaviour, marking/assessment or homework.

The three aspects of culture or ethos we find to be most significant in determining academic effectiveness at GCSE demonstrate how an effective school manages to achieve an optimal balance between the social control task achievement and the expressive social cohesion domains identified by Hargreaves (1995). Both behaviour policy and practice, leading to a safe orderly working environment and an academic emphasis are necessary for task achievement (effective teaching and learning and thus students' academic progress), while the student-focused environment concerns social cohesion and creates a positive climate for learning. The school experience thus becomes both demanding and enjoyable for teachers and students alike. Thus, while it is not possible to claim that school culture necessarily has a direct impact on student learning and achievement, its indirect effects may be profound.

The importance of school culture or ethos has implications for management at both the school and the department level. Effective leadership by both the HT, the SMT and the HoDs will be necessary to foster or maintain a positive and consistent culture. Reynolds (1995) has drawn attention to the 'increased value heterogeneity within societies, and the lack of consistency between socialising agencies of the family, the society and the community, [which] may combine to potentially elevate the consistency of the school's socialisation process to an importance it would not formerly have had' (p. 65). In an inner city context this *socialisation* process may be of especial relevance. The achievement of a positive and consistent school culture thus appears to be crucial for academic effectiveness at the secondary level.

Summary

In seeking to examine the relationship between the results of our Differential School Effectiveness project and theoretical accounts of educational effectiveness, we have outlined existing models of school effectiveness and explored how our findings relate to these, stressing the need to incorporate the departmental level into current models of school effectiveness. We argue that such models help to improve our understanding of the ways schools influence their students' academic outcomes. In particular we focus on the concept of school culture or ethos and identify three strands which, in combination, are highly significant in accounting for variations in academic effectiveness. These findings have important messages for schools involved in the processes of self-evaluation and review and concerned with school improvement. In the last section of Forging Links we move on to discuss in more detail the implications of our results for policy makers, practitioners and other players in the educational process.

Part 3

Using School Effectiveness Research – Implications for Raising Standards and School Improvement

The third section of this book focuses on the *application* of SER. As Reynolds (1997) argued, SER is an applied science being deeply concerned with influencing students' educational outcomes and promoting the quality of education for all students. Indeed this is one of the features of the field which initially attracted my interest and the first major school effectiveness study with which I was involved *School Matters* (Mortimore et al., 1988a), was conducted as part of an LEA funded team where the culture was very much intended to make research findings available and accessible to inform and, (we hoped!), influence policy makers. The strong equity emphasis of early SER has already been described in Part 1 of this book. The desire to use the experience of schooling to improve the life chances of socio-economically and educationally disadvantaged groups has a long history, but experience indicates that the application of research results is never easy. There are potential dangers of over-simplification, misunderstanding and misuse.

Given this, I believe that the major challenge facing SER at the present time concerns the ways in which guidance is given about how positive use can be made of its findings. For this it is vital for SER to engage with the school improvement paradigm. As I noted in the introduction of this volume, there have been many calls in the last decade for a bringing together of the best of the two fields so that a new tradition can develop. West and Hopkins (1996) called for a fundamental revision or paradigm shift. Likewise Gray, Jesson and Reynolds (1996) have focused on the challenges of school improvement and need for greater integration.

It has become increasingly evident that school improvement initiatives in the past have focused more on organisational outcomes and teachers' perspectives, seeing staff development as a key lever to effect change but generally failing to measure any impact on students' educational outcomes directly. By contrast, SER has generally neglected the study of change, concerning itself more with identifying the correlates of effectiveness without providing much guidance as to how schools actually become effective.

> 'To relate "key factors" concerning school effectiveness to school improvement one must draw on some (largely) implicit causal chains' (Gray, Jesson & Reynolds, 1996, p. 169).

These authors go on to note

> 'There are inherent difficulties in moving directly from statements about the correlates/causes of "effectiveness" to the formulation of strategies for improvement' (p. 171).

The third part of this book focuses on three practical examples of different ways of using SER. Chapter 10 provides an analysis of practitioners' views and understandings of school effectiveness. This is a neglected area although work by McGaw et al. (1992) in the Australian context sought to investigate the views of teachers, parents and students in connection with the question of what makes an effective school. It is suggested that the exploration of different perspectives is a valuable starting point for schools wishing to engage in school improvement, although it should be coupled with other information to provide a broader perspective.

Chapter 11 turns to ways of making information about effectiveness accessible to practitioners. It describes the results of a value added project conducted in collaboration with the LEA and involving over 100 volunteer primary schools. The method of feeding back information about the impact of children's background characteristics and the contribution of the school was decided on in collaboration with a working group of heads and LEA advisers. It provides an example of the positive use of SER results which can help to inform practitioners engaged in institutional self-evaluation and review.

The third example featured (Chapter 12) moves on to explore the results of an evaluation of a major school improvement initiative using a SER framework. It also provides an illustration of the positive role an LEA can play in promoting school improvement. The role of the LEA (or school board) is, of course, one not generally considered in school effectiveness studies, yet it can be very influential (see Coleman & Larocque, 1990).

While by no means exhaustive, I hope that, taken together, these three examples provide some indication of the potential use of SER methods and approaches and demonstrate the advantages of working towards a greater integration of the effectiveness and improvement traditions.

10

The Practitioners' Perspective

Introduction

Principals and their staff in schools are primarily responsible for the quality of their students' educational experience. The weight of school effectiveness research shows that even in difficult circumstances they can and do make a difference to students' attainment and development. Where learning is valued and academic matters are seen to be relevant for all students, staff recognise that they can make an important contribution. School effectiveness research in which I have been engaged over the last 18 years has consistently shown that powerful though the influences of home background are, schools can be a force for good (Mortimore, 1995).

Early definitions of school effectiveness which focus on the progress students make over a particular time period have been expanded in recent years. In Chapter 3 of this volume we suggested that effectiveness is outcome and time specific and that internal variations for different student groups require careful consideration (Sammons, 1996). My colleagues, Louise Stoll and Dean Fink have recently provided a definition of an effective school which links well with schools' interest in improvement and covers four important features:

- *promoting progress for all students beyond what would be expected given consideration of initial attainment and background factors;*
- *ensuring that each pupil achieves the highest standards possible;*
- *enhancing all aspects of pupil achievement and development;*

- *continuing to improve from year to year. (Stoll & Fink, Changing Our Schools, 1996, p. 28)*

In earlier chapters I have drawn attention to the potential value of school effectiveness research results for informing school improvement initiatives. The research which colleagues and I conducted on secondary schools for the Forging Links: Effective School and Effective Departments study (Sammons, Thomas & Mortimore, 1997) drew attention to the crucial influence of both the Senior Management Team (SMT) and the role of the Middle Managers (Heads of Department) in promoting secondary schools' academic effectiveness. Likewise, the theoretical review in Chapter 9 of this volume, linking this research with models of effectiveness indicated the importance of these groups in shaping the nature of school and departmental culture, which in turn influences students' educational outcomes.

Yet surprisingly little research has attempted to explore practitioners' perspectives and explanations for differences in performance from a school effectiveness perspective (for an exception see McGaw and colleagues in the Australian context, 1992). These perspectives and explanations, however, can be important indicators of culture and expectations. The need to examine teachers' views has been rightly endorsed by David Hopkins and colleagues in their development work with schools as part of the Improving Quality in Education Projects (IQEA). Also Southworth (1994) has drawn attention to the concept of schools as learning institutions while, more recently, Barbara MacGilchrist and colleagues (1997) have focused on the idea of the Intelligent School. An important starting point for school improvement must surely be to explore and understand practitioners' conceptions of effectiveness and their understanding of the influences at work in their own institution's context.

This chapter focuses on 'Practitioners' Views of Effectiveness' and is based on an article written with Sally Thomas, Peter Mortimore and Adrien Walker and published in a new journal Improving Schools. This journal is specifically aimed at enhancing the links between research and practice by making research results accessible to a wider audience. It is based on the results of detailed interviews conducted in six secondary schools and thirty subject departments with headteachers, deputy headteachers and heads of departments (HoDs), as part of the Forging Links secondary school research. These interviews pointed to a variety of common strands as well as the influence of unique school and departmental histories and contexts. The interview results were used to help develop a questionnaire survey for headteachers and HoDs drawn from a larger sample of secondary schools. The intention was to explore practitioners' views and understandings and establish the extent of any congruence with existing evidence concerning the key characteristics of effective schools. The questionnaire also sought to explore staff views about their own school's current goals and effectiveness. In the article we suggest that questions such as those included in our questionnaire survey can be used as prompts to initiate discussion and reflection amongst the SMT and staff as a whole. By establishing the extent of shared vision/goals and common perceptions about areas of strength or by contrast weakness, schools' staff may be stimulated to engage in the reflective practice necessary for the process of institutional self-evaluation and review. In this process we believe it is important to encourage staff to focus on areas of success and achievement and what can be learnt from these, as well as on challenges or problems much in the way that good systems appraisals and formative assessments do. Exploring staff views of changes over a given time period (for example, the last five years) can prove a useful way of identifying achievements as well as areas of difficulty. Open discussion of the results of such a survey can help to identify areas of agreement as well as difference. Feeding the results of such surveys into the school development planning process can identify specific foci for action and can help in the cycle of self-evaluation/review – action – reflection – evaluation/review. Of course, the results of questionnaire surveys of teachers' views may prove challenging for schools in serious difficulties. As part of the Improving School Effectiveness Project we conducted in Scotland in conjunction with the University of Strathclyde (Robertson & Sammons, 1997a & b) the feedback of questionnaires results on two occasions proved highly informative, but, in some schools, provided disturbing information for management (Smith et al., 1998). Very deep divisions may exist in some schools and open discussion may appear to exacerbate some of these at first. Nonetheless, encouraging the discussion of practitioners' views may identify some unanticipated areas of agreement and highlight the most important foci for early action.

Of course, it is not just practitioners' views which can inform school improvement initiatives. In the ISEP research in Scotland, and in the Northern Ireland Making Belfast Work evaluation of RSS, questionnaire surveys of other key stakeholders in the educational process were incorporated (see Robertson & Sammons, 1997a & b; Sammons, Taggart & Thomas, 1998). The triad of parents', students' and teachers' perspectives provides a more complete picture of the school and its culture than the perspective of any one group in isolation (Coleman, 1998). Developing competence to handle and make use of increasing quantities of information about student performance as well as the results of surveys is an increasingly important challenge for school managers as we move into the 21st century. Who has access to information and how it can be best used to inform action planning and evaluation are questions which, I believe, will distinguish successful schools in the future. The ability to collect and use information appropriately and creatively while avoiding the dangers of information overload are core skills which senior and middle managers concerned to promote school improvement need to develop.

The paper which forms the subject of this chapter helps to illustrate the range in views of what contributes to school and departmental effectiveness from the perspectives of two groups – headteachers and heads of departments. Some important common features emerge but there is also evidence of differences in priorities and understandings. Encouraging collegiality and common goals may be difficult if such differences in understanding and perception are ignored.

The 'touchstone' of school effectiveness research is the impact of schools on their students' learning, measured by their progress in specified educational outcomes (Reynolds, 1995). Nonetheless, little is actually known about the relationships between practitioners' beliefs and views about the factors which influence effectiveness, and which ought to be taken into account in judging school performance, although Glover (1992) conducted some work on parents and community perceptions of school quality. Yet staff can differ markedly in their views about what constitute appropriate goals for students and for schools (Elliott, 1996; White, 1997). It is a matter of considerable practical as well as academic interest to establish how much senior and middle managers' views on such matters vary, and the extent to which practitioners' beliefs and perceptions are related to research findings about the key characteristics of effective schools.

As part of a wider study of academic effectiveness we investigated the concept of secondary school and departmental effectiveness and their determinants from the practitioners' perspective. Our study involved 94 inner London schools and used value added approaches to analyse school effects on students' GCSE results over three consecutive years. As well as quantitative analyses of examination data (using overall GCSE results and results in 5 separate subject areas) the research involved detailed qualitative case studies of six schools and 30 subject departments and a questionnaire survey of headteachers (HTs), and heads of departments (HoDs). Full details of the results of the research and their implications for policymakers and practitioners are available in *Forging Links: Effective Schools and Effective Departments* (Sammons, Thomas & Mortimore, 1997).

In this article we focus on only one aspect of our study, using the results of the questionnaire to explore HTs' and HoDs' beliefs and perceptions about school and departmental effectiveness. The questionnaire survey was informed by the results of the qualitative case studies and was piloted in four schools not involved in the main project. We adopted an *explicitly* retrospective approach in the survey asking respondents to compare the current situation in their school (spring term 1995) with that five years before. This enabled us to examine perceptions of groups of key personnel about the processes of change and/or improvement during that time period. The first half of the nineties involved considerable pressures of change linked with the application of market forces and the increased accountability of schools (DFE, 1992). Change was especially marked in inner London, where the introduction of LMS, the National Curriculum and national assessment, changes to the GCSE examination, publication of league tables and introduction of OFSTED inspections was accompanied by the abolition of the ILEA and creation of 13 new LEAs. It was thus a particularly interesting period in which to study schools and explore practitioners' experiences and views of change.

A total of 264 questionnaires (for the HT and two subject HoDs) were sent to 88 of the 94 schools included in the analysis of GCSE results. The survey was conducted in the spring term of 1995. One or more responses were received from 55 of the 88 schools (62.5 per cent). Forty-seven HTs (53.4 per cent), 40 HoDs of Mathematics (45.5 per cent) and 39 HoDs of English (44.3 per cent) submitted returns. Schools for which questionnaires were received and those which did not respond were found to be similar in terms of their academic effectiveness – as measured by the residual estimates from the multilevel value added analysis of examination data (for details of the survey see Sammons, Thomas & Mortimore, 1997).

Our questionnaires examined HTs' and HoDs' perceptions of their schools' and department's academic performance and the factors they perceived as contributing to, or inhibiting, effectiveness. Views about their school's principle educational goals and about aspects they thought important in judging school and departmental effectiveness were also sought. The questionnaire items were informed by a review of the relevant literature, and discussion of the results of the interviews conducted in case study schools with practitioners and colleagues. Questions were pre-coded but respondents invited to provide additional comments. It should be stressed that no feedback had been given to staff concerning their own institution's academic effectiveness as measured by our value added analysis of students' progress.

Schools

Judging school effectiveness

The HTs and HoDs were asked to indicate up to 10 most important factors which ought to be taken into account in judging the effectiveness of any secondary school. Table 10.1 gives details of their responses.

Our results indicate that, for HTs, good progress or value added, for students of all ability levels was the most commonly noted factor which ought to be taken into account in judging the effectiveness of any secondary school. By contrast, less than a third felt a high level of achievement in examinations ought to be taken into account. Good quality teaching and high expectations of students were the next most frequently cited factors (joint second), followed by positive inter-personal relationships for staff and students and the creation of a positive climate for learning (joint fourth). In joint sixth place were shared goals by staff and students and the encouragement of students to take responsibility for their own learning.

The list of factors which HoDs thought should be taken into account in judging the effectiveness of any school was similar to that of HTs in terms of most commonly cited factors. High quality teaching was in first place, followed by good progress, or value added, for students of all ability levels in second, positive inter-personal relationships for staff and students (third), the encouragement of students to take responsibility for their own learning (fourth) and the creation of a positive climate for learning (fifth). These results showed a fair degree of agreement in views between HTs and HoDs, the rank correlation being 0.84.

Nonetheless, there were some differences in emphasis. HoDs placed rather less stress than HTs on shared goals and values for staff and students. In addition, they were less likely to cite a good record of student attendance, good discipline and student behaviour, and high expectations of students as important criteria for judging school effectiveness.

Table 10.1 Factors Which Ought to be Taken into Account in Judging the Effectiveness of Any Secondary School.

	HTs		HoDs	
	Rank	%	Rank	%
An improved staying on rate post-16	18	25.5	18	24.1
A caring pastoral environment	16	29.8	11	44.3
Good progress (value added) for students of all ability levels	1	89.4	2	77.2
The creation of confident, articulate people	11	51.1	8	54.4
A high level of academic achievement in examinations	17	27.7	16	30.4
Positive inter-personal relationships for staff and students	5.5	66.0	3	70.9
High quality teaching	2.5	74.5	1	79.8
A good record of student attendance	9	57.4	17	27.8
Good discipline and student behaviour	4	70.2	6.5	58.2
Preparation for work	21	2.1	21	2.5
The encouragement of students to take responsibility for their own learning	7.5	61.7	4	63.3
The encouragement of a positive attitude to school (pride in school)	15	31.9	15	32.9
The provision of a good range of extra curricular activities	14	36.2	19.5	17.7
The creation of a positive climate for learning	5.59	66.0	5	60.8
Shared goals and values by staff and students	7.5	61.7	12.5	43.0
High expectations of students	2.5	74.5	6.5	58.2
High expectations of staff	10	53.2	9	51.9
Teacher motivation and commitment	12	48.9	10	50.6
High level of student motivation	13	38.3	12.5	43.0
Parent/community satisfaction	20	17.0	14	34.2
Student satisfaction	19	21.3	19.5	17.7

Note. Respondents were asked to identify up to 10 factors.

Given the priority attached to applying market forces to education by the last Government, it is perhaps surprising that few HTs or HoDs placed a strong emphasis on either parent/community satisfaction, or student satisfaction, as factors which ought to be taken into account in judging the effectiveness of any school. Nonetheless, HoDs gave more weight to the maintenance of a caring pastoral environment and parent/community satisfaction than did HTs.

Although more HTs and HoDs rated good progress as an important criterion for judging school effectiveness – and less than a third cited a high level of achievement in public examinations as important – promoting student progress was only infrequently reported as being one of the school's principle educational goals (see Chapter 6 in Sammons, Thomas & Mortimore for details). Our findings suggest, therefore, that although both HTs and HoDs viewed student progress as essential for judging school effectiveness, the majority still attached greater importance to more visible goals, such as absolute achievement in public examination results. Again this is likely to reflect the impact of league table publications and the experience or prospect of a first OFSTED inspection for schools in our sample. Given the new Labour Government's commitment to presenting value added information in school performance tables, we would anticipate that this pattern might change.

Factors contributing to the effectiveness of the respondent's school
Headteachers and HoDs were asked to identify the most important factors which contributed to the effectiveness of their school (see Table 10.2). For HTs, the factor a 'strong and cohesive SMT' was cited most frequently, followed by 'staff and students' shared belief that the school is primarily a place for teaching and learning' in second place. The 'quality of leadership provided by the HT' ranked third. 'High quality teaching in all/most departments' and 'regular monitoring of student achievement and progress' were the next most commonly noted in joint fourth place, closely followed by the 'commitment and enthusiasm of teaching staff'.

Table 10.2 Factors Which Contribute Most to the Effectiveness of Your School.

	HTs		HoDs	
	Rank	%	Rank	%
Strong support from parents/community	21	21.3	19	20.3
Teachers feel valued	15.5	25.5	20	19.0
A strong emphasis on academic matters	18.5	23.4	17.5	22.8
A good staff development programme	15.5	25.5	22	7.6
Students feel valued as people	8	53.2	3.5	49.4
Careful monitoring of attendance	13	38.2	16	25.3
Staff stability in post	14	34.0	5	48.1
Good leadership by heads of departments	11	42.6	2	69.6
A strong and cohesive senior management team	1	78.7	15	32.9
Quality of leadership provided by headteacher	3	63.8	8	43.0
Commitment and enthusiasm of teaching staff	7	57.4	1	70.9
High quality teaching in all/most departments	5	59.6	3.5	49.4
Considerable teacher involvement in decision-making	18.5	23.4	17.5	22.8
The creation of an orderly and secure working environment	5	59.6	11	38.0
Clear and consistently applied whole school approach to student behaviour and discipline	9	51.1	12.5	36.7
Encouragement of student responsibility	18.5	23.4	14	34.2
Regular marking and monitoring of homework in all/most departments	10	44.7	12.5	36.7
School-wide policies on marking/assessment	18.5	23.4	21	17.7
Regular monitoring of student achievement and progress	5	59.6	7	43.0
No shortage of experienced and well-qualified staff	22	17.0	10	41.8
Good working relationships amongst staff	12	38.3	6	44.3
Staff and students' shared belief that the school is primarily a place for teaching and learning	2	66.0	8	43.0

Note. Respondents were asked to identify up to 10 factors.

For HoDs, by contrast, 'the commitment and enthusiasm of teaching staff' was most frequently noted, followed by 'good leadership by HoDs' in second place. The factors 'students feel valued as people' and 'high quality teaching in all/most departments' were next most frequently noted (joint third), closely followed by 'staff stability in post' (fifth). HoDs placed rather less emphasis than HTs on the factors 'quality of leadership provided by the HT', 43 per cent listed this, and 'a strong cohesive SMT' was mentioned by a third. Interestingly, HTs and HoDs showed much less similarity in their opinions about what *actually* contributed to their schools' current success (rank correlation = 0.48) than they had about what factors *ought* to be taken into account when judging school effectiveness, as we showed earlier in Table 10.1. Thus HTs, for example, gave more credit to 'the creation of an orderly and secure working environment' and a 'clear and consistently applied whole school approach to student behaviour and discipline'. This suggests that they attach more importance than HoDs to the creation of a positive behavioural climate in their school.

In terms of which factor was most important, five HTs (11 per cent) and 13 HoDs (17 per cent) were unable/unwilling to identify any single factor as most influential, suggesting that a combination of different factors was important. Of those who did respond, HTs cited 'staff and students' shared belief that the school is primarily a place for teaching and learning' most frequently (24 per cent) followed by 'the quality of leadership provided by the HT' and 'high quality teaching in all/most departments' (14 per cent each). More HoDs cited 'the commitment and enthusiasm of teaching staff' as most important than any other factor (17 per cent), closely followed by 'the quality of leadership provided by the HT' (15 per cent). 'Staff and students' shared belief that the school is primarily a place for teaching and learning' came in third, cited as most important by 12 per cent of HoDs.

It is notable that relatively few respondents believed that 'a good staff development programme' contributed to their schools' current effectiveness, particularly amongst HoDs – less than 8 per cent cited this. By contrast, HoDs attached more importance that HTs to the factor 'no shortage of experienced and well- qualified staff'. Only a minority of both groups, around one in five, felt that 'strong support from parents/community' contributed to their school's current effectiveness.

Barriers to greater effectiveness
More detailed information was sought from HTs about certain aspects of the school and equivalent information from HoDs about specific departments. Again HTs were asked to indicate the most important factors which acted as barriers to the effectiveness of their school (Table 10.3).

Our results indicate that 'inadequate leadership by some HoDs' and 'too low expectations of students by some staff' were the most commonly cited items – each noted by nearly three-quarters of HTs. The 'social disadvantage of the student intake' was also very highly rated, identified as a barrier by a similar proportion of HTs. 'Poor quality teaching in some departments' was a source of concern for over half, and an 'inconsistent approach to student behaviour and discipline' was noted by 40 per cent. External factors such as 'lack of resources' and 'pressures of external change' (e.g., NC, LMS) were also noted by over a third and a quarter cited a 'poor physical environment' as a barrier to greater effectiveness.

Five HTs also commented on other specific factors. These included: 'being an all boys school'; 'union activity'; 'low student motivation' and 'low skills of intake at transfer'and 'lack of space'. However, one HT expressed great confidence, commenting that 'nothing' held the school back from being more effective at present. In all, 43 of the 47 HTs made a response concerning which factors were the most important inhibitor of greater school effectiveness. 'Too low expectations of students by some staff' was identified as most important by over a quarter (12 HTs). The 'social disadvantage of the student intake' was noted by seven respondents, followed by 'inadequate leadership by some HoDs' – cited by five, whereas 'Lack of resources' was seen as the main barrier by four HTs.

Table 10.3 Factors Which are Barriers to the Effectiveness of Your School.

	HTs	
	Rank	%
Social disadvantage of intake	3	70.2
Shortage of qualified staff in key departments	17.5	10.6
Inadequate leadership by some heads of department	1.5	72.3
Low staff morale	20.5	6.4
Pressures of external change (e.g., NC, LMS)	7	36.2
Lack of resources	6	38.3
Poor physical environment	12	25.5
High staff turnover	22	4.3
Falling student roll	17.5	10.6
Lack of consensus in staff goals for students	15	19.1
High levels of staff absence	20.5	6.4
Too little emphasis given to homework	13.5	23.4
Inconsistent approach to student assessment by staff	10	29.8
Inconsistent approach to student behaviour and discipline	5	40.4
Insufficient academic emphasis	10	29.8
Little support from parents/community	8	31.9
Too low expectations of students by some staff	1.5	72.3
Poor quality teaching in some departments	4	55.3
Lack of commitment and enthusiasm by some staff	13.5	23.4
Poor student attendance	10	29.8
Poor student behaviour	17	10.6
Conflict within senior management team	19	8.5

Note. HTs were asked to identify up to ten factors.

Departments

In this section we turn to examine the HoDs' perceptions of departmental effectiveness and its influences in more detail.

Judging departmental effectiveness

The HoDs were asked to identify which factors they though should be taken into account in judging the effectiveness of any department. Table 10.4 compares English and mathematics HoDs' responses.

As a group, HoDs of English and mathematics gave a fairly similar pattern of responses. Both groups noted quality of teaching in the department most frequently – over 85 per cent. For English HoDs, the next most commonly cited factors were the extent to which departmental staff worked together as a team and the commitment/enthusiasm of teaching staff, each noted by over 70 per cent. For mathematics, HoDs gave equal stress to departmental staff team work and the enjoyment/interest of students which were both noted by 63 per cent. Mathematics HoDs gave more stress to the prior attainment of students at intake to school and visible student progress. Factors less frequently noted as important in judging the effectiveness of any department by both groups included the development of students' social skills and student behaviour.

Table 10.4 Factors Which Ought to be Taken Into Account in Judging the Effectiveness of Any Department.

	HoD English [N=39]		HoD Maths [N = 40]	
	Rank	%	Rank	%
Enjoyment/interest of students	4.5	59.0	2.5	62.5
Uptake at GCSE and A level	14	23.1	13	25.0
Quality of teaching in the department	1	94.9	1	85.0
Prior attainment of students (at intake to school)	9	43.6	4.5	60.0
Extent to which independent student learning is fostered	8	46.2	9.5	47.5
Development of students' study skills	14	23.1	17	15.0
Development of students' social skills	18	7.7	18	5.0
Visible student progress	10	38.5	6.5	55.0
Student motivation	4.5	59.0	6.5	55.0
Student behaviour	14	23.1	16	17.5
Student self-confidence	12	25.6	9.5	47.5
Extent to which departmental staff work together as a team	2	74.4	2.5	62.5
Personal effectiveness of teaching staff	11	35.9	13	25.0
Examination results	7	48.7	11	45.0
High expectations of students	6	56.4	4.5	60.0
Commitment/enthusiasm of teaching staff	3	71.8	8	52.5
Stability of teaching staff	16.5	15.4	13	25.0
Experience of teaching staff	16.5	15.4	15	22.5
Other	19	0.0	19	2.5

Note. HoDs were asked to identify up to eight factors.

Around a quarter of HoDs thought that the development of students' study skills; the stability and experience of teaching staff, and uptake at GCSE and A-level should be used to judge departmental effectiveness – whereas nearly half cited examination results.

Of the 34 mathematics HoDs who identified one factor as most important, a third (11) chose the quality of teaching in the department, and five nominated the enjoyment/interest of students. Four noted visible student progress. For English HoDs a similar pattern emerged. Of 35 who responded to the same question, 10 rated the quality of teaching as most important, five the enjoyment/interest of students and another five visible student progress. Overall, English and mathematics HoDs were broadly in agreement, with a rank correlation of 0.87, although mathematics HoDs attached relatively less importance to the commitment/enthusiasm of teaching staff and rather more to developing student self-confidence than their English counterparts did.

Factors contributing to departmental effectiveness

The HoDs were asked to indicate up to eight most important factors which they felt contributed to their department's effectiveness (see Table 10.5).

Table 10.5 Factors Which Contribute Most to the Effectiveness of Your Department.

	HoD English		HoD Maths	
	Rank	%	Rank	%
Quality of staff	4.5	61.5	10	42.5
Curriculum enabling students to work to their strengths	7	46.2	2.5	62.5
Building students' self-confidence	17	17.9	17.5	15.0
Careful record keeping and marking for GCSE coursework	14	23.1	11	40.0
Departmental staff working as a team	2	71.8	1	77.5
Teaching staff stability	15	20.5	6.5	50.0
Experienced senior staff	12.5	25.6	19	10.0
Teacher commitment/effort	6	59.0	4	60.0
Collective running of department/teacher involvement	10	35.9	8.5	45.0
Congruence of educational views and goals amongst teaching staff	10	35.9	17.5	15.0
Good organisation of resources	8	43.6	6.5	50.0
High quality teaching in most lessons	2	71.8	8.5	45.0
Strong academic emphasis in department	17	17.9	15.5	17.5
High expectations of students by most staff	2	71.8	5	55.0
Strong emphasis on homework by most staff	19	12.8	13.5	20.0
Good classroom management in most lessons	4.5	61.5	2.5	62.5
Effective assessment by most staff	10	35.9	15.5	17.5
Strong emphasis on examination entry	17	17.9	13.5	20.0
Encouragement of independent learning by students	12.5	25.6	12	32.5
Other	20	5.1	20	0.0

Note. HoDs were asked to indicate up to eight factors.

For both groups of HoDs, 'departmental staff working as a team' was cited most frequently as contributing to effectiveness. However, 'high quality teaching in most lessons' was identified less commonly by mathematics HoDs, only 45 per cent compared with 72 per cent for English. 'Good classroom management in most lessons' was reported by very similar numbers of HoDs in each group, as was 'teacher commitment/effort' – 60 per cent. By contrast, 'high expectations of students by most staff' was noted more frequently by English HoDs (72 per cent compared with 55 per cent). The 'quality of staff' was seen to be more important by English (62 per cent) than mathematics HoDs (43 per cent), whereas more weight was attached to a 'curriculum enabling students to work to their strengths' by mathematics HoDs.

In terms of factors seen as less influential, 'a strong emphasis on homework', emphasis 'on examination entry' and 'building students' self-confidence' were infrequently cited as contributing to success by both English and mathematics HoDs. English HoDs also gave much less weight to 'teaching staff stability' than their mathematics counterparts, only one in five compared with half. However, for English HoDs 'effective assessment by most staff' was more commonly seen to have a positive impact on effectiveness, more than a third of English HoDs compared with less than one in five mathematics HoDs noting this.

In all, 35 English and 38 mathematics HoDs made a response about which factor was most important. For English HoDs, 'high expectations of students' and 'high quality teaching' in most lessons were identified as most influential (by 23 per cent of English respondents, followed by teacher commitment/effort at 11 per cent). For mathematics HoDs, 'teamwork' was seen to be most important by almost a third of respondents followed by 'teacher commitment/effort, high quality teaching' and 'high expectations'.

English and mathematics HoDs shared a fair degree of similarity in their views regarding factors which contributed to departmental effectiveness, with a rank correlation of 0.70. English HoDs were likely to cite the following: 'high quality teaching', 'effective assessment by staff', 'congruence of views and goals' and 'high expectations of students'. By comparison mathematics HoDs gave rather more emphasis to a 'curriculum geared to students' strengths', 'careful record-keeping and marking' and 'staff stability'.

Barriers to greater departmental effectiveness
The HoDs were also questioned about 'what, if anything, holds your department back from being more effective?'. It is clear that, for the vast majority of both groups of HoDs, heavy workload was perceived as a significant barrier to greater effectiveness (see Table 10.6). Large classes were also noted by a substantial proportion (44 per cent of English HoDs and 55 per cent of the mathematics HoDs). Excess administration was also reported by over half of each group. Rather more of the mathematics than the English HoDs noted 'inadequate resources and large classes' as barriers to greater effectiveness. 'Inadequate emphasis on student assessment' and 'insufficient high quality teaching' in some classes were also noted by a sizeable minority. However, 'Low expectations of

Table 10.6 Factors Which Hold Your Department Back From Being More Effective.

	HoD English		HoD Maths	
	Rank	%	Rank	%
Differences between this department and other departments in school in teaching style	7	25.6	7.5	27.5
SMT support not always present	9.5	20.5	10.5	22.5
Inadequate resources	4	38.5	4	50.0
Heavy workload	1	82.1	1	85.0
Large classes	3	43.6	2	55.0
Insufficient high quality teaching in some classes	6	28.2	6	35.0
Low staff morale	12.5	10.3	16	10.0
Low expectations of students by some teachers	8	23.1	5	37.5
Lack of fresh ideas (static department)	21.5	5.1	12	17.5
Lack of stability in teaching staff	16.5	7.7	8.5	5.0
Lack of consensus in educational philosophy and goals in departments	12.5	10.3	14	15.0
Too much teacher autonomy	21.5	5.1	20.5	2.5
Excess administration	2	51.3	3	52.5
High teaching staff absence rates	16.5	7.7	22.5	0.0
Too little emphasis on examination entry by some staff	21.5	5.1	22.5	0.0
Too little emphasis on homework by staff	16.5	7.7	18.5	5.0
Insufficient challenge for high ability students	9.5	20.5	10.5	22.5
Insufficient challenge for low ability students	16.5	7.7	17	7.5
Poor organisation of resources	16.5	7.7	14	15.0
Insufficient feedback on work to students	21.5	5.1	9	25.0
Inadequate emphasis on student assessment	14.5	7.7	14	15.0
Little parental/community support for school	5	30.8	7.5	27.5
Other	11	11.3	20.5	2.5

Note. HoDs were asked to indicate up to eight factors.

students by some teachers' and 'insufficient feedback on work for students' were more likely to be cited by heads of mathematics than their English counterparts. By contrast, 'Too much teacher autonomy' was identified by very few HoDs as a problem area, as were 'high teaching staff absence rates' and 'low staff morale', this last despite the heavy workload. 'Insufficient challenge for low ability students' was likewise rarely rated, although just over one in five from each group identified 'insufficient challenge for high ability students' as a barrier to greater effectiveness.

A problem of 'Insufficient feedback on work to students' was noted by a quarter of mathematics HoDs, but only two out of 39 English HoDs highlighted this area. Very few in either group of HoDs identified too little emphasis on examination entry or on homework as factors holding back their department from greater effectiveness.

The HoDs were in somewhat greater agreement about barriers to departmental effectiveness than factors contributing to greater effectiveness, with a rank correlation of 0.83 vs 0.70.

Conclusions

In this paper we have reported on only one aspect of a larger study. Little research has been conducted into the topic of departmental differences in effectiveness, although work by Harris, Jamieson and Russ (1995) and Brown and Rutherford (1995) has highlighted the importance of this topic.

Broadly speaking, as far as HTs were concerned, five years previously (1990) the five main educational objectives of their school were centred around personal qualities or social mores, rather than overall academic achievement, but by 1995 both curriculum and examination results were included in the top five objectives. Nonetheless, promoting a sense of achievement, good behaviour and a pleasant environment were still seen as vital to the effectiveness of the school. Perhaps predictably, unlike the HTs, HoDs laid less weight on behaviour and more on the curriculum. Although not rated highly amongst their school's principle goals, judging school effectiveness, good progress or value added by the students was considered by HTs to be the single most important factor. This was also the opinion of the HoDs.

Asked about factors contributing to effectiveness, HTs most frequently mentioned a strong SMT, closely followed by an aspect of shared school culture or vision, staff and students' shared belief that the school is essentially a place for learning. On the other hand, HoDs were more likely to cite commitment by staff and strong leadership from themselves as important.

Regarding barriers to effectiveness, inadequate leadership by some HoDs and low expectations of the students by some staff were the factors most mentioned by HTs, although a socially disadvantaged intake was also identified as a major problem by many in our sample.

In connection with criteria which should be used to judge departmental effectiveness, heads of English and mathematics, respectively, displayed a fairly high level of agreement – 'quality of teaching' being the factor noted most frequently by both groups, followed by the 'extent to which departmental staff work together as a team'. Of the factors contributing to their own department's effectiveness, teamwork was the most mentioned element, with similarity of views held by heads of English and mathematics. A heavy workload was seen as the most unyielding barrier to departmental effectiveness. Again, this probably reflected the impact of implementing the NC and its subsequent revisions, compounded simultaneously by LMS and GCSE changes.

By and large, these results for both schools and departments indicate that HTs and HoDs tend to hold broadly similar views, as do the heads of English and mathematics departments. There were indications that the HTs laid somewhat more emphasis on aspects which may be viewed as highly visible criteria – academic achievement levels, behaviour and attendance – which tend to affect a school's public image. In the light of the last Government's policy emphasis on applying market mechanisms in education, it is notable that student and parent satisfaction were not rated highly by most of the sample in connection with judging school effectiveness.

Implications for school improvement

As we have argued elsewhere (Mortimore et al., 1988a, Sammons, Hillman & Mortimore, 1995) school effectiveness research should not be regarded as prescriptive and there are no easy formulas for school improvement. Nonetheless, we believe that there are promising signs that the closer links developing between school effectiveness researchers and those engaged with practitioners in the challenging task of improving the quality of the education offered to young people will benefit both fields (Reynolds et al., 1996; Stoll & Fink, 1996; Robertson & Sammons, 1997).

The findings of school effectiveness research can provide a basis for school self-evaluation and review (Sammons, Hillman & Mortimore, 1995). As Hopkins (1994) has argued in connection with school improvement *'the knowledge is not there to control, but to inform and discipline practice'* (p. 89). To date, however, school effectiveness researchers have paid too little attention to practitioners' perspectives and explanations of effectiveness. Our research demonstrates the existence of both important *communalities* in understanding and explanations and also variations. In *Forging Links* (Sammons, Thomas & Mortimore, 1997) we examine the practical implications of the research for different groups including the senior management team, middle managers and classroom teachers. One of our suggestions for those interested in school improvement is to use the kinds of questions reported in this article as a starting point for school self-evaluation and review. Whilst we do not suggest that these questions offer any magic solutions we consider they could be valuable in helping schools to analyse the extent to which different groups of staff (SMT, HoDs etc.) share common perceptions about strengths and weaknesses. Encouraging whole-staff discussion of such issues may well reveal multiple perspectives and could be painful. Nonetheless, it offers one way of attempting to analyse school and departmental culture and of stimulating the discussions necessary to develop a shared vision and commitment to common goals.

11

Using Value Added Research

Introduction

This chapter provides an illustration of the way value added analyses of school effectiveness have been used in one large English LEA (School Board) to support its interest in equal opportunities and raising school standards. Surrey Education Service was one of the first LEAs to introduce reception assessment of children at entry to school in 1993. This policy was intended first and foremost to assist teachers in their educational planning to meet the range of needs of children as identified in individual assessments made in the term of their fifth birthday. Teachers were encouraged to develop individual education action plans for children who obtained low scores on these baseline assessments and to develop strategies to monitor children's subsequent progress and development. Additional training for teachers, guidance documents and extra resources for schools having higher proportions of children with low scores at entry were all part of the reception assessment package and undoubtedly contributed to the successful implementation of the policy in schools. Parents received leaflets about the entry to school assessments and teachers were expected to discuss each child's results with his or her parents. Teachers were also encouraged to suggest ways that parents could help their children at home and the underlying rationale was that this would help to foster parental partnership.

The LEA also wished to use the reception baseline measures to help it explore equity in primary education for different groups of children. There was consid-

erable interest in establishing whether different profiles of prior achievement could be identified for girls as a group compared with boys, for those children young for their year (the summer born in the UK system) compared with older children in the cohort, for children for whom English was an Additional Language and for those from socio-economically disadvantaged (low income), families as identified by eligibility for free school meals. The LEA wished to use reception screening to provide a baseline against which the subsequent progress of individual children, and different groups of children could be compared in National Assessments in the three core curriculum subjects of English, mathematics and science conducted at the end of Key Stage 1 (age 7 years plus).

One feature of the introduction of baseline reception assessment was the opportunity it afforded for value added analyses of children's progress across KS1. The LEA was aware of the limitations of using 'raw' National Assessment results to compare schools and therefore convened a Value Added Working Group of headteachers and LEA officers to develop a strategy for such analyses which would provide useful information for the LEA and individual schools while avoiding potentially invidious public comparisons of individual institutions.

A key feature of the Surrey Value Added Strategy was the development of a Code of Conduct by which schools agreed not to publish or use their value added results for the purposes of marketing and publicity, and the LEA agreed it would not publish any league tables of schools' value added results.

Participation in the value added analyses by schools was voluntary, individual institutions opted in rather than opted out of the system. In return, schools were promised feedback in an accessible (non-statistical) format about their value added results in four areas of curricula: reading, writing, mathematics and science. In addition, a report giving overall results and those of their own schools, presentations and discussions were made to headteachers. The LEA's Advisory Service also made visits to schools to discuss the results and their interpretation on a one-to-one basis.

The Surrey project provides an example of collaborative research involving schools, the LEA and a Higher Education Institution. The multilevel analyses and feedback were conducted by myself and an ISEIC colleague, Rebecca Smees, in the first (pilot) year. As part of the project we trained LEA personnel in the statistical analysis and its interpretation so that in subsequent years they could conduct the multilevel analyses of their schools' data. The LEA conducted an evaluation of the pilot project to investigate headteachers' views about the value of the exercise.

Initially 107 schools (just over a third of the total in Surrey) opted in to the pilot project which analysed 1996 National Assessment results. The evaluation of the value added feedback proved highly positive and in 1997 all primary schools decided to participate in Surrey's value added project. Early indications suggest that, in combination with changes in national policy related to the enhanced priority attached to literacy and numeracy and associated target setting, a climate which gives greater emphasis to monitoring children's progress *has been created. The feedback from participants suggests that the value added project has enabled the LEA to provide useful feedback to schools on their effectiveness in promoting pupils' progress during their early years in primary school and encouraged schools to focus on the equity implications of lower attainment and subsequent poorer progress of certain groups (e.g., boys and those from socio-economically disadvantaged families). It is too early as yet to establish the long term impact on educational standards of the feedback of value added results to schools in Surrey, but the early indications are that schools have welcomed the additional information and are using it in a variety of ways to assist in evaluating their institution's performance.*

Baseline assessment is being introduced from September 1998 as part of a national policy of assessing children at the start of primary school. Although initiated under the previous Government, the new Labour administration has followed up this policy, the stated rationale being that

> 'The assessment will help in planning children's learning and the measurement of their future progress, and will ensure an equal entitlement for all children to be assessed on entry to school' (QCA, 1997, p. 3).

A number of LEAs and individual schools have used baseline assessment for several years. Blatchford and Cline (1992) provide an analysis of a number of schemes and argue that issues such as theoretical integrity, practical efficacy, equity and accountability are relevant for both practitioners and policy-makers in assessing the value of different baseline schemes. They also question whether the same assessments should be used for the range of different purposes baseline schemes are hoped to meet (formative, summative, evaluative and informative).

This paper presents the baseline results of an analysis of an LEA initiative in this area. It examines both the extent to which measures relate to children's later cognitive performance at KS1 and their usefulness in measuring the value added by schools.

Surrey Education Services introduced baseline reception screening for all pupils on entry in September 1993. The baseline screening was intended to fulfil four purposes.

1 To assist schools in identifying pupils 'at risk' so that action can be taken early to address need.
2 To enable the LEA to distribute a proportion of school funding on the basis of need.
3 To allow for overall monitoring of pupil progress in Surrey schools.
4 If desired, to provide a baseline against which future performance can be judged.

The Year R screening is conducted in the term pupils enter reception classes and covers a range of areas considered to be important for effective learning. The assessment comprises nine sub-scales which cover five main areas: language, early literacy, mathematics, drawing skills and social skills. The Year R screening results in the production of an individual profile for each child. The LEA has published guidelines to assist schools and teachers in interpreting these profiles and devise individual education plans to overcome weaknesses and develop strengths. In common with most baseline assessments the prime purpose of Surrey reception screening was to assist teachers in planning their work and thus to promote pupil progress.

Statistical analysis of pupils' matched baseline and 1996 KS1 results were conducted. Both simple descriptive analyses and more complex multilevel analyses (Goldstein, 1995) were used to:

- examine the links between pupils' reception screening results and their later performance at KS1;
- to examine the impact of pupil background factors including age, gender, eligibility for free school meals, and English as an Additional Language (EAL) status on later KS1 results;
- to establish whether pupils' previous educational experiences (type of pre-school provision and number of terms in nursery) were related to later performance at KS1.
- to provide schools with simple feedback about the 'value added' to their pupils' KS1 results after controlling for differences between schools in pupils' prior attainments and background factors at intake.

Code of conduct

Surrey introduced a code of conduct which was written in collaboration with schools concerning the use of information about schools' performance produced by any value added analysis. Participation was voluntary (schools 'opted in') and confidentiality of individual schools' results was assured. Surrey Education Service stated that no ranked 'league tables' would be produced from the value added analysis and participating schools agreed not to use their value-added results for marketing and publicity purposes. Just over a third of Surrey primary schools decided to take part in the value added survey in 1996 and, after the feedback and discussion of results, the Authority asked these schools to complete questionnaires to assist in evaluating the value added project's impact.

Results

Data were available for a total of 3703 pupils and 107 schools which had volunteered to join the value added project. Only pupils for whom full data were available were included in the analyses. In all, 49.3 per cent of the sample were girls, 12.7 per cent were eligible for free school meals (FSM) and 2.6 per cent used English as an Additional Language (EAL).

Details concerning the range of pupils' scores in each of the nine sub-scales used for reception screening are shown in *Appendix 11.1*. It should be noted that both Test/Task and Teacher Assessment (TA) results were available for the KS1 analyses of English and mathematics. For English attainment, levels in reading and writing were analysed separately. For the Test/Task results details of level 2 performance were also recorded (2C, 2B, 2A).

It was found that the Test/Task results showed a closer relationship with the reception screening sub-scales than the TA (probably a reflection of the more detailed information about level 2 results recorded by these measures). Given this, the analyses reported here focus on Test/Task results at KS1 for both English and mathematics. For science only TA data were available.

Table 11.1 shows the number and percentage of seven year old (Year 2) pupils at each level at KS1 for four KS1 outcomes – reading, writing, mathematics and science.

Table 11.2 shows the relationships between pupil performance across these KS1 measures. As might be expected, pupils who achieve well in one area tend to achieve well in others. Pupils' KS1 reading and writing results showed the strongest association ($r = 0.79$). Mathematics was more strongly correlated with reading and writing performance (0.67 in both cases) than with science results ($r = 0.61$). The weakest association was between science and reading (0.59).

The same children had varied quite markedly in their 1993 baseline attainments at reception (see Surrey Education Service 1994). As might be expected children's scores in the different baseline scales were associated. Table 11.3 shows the relationships between pupils' earlier performance in the nine reception screening sub-scales. It can be seen that the 'language' sub-scale shows the highest correlation with 'behaviour' (0.62) followed by 'sound' (0.57) and 'mathematics' (0.51). After 'language', 'behaviour' showed the strongest correlation with 'writing' (0.50).

The sub-scale 'book' was most closely associated with 'mathematics' (0.56), followed by 'letter' (0.55). 'Sound' was most closely related to language (as noted above), followed by 'mathematics' (0.48). The sub-scale 'letter' was most strongly related to 'mathematics' (0.63) followed by 'book' (0.55) and 'writing' (0.54).

Table 11.4 shows the associations between each of the reception screening scales and the four outcome measures at KS1. As might be expected given the time lapse between reception screening and KS1 assessment, the associations between performance in the nine sub-scales and KS1 results are somewhat weaker. Although all correlations are statistically significant and positive, the range is between 0.54 (maximum) and 0.15 (minimum).

As shown in Table 11.4 the nine baseline measures do show significant associations with pupils' later (KS1) academic achievement. These correlations are by no means perfect, however. Therefore, low scores on the reception screening sub-scales, whilst useful to identify potentially 'at risk' pupils, should not be allowed to lower teachers' or parents' expectations. The mathematics sub-scale shows the strongest association with all four outcome measures at KS1, followed by the 'letter', 'language' and 'book' sub-scales.

It should be borne in mind that many reception teachers will have used the reception screening results for diagnostic and formative purposes and this was encouraged by Surrey which had issued guidance on individual action planning. The early identification of low scores at entry on specific scales may have prompted teachers to focus on these topics with individual children. This action would be expected to reduce the overall correlations with later performance at KS1.

The value added analyses
As has been shown, children's scores on the nine reception screening sub-scales were linked as well as being associated with their later KS1 achievement. Multilevel modelling techniques (a form of hierarchical regression analysis) were therefore used to explore in more detail the relationships between children's achievement at intake to infant school and their later KS1 performance (Goldstein, 1995). These analyses give the best estimates of the impact of prior attainment measures and pupils' background characteristics on later attainment, as well as allowing the identification of any systematic differences related to school attended. Panel 1 gives a brief exploration of some of the technical terms commonly used in value added analyses.

Panel 1 – glossary of terms
- *correlation* – a statistical measure of the strength of the association between two measures (e.g., prior attainment and outcome variables)
- *hierarchical nature of the data* – data that clusters into pre-defined sub-groups or levels within a system (e.g., pupils, classrooms, schools, LEAs)
- *measurement error* – the error associated with trying to obtain a 'true' measure of pupil attainment from an 'observed' measure of attainment
- *multilevel modelling* – a methodology that allows data to be examined simultaneously at different levels within a system (e.g., pupils, classrooms, schools, LEAs). Essentially a generalisation of multiple regression
- *multiple regression* – a method of predicting outcome scores on the basis of the statistical relationship between observed outcome scores and one or more predictor variables
- *95 per cent confidence interval* – a range of values which can be expected to include the 'true' value in 95 out of 100 samples (i.e., if the calculation was repeated using 100 random samples)
- *prior attainment factors* – measures which describe pupils' academic attainment at the beginning of the phase or period under investigation (e.g., taken on entry to primary or secondary school)
- *pupil background factors* – pupil characteristics such as age, gender, socio-economic context, ethnicity
- *pupil level data* – data, such as attainment scores, collected directly from individual pupils
- *residuals* – the residual score is an estimate of the difference between an actual observed score and a score predicted from previous attainment – used as the basis of school effectiveness measures
- *retrospective nature of data* – data obtained about a previous period of a pupil's or school's history

Table 11.1 Percentage of Pupils at Each Level for Reading, Writing, Mathematics and Science 1996 KS1 Results.

Level	Maths %	Writing %	Reading %	Science %
Working towards Level 1	0.9	2.3	1.4	0.6
1	11.7	13.1	12.8	10.8
2A	25.5	20.1	13.2	
2B	19.2	28.4	18.9	69.4
2C	18.7	31.3	16.2	
3	23.8	4.6	37.3	19.2
4+	0 (n= 1)	0.1 (n= 3)	0.1 (n= 3)	0
Mean	2.10	1.87	2.22	2.07
SD	0.62	0.51	0.72	0.57

Note. $n = 3708$. Pupils with any missing KS1 data including those with a Temporary Disapplication (T) were excluded from further analyses leaving a sample of 3703.

Table 11.2 Correlations Between Pupils' 1996 KS1 Results in Reading, Writing, Mathematics and Science.

	Reading	Writing	Science	Mathematics
Reading	1.00			
Writing	0.79	1.00		
Science	0.59	0.61	1.00	
Mathematics	0.67	0.67	0.61	1.00

Note. n of pupils = 3703.

- *school level data* – pupil data aggregated or grouped at the level of the school
- *statistical uncertainty* – attached to an individual school's results and preventing fine distinctions to be made between individual schools, e.g., league table rankings (see 95 per cent confidence interval)
- *variance* – a statistical measure of the extent to which individual scores (e.g., pupils' test or assessment results) differ from the average (mean) score for a sample.

The impact of prior attainment

The first set of models were used to establish the extent to which, taken together, the nine sub-scales could be used to account for the variation in pupils' scores for each of the four outcome measures at KS1. A non-statistical summary is given in Table 11.5.

Table 11.3 Correlations Between Pupils' Baseline Reception Screening Results Across Nine Sub-Scales.

	Lang	Sound	Motor	Draw	Behav	Book	Letter	Writing	Maths
Lang	1.00	0.57	0.37	0.36	0.62	0.49	0.40	0.46	0.51
Sound		1.00	0.29	0.30	0.44	0.43	0.40	0.41	0.48
Motor			1.00	0.20	0.36	0.22	0.13	0.28	0.21
Draw				1.00	0.39	0.35	0.32	0.43	0.37
Behav					1.00	0.43	0.36	0.50	0.44
Book						1.00	0.55	0.49	0.56
Letter							1.00	0.54	0.63
Writing								1.00	0.50
Maths									1.00

Note. n = 3703.

Table 11.4 Correlations Between Pupils' Reception Screening Sub-Scales and 1996 KS1 Performance.

	KS1			
	Reading r	Writing r	Mathematics r	Science r
Reception Screening				
Language	0.44	0.40	0.39	0.40
Sound	0.41	0.37	0.34	0.35
Motor	0.17	0.19	0.18	0.15
Draw	0.30	0.30	0.28	0.26
Behaviour	0.39	0.36	0.36	0.34
Book	0.45	0.35	0.41	0.38
Letter	0.54	0.41	0.41	0.41
Writing	0.43	0.39	0.38	0.36
Mathematics	0.54	0.49	0.50	0.44

Note. n of pupils = 3703.

The results in Table 11.5 indicate that, although not showing the strongest association when correlated separately, taken together with the other baseline scales, the sub-scale 'sound' was the best predictor of later KS1 results. This may point to the value of developing children's phonological knowledge at a young age. By contrast, the sub-scale 'motor' was not found to be important, when the influence of the other baseline sub-scales were taken into account.[1] As the

---

[1] The sub-scale 'language' (which was fairly strongly correlated 0.57 with 'sound') was not significant for two of the outcome measures. Although it just reached statistical significance in the others, its sign changed, probably a reflection of its links with performance on the 'sound' sub-scale.

majority of scales were found to help predict later KS1 results, there is no strong case for eliminating any from the baseline screening assessment (except motor which may be relevant for other purposes) on the basis of this analysis.

Overall, controlling only for prior attainment the models accounted for a greater proportion of the total pupil level variance in reading (43 per cent) and a markedly lower proportion (only 26 per cent) for science. It can be seen that controlling for prior attainment accounts for a greater percentage of the school level variance in pupils' KS1 results for reading and writing than for the other outcomes.

The influence of pupil background
Many studies have demonstrated that pupils' background characteristics are related systematically to measures of cognitive attainment by age seven years (e.g., see Mortimore et al., 1988a; Tizard et al., 1988; Sammons, West & Hind, 1997). Surrey has already conducted analyses of such links for the reception screening sub-scales and these are not replicated here (see Surrey Screening Report 1993–94). Few English studies, however, have examined the impact of

Table 11.5 Summary of Multilevel Analysis of Reception Screening Measures as Predictors of Later 1996 KS1 Results (Simple Value Added Model).

	KS1			
	Reading	Writing	Mathematics	Science
Reception Screening				
Language	ns	*	*	ns
Sound	+*	+*	+*	+*
Motor	ns	ns	ns	ns
Draw	*	*	*	*
Behaviour	*	*	*	*
Book	*	*	*	*
Letter	*	*	*	*
Writing	*	*	*	*
Mathematics	*	*	*	*

Note. + = Sub-scale with largest effect.
  * = Measure significant ($p < 0.05$) in combined analysis.
  ns = not significant ($p > 0.05$).

Model Fit	% Total variance accounted for	% School level variance accounted for	% Unexplained variance attributed to school
Reading	43.0	43.0	5.3
Writing	40.5	34.4	9.8
Mathematics	34.1	12.1	11.6
Science	26.3	12.8	14.0

pre-school experiences on KS1 results although Davies and Brember (1997) have reported a cross sectional study which indicates positive effects for some pre-school experience.

Multilevel models were used to explore the links between KS1 results in the four outcome measures and the measures for which pupil level background data were available. These contextualised models are cross sectional, unlike the simple value added models shown in Table 11.5 which control for prior attainment and measure pupil progress over KS1.

Information about age (in months), gender, eligibility for free school meals, EAL status and type and length of pre-school provision and term of assessment for the baseline measures were available. However, due to the number of pupils for whom length of nursery provision could not be ascertained accurately this variable was not tested for the full sample. Term of assessment (entry to school) was highly correlated with children's age in months ($r = 0.91$) and because of this it was not possible to separate effects on later performance. Therefore, only age was included in the analyses.

Table 11.6 summarises the results of the analysis of the impact of background factors. It should be noted that for the type of pre-school experience measures, the contrast is in each case against pupils known to have had no experience of any kind of pre-school provision.

The results from the contextualised multilevel analysis show evidence of significant relationships between pupils' background characteristics and KS1 outcomes. These have important equity implications for practitioners and policy-makers.

The following factors were consistently associated with better performance at KS1:

- age (older members of the year group doing relatively better);
- all forms of pre-school experience had a positive impact in comparison with no experience. However, early school entry at rising rising five (i.e., before the term of their fifth birthday – the RISE category) showed the strongest relationship, followed by the category private nursery. (In the case of private nursery, socio-economic factors may also be linked as parents have to pay for the latter type of provision.)

By contrast, two factors were consistently associated with poorer performance at KS1:

- eligibility for free school meals (FSM); and
- English as an additional language (EAL).

Gender (girls versus boys) showed a strong and significant positive relationship with both reading and writing at KS1. However, whilst weaker, there was a significant negative relationship between gender (girls) and performance in mathematics and science at KS1 controlling for other factors.

It is well know that schools differ in the characteristics of their pupil intakes in terms of these factors. The next stage of the analyses, therefore, explored the

Table 11.6 Summary of Multilevel Analysis of Pupil Background Factors as Predictors of 1996 KS1 Results (Contextualised Model).

	KS1 Outcomes			
	Reading	Writing	Mathematics	Science
Background Factors				
Gender [girls]	[+] *	[+] *	[–] *	[–] *
EAL	[–] *	[–] *	[–] *	[–] *
FSM [eligible]	[–] *	[–] *	[–] *	[–] *
Age [in months]	[+] *	[+] *	[+] *	[+] *
Model Fit				
	% Total variance accounted for	% School level variance accounted for	% Unexplained variance attributed to school	
Reading	11.7	38.3	5.7	
Writing	12.0	28.8	10.6	
Mathematics	9.3	20.4	10.5	
Science	8.1	15.6	13.5	
Type of Pre-School Experience				
LEA	ns	ns	[+] *	[+] *
Play Group	[+] *	[+] *	[+] *	[+] *
Private Nursery	[+] *	[+] *	[+] *	[+] *
RISE	[+] *	[+] *	[+] *	[+] *
Mix	[+] *	[+] *	[+] *	[+] *
UNK	ns	ns	[+] *	[+] *

Note. * = factor significant ($p < 0.05$).
[–] = negative relationship.
[+] = positive relationship.
ns = not significant ($p > 0.05$).
UNK = pre-school experience unknown.

combined impact of controlling for both baseline attainments at reception and pupils' background characteristics.

Table 11.7 summarises the results of the multilevel analysis of the extent to which both background and prior attainment subscales can predict pupils' KS1 outcomes. Only measures with a statistically significant relationship ($p < 0.05$) were retained in the model.

The results indicate that only three pupil-level background characteristics were important in accounting for pupils' KS1 results when account is taken of baseline attainment. These are age (in months), gender and eligibility for free school meals. English as an additional language was no longer influential when prior attainment is controlled. This shows that such pupils made similar cognitive progress across KS1, in other words the attainment gap evident at entry did not alter.

In addition to the impact of individual pupil background characteristics, there was evidence of small but significant contextual effects related to the percentage of pupils eligible for free school meals in the pupil sample at the school level.

Table 11.7 Summary of Multilevel Analyses of the Impact of Reception Screening Sub-Scales and Pupil Background Characteristics as Predictors of Later KS1 Results (Full Value Added Model).

	KS1 Results			
	Reading	Writing	Mathematics	Science
Pupil background factors				
Age [in months]	[+]*	[+]*	[+]*	[+]*
Gender [girls]	[+]*	[+]*	[−]*	[−]*
FSM [eligible]	[−]*	[−]*	[−]*	[-]*
% pupils in school sample eligible for free school meals	[−]*	[−]*	[−]*	[−1]*
Prior attainment measures				
Book	[+]*	[+]*	[+]*	[+]*
Writing	[+]*	[+]*	[+]*	[+]*
Maths	[+]*	[+]*	[+]*	[+]*
Sound	[+]*	[+]*	[+]*	[+]*
Draw	[+]*	[+]*	[+]*	[+]*
Behaviour	[+]*	[+]*	[+]*	[+]*
Letter	[+]*	[+]*	[+]*	[+]*

Note. $p < 0.05$.

In schools with higher proportions of disadvantaged pupils, all pupils' KS1 results tended to be lower.[2] This finding is in accord with school effectiveness studies conducted at the secondary level (Willms, 1985; Sammons & Smees, 1997; Thomas et al., 1997a & b).

The majority of the reception screening sub-scales remained significant predictors of KS1 performance. The sub-scale 'sound' was the best predictor for reading, writing and mathematics but interestingly not science.

Older pupils did better in all areas than younger members of the year group. Because prior attainment is controlled, this means that older pupils made more progress over the infant years, as well as having higher initial attainments at reception.

Pupils eligible for free school meals (an indicator of low family income) scored less well at KS1 than others, after taking account of baseline attainment. In other words these pupils made significantly less progress over the infant years. The gap between disadvantaged pupils and others increased over time. Also the percentage of pupils eligible for free school meals was significant, indicating a contextual effect for all four outcome measures.

---

[2] Contextual effects were also tested for the percentage of pupils scoring highly on the sound sub-scale at reception but the results indicated that, when the percentage of pupils eligible for free school meals was controlled the factor did not improve the model fit.

Gender effects varied for different KS1 measures. Taking account of prior attainment (as measured by reception screening) girls did significantly better at KS1 in reading and writing. Thus girls made more progress over the infant years in these skills – the gender gap widened. By contrast, taking account of prior attainment levels (in which girls tend to outperform boys) girls made significantly less progress in mathematics and science than boys.

These findings have clear equal opportunities implications for the education of both sexes during the Key Stage 1. They also show that children who are young for their year (summer born children) and those eligible for free school meals are at risk of poor later performance. The influence of free school meals was strongest for later reading performance and again points to the important and continuing impact of socio-economic disadvantage. These findings demonstrate that the emphasis given by the SCAA (now QCA, 1997) to developing simple value added models which do not incorporate information about the background characteristics of pupils will provide a misleading picture of effectiveness for some schools, especially those serving socio-economically disadvantaged communities.

Table 11.8 shows the percentage of total variance (the amount of variation in pupils' KS1 scores in comparison with the mean score for the sample) accounted for by the model. It also gives an indication of the percentage of unexplained variation (i.e., that not related to prior attainment or background factors) attributable to the school.

The results indicate that background factors and prior attainment measures at reception are better predictors of later performance at KS1 for reading than for other outcomes. The level of statistical explanation is significantly lower for science than for other measures. (It should be remembered that the TA data available for science is cruder than the Task/Test data for other KS1 outcomes because there is no differentiation of pupils at Level 2.) In all, 51.3 per cent of the total variation in reading performance at KS1 is accounted for, but only 30.9 per cent for science.

A greater proportion of the variation in children's performance was found to be due to school attended for science, mathematics and writing attainment at KS1 than for reading. This is in line with findings from other studies (e.g., Brandsma & Knuver, 1989; Sammons et al., 1993; Sammons, 1996). We can conclude that schools are able to exert a proportionally greater influence over

Table 11.8 The Percentage of Total Variance in Pupils' 1996 KS1 Results Attributable to School Attended.

	% Total variance accounted for	% School level variance accounted for	% Total variance attributed to school
Reading	51.3	48.3	4.8
Writing	45.0	43.7	8.4
Mathematics	39.4	14.1	10.9
Science	30.9	15.7	12.8

subjects primarily taught at school than over reading which appears to be relatively more susceptible to home background influences (including pre-school learning). In all only five per cent of the total variance in reading was attributable to school, compared with over ten per cent for mathematics and science measures. In terms of unexplained variance the intra school correlations were 0.099 (reading), 0.153 (writing), 0.179 (mathematics) and 0.186 (science).

These results demonstrate that raw league table presentation of schools' KS1 results are liable to be particularly misleading for reading due to the greater impact of prior attainment and background factors on reading results. Although analyses reported here concern KS1 results, the same difficulty applies to 'league table' publication of raw KS2 performance tables. These results also have particular implications for the policy of target setting in literacy and lend some support to the choice made by some LEAs of using free school meals as an important though inevitably crude contextual indicator.

Are some schools performing better than others?
The multilevel analyses clearly demonstrated that a statistically and educationally significant percentage of the variation in pupils' KS1 results which cannot be accounted for by prior attainment and background factors remains attributable to school attended. In other words, of the unexplained variance a significant proportion relates to school attended rather than to individual pupil differences. In order to explore the issue of the value added by schools further residual estimates were calculated for each school in the sample for the four KS1 outcome measures.

These residual estimates enable 'like with like' comparisons of schools' results, taking into account relevant intake factors measuring individual pupils' prior attainment and background characteristics. These provide indicators of school effectiveness or of the value added by the school to pupil progress over the infant years. However, it is most important that account is taken of the statistical significance of any differences in the residual estimates (Sammons, Nuttall & Cuttance, 1993; Thomas & Mortimore, 1996). Taking account of the probability that differences might have arisen by chance (via the calculation of confidence limits for each school), means that it is not valid to make rank order distinctions (i.e., a league table presentation) for value added estimates. Only schools in which, taking account of intake, results are significantly ($p < 0.05$) better or significantly worse than predicted on the basis of intake relationships calculated for the whole sample, can be distinguished.

The vast majority of schools' value added results were based on pupil samples exceeding 15 in size (104 of 107). In calculating confidence limits the number of pupils in each school in the sample is taken into account (confidence limits are larger for small samples). The school effectiveness estimates (residuals) are also calculated with reference to the number of pupils at the school level. Thus multilevel techniques provide conservative estimates of the significance of any differences in performance identified.

Stability over time and subject differences
It is important to remember that only one year's KS1 results (1996) have been analysed to date. Therefore we cannot draw any conclusions about stability over time in schools' effectiveness. As future results are analysed it will be possible to examine trends over time and this will provide helpful indicators concerning improvement.

As might be expected given the impact of background and prior attainment factors, schools varied more in terms of their value added results for mathematics and science, and least for reading. Nonetheless, evidence that in some schools pupils were performing significantly better or worse than predicted was found for each KS1 outcome measure. For 30 (28 per cent) of the 107 schools reading results were significantly better or worse than predicted. For writing, 39 schools showed significantly better or worse results (36.4 per cent). For mathematics the significant differences were found for 39 schools (36.4 per cent) and for science the figure was 44 schools (41.1 per cent).

Consistency in schools' effectiveness at KS1
It should be noted that schools showed some marked internal variations in their effectiveness. Only a small minority 7 (6.5 per cent) had a significantly better or worse impact on all four measures at KS1. By contrast, 27 schools (25.2 per cent), almost exactly a quarter of the sample, were performing broadly as expected given their intakes in all areas. For three-quarters of schools, however, KS1 performance was either significantly above or significantly below that expected on the basis of intake in one or more areas.

In discussion with Surrey's Value Added Working Group it was decided that the feedback of value added results in terms of simple divisions (significantly above expected, as expected, or significantly below expected) would prove helpful as a starting point for evaluating current performance and identifying apparent areas of strength and weakness. This strategy was thought to be relevant in encouraging school target setting.

In this example, the school was performing above expectation in reading, but below expectation for its intake in science.

Table 11.9 Example of Method Used for Feeding Back Value Added Results to an Individual School.

	Reading	Writing	Maths	Science
Above expected	×			
As expected		×	×	
Below expected				×

Note. The above and below expected categories are statistically significant ($p < 0.05$).

Differential effectiveness
In order to establish whether some schools were more effective in promoting the progress of particular pupils groups, further multilevel analyses were conducted to test for differential effectiveness. Differential effects were tested for three intake measures for each of the four KS1 outcome measures. These examined groups of pupils based on prior attainment, gender and eligibility for free school meals specifically.

The results provided only limited evidence of statistically significant differential effects. Out of the 12 models tested, four (a third) indicated the evidence of differential effects for specified groups.

For reading no significant differential effects were identified. For writing, however, gender was significant suggesting that some schools were relatively better at promoting the progress of either girls or boys. For mathematics significant differential effects were found for the 'sound' baseline measure, suggesting that some schools were better or worse at promoting the progress of pupils with low or with high scores on this screening sub-scale. Finally, for science at KS1 significant differential effects were identified for two measures, eligibility for free school meals and 'sound'.

Although some evidence of differential effects was identified, in the main such differences were modest. On the basis of the results it was considered appropriate to calculate separate residuals for only two pupil groups (girls and boys) for writing. It is suggested that such results should only be fed back to schools where the pupil sample is large enough (more than 20 pupils in the analysis) and where a school's results were significantly better or worse for one pupil group. In practice differences were only found for a small number of schools, and in no cases were schools found to have significantly positive effects on progress for one sex but negative effects for the other. Nonetheless, providing indicators of differential effects may prove helpful in future for those schools where girls or boys respectively appear to have made significantly more or less progress, where a clear trend is identified over several years.

Summary and conclusions

The analysis of baseline reception screening measures in relation to KS1 results to provide indicators of variations in school effectiveness and can offer schools valuable additional information for the purposes of self-evaluation and review. Similar initiatives have been reported in Birmingham (Jesson, Bartlett & Machon, 1997), Wandsworth (Strand, 1997) and Hampshire (Yang & Goldstein, 1997).

Schools need to be cautious in interpreting value added analyses given that they are based on only one year's data. When further work is conducted on subsequent years' results greater confidence will be possible as trends over time can be examined. Year-on-year analyses should prove valuable for the analysis of improvement and in relation to monitoring the achievement of specific school targets.

The numbers of pupils in the sample at the level of the individual school used in calculating the value added measures should be referred to in interpreting the results. If the numbers for whom full data were available (and therefore who could be included in the analysis) in the 1996 Year 2 age group was much smaller than the total in that year group, the value added estimates will provide only a partial picture of school effectiveness. They will be accurate (at the 95 per cent confidence level) for the pupils included in the analysis, but not necessarily for the wider year group. However, if data were available for 90 per cent or more of the Year 2 age group, they should provide a fair representation of value added for the school at KS1 in 1996 in each of the four outcomes reported.

The value of linking school effectiveness and school improvement work has been noted by Creemers and Reezigt (1997) who argue that school improvement initiatives offer a valuable way of testing the validity of effectiveness research. It is advised that the value added feedback may be best used to encourage schools to identify apparent areas of strength and/or weakness to formulate provisional hypotheses about the factors which may have influenced pupil performance in specific areas for discussion with staff and primary advisers rather than as a mechanism for accountability. In other words, for raising questions rather than making judgements. The need for caution in using value added methods to make statements about school effectiveness has been highlighted by Goldstein (1997).

After such discussion, targets and possible strategies for action to promote improvement may emerge which can be incorporated into school's development plan for promoting performance in particular outcomes for specific pupil groups.

The possibilities of networking amongst schools to share strategies may also be worth exploring if individual schools would find this helpful, and if confidentiality can be maintained. An important feature of the Surrey value added initiative was the development of a clear code of conduct regarding the use of value added results.

Baseline assessment clearly provides opportunities to identify individual pupils' strengths and weaknesses in a variety of areas. Strengths of the Surrey scheme are the range of areas covered, the investment made in training teachers in their use and the guidance given on developing individual education plans for pupils. Parents were informed about the baseline assessment and suggestions about ways of helping their child at home were also published. The authority's policy of linking of resource distribution to low baseline assessment scores is beyond the scope of this paper but was clearly important in the successful implementation in Surrey and the generally positive views of schools towards the initiative.

The majority (three-quarters) of schools involved in the value added project differed in their performance from that expected on the basis of intake for at least one KS1 outcome. However, most varied being above or below expected in specific areas. Only a very small number of schools (under 5 per cent) were performing significantly below expected in all four measures at KS1 after controlling for the influence of intake, and equally small numbers (under 5 per cent) were performing significantly above expected levels. For most schools, therefore, the results indicate that internal variations in effectiveness exist. Overall, the school's contribution to pupil progress over the period Year 1 and 2 was greater for science and mathematics and smaller for reading which showed strong home influences. Given this, parental involvement in reading (already encouraged by many schools) should perhaps receive even further attention in all schools if reading standards are to be raised.

The lower rates of progress of boys in reading and writing, and of girls in mathematics and science apparent across the whole sample, suggest that equal opportunities issues remain highly relevant in considering whether differences in the nature of infant school experiences of the two sexes, or of teacher (or parent) expectations may play a part. The under-performance of girls in mathematics and science should not be forgotten given the high priority currently focusing on boys' poorer performance at later ages, especially in English.

The greater progress made by older pupils in the year group (even given initially higher attainment at reception) and the poorer progress of those from socio-economically disadvantaged backgrounds is also in line with other school effectiveness research findings conducted in different contexts and again has equal opportunities implications. It is worth remembering that, in contrast to standardised tests, national assessments do not explicitly take account of age related differences in attainment at KS1. Informal and formal methods of teacher assessment also do not make reference to age. It is possible that younger pupils are disadvantaged by being compared with their older peers rather than similar age pupils and that teachers may wrongly interpret lower results as a reflection of lower ability rather than due to age (Mortimore et al., 1988a). Given this, there may be particular advantages for the use of some age standardised tests to give a better picture of such children's performance, without losing sight of the other benefits of teacher assessment.

Surrey Education Service provided advice to schools as part of its feedback of value added results early in 1997. Examples of some of the questions raised by its consultants are shown in Box 1.

Surrey's evaluation of the VA project revealed that most schools (over 70 per cent) found the document and meetings giving their results useful and helpful and the process of providing the data straightforward. Surrey is now working on the 1997 KS1 results so that schools will be able to look at their pattern of results across the years. Some examples of the range of comments drawn from the evaluation questionnaires are shown below.

> 'We were really pleased we took part and are looking forward to comparing the outcome this year.'

> 'I find it all rather worrying that we are asked to achieve more and more – all the time raising standards even though they are good at the moment.'

> 'Thank you for the work carried out by the LEA on behalf of schools – very useful.'

> **Box 1** *Questions about value added feedback for schools suggested by Surrey curriculum & management consultancy*
>
> *Were the results as expected?*
> *Are there any marked differences between English, maths and science?*
> *Has the school any reason for the results?*
> * e.g., special circumstances staff/pupils*
> *Do the results reflect what the school values?*
> *Are there key implications for some pupil groups?*
> * e.g., those eligible for free school meals*
> * those young for their year*
> * boys or girls*
> *What has been your approach to*
> * grouping/setting*
> * provision for more able*
> * SEN provision?*
> *What schemes of work and resources are used?*
> *What is the recent pattern of teacher continuity?*
> *What is the level of teacher knowledge/expertise?*
> *Has the school set or planned benchmarks in these subject areas?*

'For a school such as this value added is the only way in which achievements of staff and children will even gain recognition in any league tables. Leagues tables as they are demotivate. Value added offers the possibility for improvement.'

The introduction of baseline screening in Surrey provided an opportunity to conduct an analysis addressing issues relevant to practitioners and policy-makers concerned with raising standards and is particularly relevant to the current debate about using target setting as a means to foster school improvement. It demonstrates that baseline measures are predictive of children's later cognitive performance at the end of KS1. However, although useful for identifying 'at risk' groups, performance is by no means set in stone and should not reduce teachers' or parents' expectations of low attaining children. Considerable attention was paid to encouraging individual education planning as a result of the screening results. Baseline screening can provide a basis for measuring pupil progress over KS1. Multilevel analyses allow the value added by schools to be calculated. The feedback of results using a code of conduct to protect schools' confidentiality proved popular with headteachers and it is suggested provides a useful starting point for school self-evaluation and review.

Appendix 11.1

Surrey baseline screening – pupils' sub-scale scores 1993.

	Mean	SD
Behaviour	32.3	5.17
Language	20.2	3.30
Book	4.1	1.70
Writing	9.4	3.13
Sound	15.0	3.06
Letter	25.2	17.77
Mathematics	17.9	5.51
Gross Motor Control	4.7	0.77
Drawing	11.7	4.04

Note. $N = 3708$.

12

Evaluating School Improvement

Introduction

School improvement initiatives have burgeoned in the last decade in many different countries and contexts. In the UK indeed the slogan 'improvement through inspection' was adopted by OFSTED, the national inspection body responsible for judging the performance of schools and quality of pupils' education in England and Wales, although this approach to improvement has proved highly controversial and has been challenged by academics and practitioners alike (see Earley et al., 1996 or Wilcox & Gray, 1996).

In part, the recent interest in school improvement has stemmed from the high profile accorded to school effectiveness research, which provided evidence of significant difference in the value added to students' educational outcomes between institutions with similar intakes. The successful challenge to the widespread acceptance of interpretations of mainstream educational research (both from sociologists of education, with their emphasis on structural determination and educational psychologists who stressed the primacy of early, family-based experience), that schools were of marginal importance as influences on educational outcomes led to an increased interest in the potential of school improvement programmes. The

> *'historic lack of any interface between school effectiveness research and school improvement practice, reflecting the very different intellectual ancestries of the two paradigms' (Reynolds, 1997, p. 104)*

has begun to change markedly during the last few years (Hopkins, Ainscow & West, 1994; Stoll & Fink, 1996) with increased recognition of the potential for synergy between the two fields. The benefits of grounding school improvement initiatives in the knowledge base of school effectiveness have been noted earlier in this volume. As described in Chapter 8 there is a growing body of evidence about the characteristics of effectiveness, for example, documented in the Key Characteristics Review, but less research attention has been paid to the ineffective end of the spectrum (Reynolds, 1996). Moreover, descriptive knowledge about the features of effective institutions does not of itself provide a clear strategy to guide policy makers or practitioners engaged in the dynamic process of attempting to change institutional culture or classroom practices. Indeed, the study of schools in difficulty suggests that turning around the 'failing' school is both difficult and time-consuming and that success is by no means predictable. There are 'no quick fixes' as my colleagues Louise Stoll and Kate Myers (1998) have succinctly concluded, and there is a need for vigilance to avoid 'peddling snake oil' – simplistic and prescriptive interpretations about 'what works'. The powerful influences of context, historical, socio-economic and cultural, have been rightly stressed by leading authorities (Hopkins, Ainscow & West, 1994; Mortimore, 1995; Reynolds et al., 1996; Teddlie & Reynolds, 1998).

Important questions, therefore, remain of how the two fields – school effectiveness and school improvement – can best inform research and practice in each? Is a marriage of minds possible and can the dangers of merger or take over be avoided? These are questions which lie at the heart of recent work conducted with Scottish schools on behalf of the Improving School Effectiveness Project funded by the Scottish Office Education & Industry Department (SOEID) which sought to explore issues of school effectiveness and improvement and bring together quantitative and qualitative methodologies and evidence to examine three major themes teaching and learning, ethos and school development (Robertson & Sammons, 1997a,b). This innovative study adopted both an effectiveness (value added) framework for the analysis of student outcomes using measures of progress over two years, and work with case study schools involving a critical friend and a researcher to explore the improvement process (MacBeath & Mortimore, 1994).

One of the most fruitful illustrations of the ways school effectiveness and improvement research can be linked is provided by the evaluation of school improvement initiatives. A recent example of an evaluation which explicitly sought to make such a link has been chosen for the focus of this chapter. The Making Belfast Work (MBW): Raising School Standards (RSS) evaluation was itself informed by the experience of working creatively with a large team of researchers drawn from the Quality in Education Centre at the University of Strathclyde and from the International School Effectiveness and Improvement Centre at the London Institute of Education on the Improving School Effectiveness (ISEP) project (described above) over four years from 1994 to 1998. Both conceptually and practically the ISEP research provided a valuable *grounding for the design of the MBW RSS evaluation, particularly in relation to the definition and measurement of improvement and process of linking qualitative and quantitative evidence. It also proved possible to adapt a number of instruments used in the ISEP study for the exploration of teachers', students' and parents' views in the MBW evaluation*

The need for rigorous evaluations of the impact of different approaches to school improvement is increasingly recognised, yet it remains true to say that few programmes have been evaluated in ways which explicitly focus on the impact on students' educational outcomes, rather than on purely process indicators (see the review by Barber & Dann, 1996). Yet much remains to be learnt about the processes of change which can improve our understanding of schools and teaching. Carefully designed evaluations offer valuable information for the development of school improvement theory and have the potential to inform policy makers and practitioners and refine future improvement programmes, as work by Bollen (1989) has illustrated in the Dutch context. School improvement strategies based on research evidence such as Slavin's (1996) Success for All literacy programme or the Dutch School Improvement Project (Reynolds et al., 1996) are unusual in their clear focus on student outcomes and have incorporated evaluation mechanisms which have demonstrated a clear impact on raising attainment levels.

This chapter is based on the results of a detailed evaluation of the impact of a recent UK initiative conducted in a highly disadvantaged urban area. It is based on a paper written with my colleague Brenda Taggart who played a leading role in conducting the fieldwork and drafting our field report. In 1994 a high profile (£3 million) three year school improvement initiative was launched in Belfast, Northern Ireland, funded by the Making Belfast Work programme, a body set up to help promote urban regeneration. The project was a collaborative venture involving 14 schools (both primary and secondary), the local education authority Belfast Education and Library Board (BELB), and the Department of Education Northern Ireland (DENI). In 1996, during the second year of the project, the BELB commissioned a 16 month evaluation of the impact of the MBW RSS Project.

The main aim of the initiative was to raise educational standards by providing additional support and resources to help the 14 participating schools address significant disadvantage and under-achievement amongst their pupils. Four secondary schools were included. These were chosen on the basis of their low public examination results – based on a three year period 1991–1993 – by inspectors. All had very high levels of socio-economic disadvantage amongst their student intakes, as measured by the percentage of students entitled to free school meals, and attendance levels were also poor. A number of feeder primaries were then identified for each secondary school. The selected primaries (10 in all) showed greater variation in their students' academic achievement results and levels of disadvantage but the rationale was to promote the attainments of future intakes to the four secondary schools at the heart of the initiative.

School development planning was given a high priority in the MBW RSS design. Participating schools produced annual Action Plans over the three years to help focus their work around the key foci of the initiative:

- *the quality of management, teaching and learning within the school*
- *the standards of literacy and numeracy*
- *levels of qualifications (secondary schools)*
- *links with local industry*
- *parental involvement*
- *discipline*
- *attendance and punctuality.*

The evaluation included the analysis of documents (such as school development plans, Action Plans, policy documents etc), schools' statistics (related to attainment, attendance and behaviour), and participants' views (key personnel in the BELB and in schools were interviewed and questionnaires were used to explore the views of three groups of stakeholders – teachers, parents and pupils).

The paper included in this chapter focuses on the design of the evaluation, and the impact of the initiative on three of the main foci: the quality of school management, teaching and learning, and educational standards. It was written by Brenda Taggart and myself and originally presented at the American Education Research Association Conference in San Diego 1998 (Taggart & Sammons, 1998). A full account of the evaluation is provided in the final report (Sammons, Taggart & Thomas, 1998).

As well as assessing the impact of the MBW RSS project, the evaluation sought to explore its legacy for the future and wider implications for those engaged in designing and implementing school improvement programmes.

Background

Concern about the quality of schooling in disadvantaged urban areas has been especially marked throughout the U.K. during the 1990s. This led, in part, to the then Minister for Education in Northern Ireland announcing in May 1994 a major school improvement initiative under the Making Belfast Work programme. The Raising School Standards (RSS) initiative was intended to 'help schools address significant disadvantage and under achievement amongst their pupils'. The project was designed to target a small number of secondary schools and their main contributory primary schools. A major difference between this initiative and other school improvement projects was the substantial financial investment (£3 million) agreed over a three year period and the strong emphasis placed on pupil outcomes.

In evaluating the MBW RSS initiative it is important to acknowledge the context of civil unrest which has been an everyday fact of life for people living in the city for over a generation. Recent political initiatives to move forward the peace process have been welcomed by all who are concerned about the quality of life in Belfast, although uncertainty about the future remains evident. Whilst outside there has been great tragedy, the schools themselves have been seen by many as 'peace havens' (BELB advisor 1996) although the impact of 'the Troubles' has been seen to have had a direct effect on some students' behaviour.

During the last fifteen years there has been a growth in public and policy concern about educational standards, particularly in literacy and numeracy, and widespread recognition of the importance of raising standards to meet the increasingly complex economic and social needs of society in the 21st century. This, in turn, has led to the demand for greater accountability of schools and teachers from policy makers, inspector and parents. Adding to the debate has been increasing evidence from school effectiveness research (Rutter et al. 1979, Reynolds, 1985, Mortimore et al. 1988a, Sammons et al. 1994a) and more recently school improvement (Hopkins, Ainscow & West 1994, Myers K 1995, Stoll & Fink 1996, Barber & Dann 1996, Sammons et al. 1997) about what makes an effective school. This research evidence has highlighted the existence of 11 key characteristics of effective schools (Sammons, Hillman & Mortimore 1995).

Since the introduction of the Educational Reform Act (ERA) in 1988 in the UK, schools have experienced a period of rapid change (the introduction of the National Curriculum, LMS [1992], School Development Planning, Appraisal, statutory testing at the end of Key Stages etc.). The statutory publication of examination results in 1992 resulting in league table rankings in the press added yet another dimension to the debate about standards. The argument as to whether raw results tell us more about the catchments served by schools rather than the quality of education in schools in socio-economically disadvantaged areas, has remained a source of controversy (Mortimore, Sammons & Thomas 1994).

The school effectiveness research base provides evidence of the need to develop value added approaches for the evaluation of school performance. Such approaches are becoming increasingly widespread and a number of Local Education Authorities have introduced innovative schemes (for a review see Thomas & Sammons 1997). The School Curriculum and Assessment Authority (SCAA now QCA) has undertaken work in an attempt to develop value added approaches for the reporting of schools' national assessment results at Key Stage (KS) 2 and KS 3 in England and Wales, while in the Northern Ireland context the Department of Education Northern Ireland (DENI) has funded research to develop a comprehensive value added framework for the evaluation of school performance as part of the expanded Raising School Standards Initiative.

In evaluating the success of the original MBW RSS initiative it is important to remember the highly selective nature of secondary education in Northern Ireland as well as the high levels of unemployment and socio-economic disadvantage evident amongst many schools in Belfast. Pupil choice at the secondary phase in particular is complicated by: selection by aptitude (11+), geographical boundaries and religious background. It is known that there is a relationship between the concentration of socio-economically disadvantaged pupils (those eligible for free school meals and those with low attainment at entry) and general levels of pupil achievement. Contextual effects on secondary school pupils' achievement have been identified by a number of school effectiveness researchers (Willms, 1986; Goldstein et al., 1993; Sammons et al., 1994; Thomas et al., 1997). Inevitably, a selective educational system is likely to exaggerate such contextual effects amongst the least popular and most disadvantaged secondary schools.

Making Belfast work

The funder of the initiative, Making Belfast Work, a programme within the Department of the Environment, was launched in July 1988 to 'strengthen and target more effectively the efforts being made by the community, the private sector and the Government in addressing the economic, educational, social health and environmental problems facing people living in the most disadvantaged areas of Belfast' (March 1995: MBW Strategy Statement). As MBW is concerned with urban regeneration it has an important role in the delivery of the Government's Targeting Social Need (TSN) strategy designed to provide financial assistance to overcome social disadvantage.

A recent paper by MBW (1997) made the following comments, 'education is pivotal to the Making Belfast Work strategy for urban regeneration within the city of Belfast. MBW's focus on education reflects one of its core aims to 'improve the ability of residents to compete for jobs' and is a considered response to a long standing awareness of the link between socio-economic disadvantage and educational attainment.' Their intervention is focused upon their own core wards, and recognises the importance of including all parties; The Department of Education, the Education Library Boards, the Council for Catholic Maintained Schools, and the individual schools and parents, in the process of improving performance. That their focus on improving attainment at primary and secondary level is set clearly in the context of other social objectives reflects their holistic approach to the problems (Hutson forthcoming 1998).

The aims of the project

The project focused on four secondary schools (two maintained Catholic and two controlled Protestant) identified as having high levels of under achievement and their ten main contributory primary schools. The primary aim of the project was to provide additional support and resources to schools with the overall objective of accelerating an improvement in the performance and employability of school leavers by:

1. improving where appropriate the *quality of management*, teaching and learning within the school;
2. improving the *standards of literacy and numeracy* and overall standards at both primary and secondary level;
3. improving the *level of qualifications* which they achieve at school and improving access to further education;
4. improving *links with local industry*

5   increasing *parental involvement*
6   dealing with problems of *discipline* in class
7   improving *attendance and punctuality*

It should be noted that of these seven foci, three contain measurable pupil outcomes (2, 3 and 7) whereas two focus on school and classroom processes which influence the quality of the educational experience (1 and 6). By contrast, 4 and 5 concern aspects of links with the broader community.

The external evaluation of the MBW RSS initiative

In May 1995 the Belfast Education and Library Board (BELB) commissioned the International School Effectiveness and Improvement Centre (ISEIC) at the Institute of Education, University of London to undertake an external evaluation of the initiative. As well as an investigation of the impact of the overall project it was thought desirable to examine the experience of implementing the project in individual schools in order to identify the factors which facilitate improvement and barriers to success. The evaluation was also intended to examine the *general* implications of the project for school improvement projects elsewhere and to identify examples of good practice for dissemination to other schools.

Individual schools involved in the RSS initiative were required to submit Action Plans for approval by the Central Management Committee (CMC) demonstrating how they proposed to meet the stated aims and objectives of the initiative. Schools were also required to engage in self-evaluation and to collect relevant information to enable BELB to review the impact of the initiative, including the identification of a range of targets and baseline measures. A key feature of the initiative was a focus on the use of measures of pupil achievement to establish the extent to which improvements in standards occur.

A number of commentators have pointed to the value of a case study approaches to increase understanding of the processes which foster school effectiveness and positive change and provide the 'rich description' of processes needed by practitioners concerned with school improvement. The National Commission on Education's (1996) *Success Against the Odds* study of 11 schools in disadvantaged areas provides a source of evidence relevant to policy makers and practitioners concerned with school improvement. Other school effectiveness research, which combined detailed qualitative case studies of both more and less effective secondary schools and departments with quantitative value added analyses of academic effectiveness at GCSE, of relevance to the evaluation of the project is provided by Sammons, Thomas and Mortimore (1997).

The results of research on the impact of School Development Planning (MacGilchrist et al., 1995) also has implications for the evaluation of the project. This drew particular attention to the need to examine the extent to which activities included in schools' written plans actually affect classroom practice and the extent of staff ownership of, and involvement in, preparing the plan.

The extended MBW RSS evaluation adopted a case study approach in considering the four secondary and ten feeder primaries involved in the BELB project. The case studies recognised both the overall aims of the initiative *and* the differing aims of individual schools expressed in their individual Action Plans. A variety of sources of data and evidence were considered:

- document analysis
  school and authority (BELB and DENI) level e.g., school development plans, RSS Action Plans, school policy documents, Inspection reports
- schools' statistics
  pupil outcome data, attendance, expulsions and suspensions as well spending patterns of the additional resources
- interviews to explore the views of key personnel BELB advisors, Principals, Co-ordinators, Field Officers and Chairs of Governors (COGs)
- questionnaires to explore perceptions of the school amongst pupils/parents (Primary = Year 5 and 7, Secondary Year 9 and 11) and all teachers.

This paper examines a number of aspects pertaining to the management of a school improvement initiative, including the launch and selection of schools. It also analyses the process of action planning in schools and addresses three of the main aims of the overall initiative: the quality of school management; teaching and learning (1) and educational standards (2 & 3). Barriers to the implementation of the project and factors which facilitated its success are identified. The evaluation was commissioned to explore the impact of an *educational initiative* and the factors which influenced its implementation and success. It does not attempt to evaluate the success of individual schools. For full details of the evaluation see Sammons, Taggart and Thomas, 1998.

Launching the initiative
It was reported that the idea for the initiative came originally from DENI, set against a backdrop of concerns about under achievement in the Province as a whole. Interviewees said it took some two years from inception to the press announcement by Education Minister, Michael Ancram MP on 11 May 1994.

How any project is launched is important in terms of setting the tone for what is to come. The Minister's press release referred to schools 'who are presently achieving no or only very poor qualifications' but did not publicly name those schools. The press coverage (Ulster Newsletter and Belfast Telegraph) factually reported the main points from the press release. The Irish News (12.5.94) however published names of the four secondary schools and called them 'schools identified as having the highest levels of under-achievement.' This high profile public naming was not well received by school staff who were mindful of their institution's reputation in the community.

> 'Before any public announcement a great deal of work has to be done with schools to ensure they are prepared for critical publicity.'

Clearly, the same issues have been debated elsewhere in the UK concerning the relative merits of the popularly coined 'naming and shaming' policy concerning failing schools. Public naming is seen by some to be a necessary stimulus for improvement for schools with serious weaknesses, but others believe it can exacerbate schools' difficulties, lowering staff morale, making recruitment harder and exacerbating problems due to falling rolls (Mortimore & Whitty 1997).

> 'The approaches made to schools to invite them to participate in an initiative must be given careful consideration and procedures must be uniform. It is important that schools are given sufficient notice of inclusion in any developmental project work to enable them to think through the implications of this with their Chair of Governors and key personnel in school.'

The reaction to being included

Despite the controversy generated by the launch, only two Chairs of Governors (CoGs) thought their school reacted badly from being included in the initiative. In one school the CoG perceived the staff were 'alarmed' about inclusion as some stigma might be attached to the school. In the second school the CoG thought the Principal and staff were 'apprehensive' because of the extra work the initiative would generate. All of the remaining CoGs thought their schools reacted very positively to being included in the initiative. One CoG of a secondary school said he thought the staff accepted the criteria for selection as there seemed 'no point in arguing about it,' and the additional resources would help teachers 'tackle problems we knew existed.'

The reaction of secondary Principals to being included in the initiative was on the whole more negative than positive. In one school, it was reported that the senior management had to work hard to prevent a 'negative feeling' from being labelled a failing school from prevailing. They deliberately presented RSS to the staff as 'recognition of the advances we've made' with the funding being seen as 'help to enable us to continue.' The challenge for this Principal was to 'turn it around': to foster a 'no blame culture' so that staff would view the initiative positively. He reported some success in this area. A second Principal reported how betrayed and angry he and his staff were, feeling the school had being unfairly stigmatised. A third Principal thought staff saw RSS as just 'another initiative' (one of many the school had been involved in). One Principal was unable to comment as he was not in post at the time.

Whilst some secondary school Principals may have felt uneasy by press coverage of the initiative this did not affect their primary colleagues in the same say. Primary school Principals saw the criteria for selection (low achievement) being unique to secondary schools with feeder primaries chosen to support the four secondary schools. The most common reaction to inclusion in the initiative in the primary sector was excitement about the additional funding. This was not however all that primary Principals considered. In three schools there were concerns about work load issues.

> 'If initiatives are sprung on teachers their reactions could delay progress in the initial stages and deflect their focus away from the aims of the project.'

Selection of schools

It was considered important when setting up the initiative that the selection of schools was based on objective criteria. In addition, a stated pre-requisite for inclusion, regardless of other factors, 'was the commitment within the management team to achieve successful change.' The final list of schools included in the initiative followed a 'catchment area' approach to identify feeder primaries for the four selected secondaries. Once schools were selected they were asked to sign a formal contract.

The initiative had a clear focus on raising school standards and the secondary schools included in the project were selected after consideration of DENI statistical information using the following criteria:

- a performance index which took account of both the number and level of GCSE and other vocational qualifications;
- the percentage of pupils who achieved A–C in at least five subjects;
- the percentage of pupils with no GCSEs.

An AVOWQI (Average Weighted Qualification Index) was considered the most appropriate indicator for selection purposes. Points were assigned according to grades obtained by school leavers. The final AVOWQI was calculated by summing the total of grade points for all pupils in each school across the three year period 1989–90 to 1991–92 and dividing this by the total number of school leavers for that school across the same three year period.

The primary schools invited to participate had to satisfy the following criteria over three years:

- the primary school must be a feeder for one of the participating secondary schools
- 25 per cent or more of P7 pupils must transfer to one of the participating secondary schools
- the number of pupils who transfer must constitute 10 per cent or more of the total number transferring to one of the participating secondary schools;
- the results of transfer tests.

Although the selection criteria for inclusion in the MBW RSS initiative were clearly identified, there was some controversy about the selection of schools. Much of this debate focused on the following issues:-

- Targeting Social Need (TSN)
- selection of the four secondary schools in relation to under achievement
- selection of the feeder primary schools

Targeting social need

The controversy about the amounts of money allocated to any one individual school must be seem within the context of a much larger debate taking place in Northern Ireland at present about the efficacy of the 'Targeting Social Need' policy.

In order to help combat 'social deprivation' and 'need' the policy of Targeting Social Need was introduced as a Government Public Expenditure priority for all Northern Ireland Departments in 1991. The operation of the TSN policy in DENI since 1994, enabled five per cent 'top slicing' (approximately, £40 million) of the education budget to be redistributed across the five Education and Library Boards. This money is *additional* to that distributed to schools by the normal Local Management of Schools (LMS) formula. The Boards distribute TSN money to schools using the number of children on role entitled to free schools meals as one indicator of social need. Some boards also include in their distribution formula to schools other indicators of disadvantage i.e., the special needs of pupils etc., which has resulted in some anomalies in the amounts individual schools receive in TSN allowances between the different Education and Library Boards. The Northern Ireland Affairs Committee, Second Report Feb. 1997 on Under achievement in Northern Ireland Secondary Schools came to the conclusion that 'there were no clear mechanism for ensuring an equitable allocation of resources to schools with similar proportions of deprived children in different Board Areas' (p. xix). It would appear that the impact of the TSN policy in tackling under achievement has been little researched and the Affairs Committee concluded that

> 'DENI should monitor the use of TSN money more closely to ensure adequate and equitable funding for social need and to avoid the problems of a geographical lottery may require consideration of how a common funding formulae could be used to target funds. The aim should be to meet the needs of the most disadvantaged pupils without rewarding schools with low standards. This would only be possible if information about pupil disadvantage and low performance were collected concerning pupil *intakes* to schools. Any formulae would need to be transparent and take account of existing knowledge of the relationships between social disadvantage and educational attainment.' (p. xxi ibid.)

The TSN policy has a bearing on the initiative given that the funders, MBW, saw themselves as being at the forefront of the Government's TSN strategy. Critics of the initiative said they thought MBW's involvement with RSS ran counter to its remit in delivering TSN money (particularly in the selection of primary schools, which will be addressed later). It is clear that the use of TSN money is a matter of interest and concern for those working within the Northern Ireland educational system, not least those working in schools who are ever mindful of levels of resourcing in an increasingly competitive climate.

The situation in Northern Ireland is not different in this respect to that operating in England during this period. For example an analysis of LMS funding focusing on the definition and funding of 'educational need' by Sammons (1993) produced very similar conclusions and indicated that a pupil with an identical score in terms of 'need characteristics' would attract different levels of funding in different Local Education Authorities (LEAs).

The links between poverty, social deprivation and under achievement continue to be a subject for debate amongst educationalists, employers and politicians (see Mortimore & Whitty, 1997, Robinson 1997, Sammons et al. 1994b; 1997). In this context it is relevant to note that, under the last Government, there was considerable reluctance to recognise any causal links between social disadvantage (as indicated by FSM) and educational achievement since this was felt to condone low expectations. Indeed methods of contextualising school performance data using such measures developed by OFSTED (Sammons et al., 1994b) were suppressed in 1995. The MBW RSS initiative was set up to help schools tackle under achievement, 'aimed specifically at improving the levels of performance ... including those at secondary schools who are presently achieving no or only very poor qualifications.' (Michael Ancram press release May 1994). It is therefore unsurprising that the four secondary schools chosen also scored highly on measures of social deprivation i.e., free school meals.

Clearly the bracketing together of TSN and RSS did result in a 'layer of confusion' but the criticism that the 'GCSE criteria were never defined precisely' or that 'they were not applied robustly,' made in press coverage (McGill, 1996) cannot be levelled at the original Making Belfast Work RSS initiative on the evidence provided for the evaluation.

> 'It is helpful when launching an initiative to consider the effects of the selection criteria on a wide audience. Where the criteria for selection is unambiguous it may be sufficient to publish this information only. Where data are more complex and could be misinterpreted then it is perhaps wise to include some exemplification material to assist in the interpretation of the selection criteria.'

The rationale behind the selection of primary schools was in keeping with the initiative's aims of improving levels of achievement in the secondary sector by placing particular emphasis on improving the basic literacy and numeracy levels of children entering those schools in Year 8. However, given the debate about TSN, the inclusion of some of the primary schools appeared anomalous. There were many other primary schools in Belfast with much higher levels of social deprivation and need, so the inclusion of some of the MBW RSS primary schools was seen as controversial given the MBW's stated policy of Targeted Social Need. However, given that the stated criteria for selection included transfer numbers alone then other aspects such as low levels of academic standards and disadvantage could not be covered fully.

To influence the levels of achievement in secondary schools it was important when setting the selection criteria that the *actual* numbers of children transferring from a given primary was considered. But in looking at the actual numbers rather than a percentage of Year 7 leavers, the selection criteria worked in the favour of large primary schools. Some smaller primary schools, not selected for RSS, sent a higher *percentage* of their pupils to RSS secondary schools. Some of the larger RSS primary schools whilst sending a larger number of *actual* children to one of the RSS secondaries could not count this as one of their more popular receiver schools in percentage terms.

Thus although the selection criteria, devised for the MBW RSS project appeared straightforward, the bias towards larger primary schools added to the controversy about the selection of the primary schools. Against this it can be argued that the selection of larger primaries ensured that larger numbers of children were covered by the project.

> 'Any criteria for the selection of schools must be relevant and evidence-based, clearly explained and applied fairly.'

The implications of being labelled an underachieving school
For BELB the choice of the secondaries was an acknowledgement that these schools were faced with problems, but given the publicity, being included on the initiative brought the public stigma of failing. Some Principals were uncomfortable with the qualifications based performance indicator used, which they felt left other areas of success in their schools unacknowledged. All of the four secondaries knew they scored poorly in terms of examination performance but staff would not necessarily have described the schools as failures. It should be noted there was also a feeling amongst colleagues in non-selected schools with high levels of disadvantage that failure was being rewarded and that more able Principals were being penalised for being effective managers who run successful schools. It is important to recognise that the RSS schools had the advantage of considerable additional resources and external support for school improvement as a result of the project. Nevertheless, the demands for accountability and the requirements for successful improvement strategies are not easily reconciled as experience elsewhere in the UK also demonstrates (TES Nov. 1997).

> 'The implications of tackling poor performance must be thought through very carefully. How this impinges on morale within a school and on public confidence outside of the school must be *managed* by the project leaders'

> 'Schools need to have some time to adjust to being labelled and stigmatised as under performing. The Principals, Staff and Governors of these schools need to be worked with sensitively in order to get over the *improvement* message.'

The schools' action plans – writing, implementing and evaluating
One of the key features of the MBW RSS was that selected schools had to submit for approval to the Central Management Committee (CMC) an Action Plan (AP) focusing on a whole school approach to meeting the objectives of the initiative. In many respects the AP was intended to be the school's blueprint for improvement. As the initiative was to extend over three years, it was considered important that schools build in improvements to their structure and working arrangements so they could maintain and further improve standards within the

normal LMS funding arrangements after the conclusion of the project. Schools also had to seek approval from their Board of Governors who were asked to give undertakings to assist in the monitoring and evaluation of the programme.

The APs were designed to broadly detail the problems which schools wished to tackle and the methods they would use to do so and in addition, schools were advised that the plans should identify:

- what can be achieved within *existing resources*, including funds to which schools are entitled under the LMS formula
- *funds from other MBW projects and from the Belfast Action teams* (schools already participating in MBW initiatives such as the Numeracy/Literacy project and the Discipline project were expected to indicate what benefits they had already derived from their participation in such projects, how lessons learned would be disseminated throughout the school and how ongoing work would be subsumed within the overall school plan) and
- *additional resources* needed from the project.

All APs had to be agreed by the CMC and any *additional funding was to be determined on their judgements of the merits of each individual proposal.*

It is widely recognised that school improvement needs to be planned, it is not therefore unreasonable to suggest that schools might be more confident in approaching Action Planning if they had experience of planning for school improvement. The process by which most schools would be familiar with planning for improvement would be through the production of a school development plan (SDP). There are no statutory requirements for a school to have a development plan in the UK. However given the evolution of school development planning (MacGilchrist et al., 1995, Hargreaves & Hopkins, 1991, 1994) the presence of a plan has become increasingly important as a developmental tool.

In England, schools have been assisted in formulating SDP by a number of Government publications (DES 1989a, 1989b, 1990, 1991 and OFSTED 1995). Mention of school development planning in the NI context is contained in *The Inspection Process in Schools (1992)* DENI. It states that 'Increasing use is being made in schools of Development Plans, in which schools outline their policies, identify aims, undertake their own evaluation of how well or how far those policies, aims and objectives are being implemented, and identify priorities and targets for review and development' (p. 7) and in *Evaluating Schools* (DENI 1992) comments under 'Management Arrangements. Organisation of the curriculum' that the organisation of the curriculum can be considered good when: the senior management and the teachers have contributed to discussions of the place of the statutory curriculum within the wider curriculum and have agreed on a strategy for its implementation; there is a costed Development Plan covering curricular needs, changes and potential improvements; this is available to all interested parties, including parents' (p. 9). Hargreaves and Hopkins (1991) argue that leadership, management and efficiency should be judged by the extent to which, 'the school, through its development planning, identifies relevant priorities and targets, takes the necessary action, and monitors and evaluates its progress towards them' (p. 21). Matthews and Smith (1995) state that 'the introduction of the inspection system, together with post-inspection action planning, is making a contribution to school improvement' (p. 29).

Thus the reliance placed by the MBW RSS initiative on Action Planning and school development planning should be seen in the wider context of the highest profile given to school inspection and reflect the importance attached to planning as a vehicle for improvement by the Inspection service in both England and NI at this time.

> If new and innovatory techniques are to be part of a project's methodology, the implications of this for school personnel must be considered *before* implementation. Where this is likely to present challenges, project leaders should ensure that practical measure are put into place to help overcome difficulties.
>
> If Action Planning is to be used as a tool for school development, those responsible for this task must be equipped to manage the process. They need to be able to build on existing school development frameworks and set realistic targets.

Elements to be considered within the Action Plan
The MBW RSS APs were intended to be specific to the needs of individual schools. Nonetheless, the programme was expected to include some or all of the following elements:

- developing effective management at all levels within the school
- the development of literacy/numeracy across the school
- raising expectations amongst pupils, parents and teachers
- increasing educational achievements
- involving parents in the support of schools' provision
- improving attendance and punctuality
- increasing and improving links between the school and employers
- developing appropriate certifiable vocational courses
- developing and implementing effective procedures to monitor and evaluate the curricular and organisational feature of the school
- preparing and planning programmes.

A proforma to help schools write up their APs was developed containing, in the first year, the following headings:

- curriculum focus
- year groups to be involved
- current position
- targets
- time-scale

- staff involvement
- resources required – including costings and
- evaluation.

The BELB perspective on action planning

The majority of BELB Advisors interviewed considered the production of an AP to be *the main challenge* for schools in the early stages of the initiative. All the BELB personnel interviewed suggested that many of the problems associated with the first year of the initiative were as a result of a lack of experience in Action Planning. One Field Officer summed up the views of many when reporting that because Action Planning was couched in terms such as 'success criteria,' 'target setting,' 'areas of planning,' it demanded from Principals a 'clarity of thinking' which took some time to develop.

BELB Advisors were familiar with the processes of development planning but they reported facing some specific challenges when considering the whole notion of Action Planning. Some Advisors considered themselves, 'ill prepared for what APs was all about in terms of the depth, detail and rigour of what was required.'

The submission of the first APs was considered by all to be very rushed. The original timetable identified January 1994 as the time when schools were to be approached and invited to become involved in the initiative. Implementation was to begin in April 1994. In the event, this time scale proved impossible to adhere to because the initiative was not officially announced until May 1994. As the Northern Ireland academic year finishes at the end of June, this left little time in which to begin any necessary consultation. Most schools began writing their plans when they returned to school in the Autumn of 1994. By November 1994 all schools had submitted their Action Plans to the CMC. As a consequence the initiative did not start in earnest for some schools until the following January (1995), almost three quarters of the way through what was meant to be the first year of the initiative.

> The time-table of a project needs to be carefully mapped out. Any time-table needs to be sympathetic to the other calendar demands on schools. Schools need time sufficient time to *plan* for developments, before embarking on implementation. Time-tables need to be *explicit* and realistic.

The lack of expertise, coupled with the very short time in which schools were asked to formulate and present plans, resulted in APs being submitted which BELB Advisors could see would be problematic when it came to implementation. The most commonly reported problem in the first APs was their tendency to be over-ambitious. It was believed, with hindsight, that schools planned to cover too much in too short a time scale. Where some schools identified strategies for improvements in all of the nine key areas, advisors had serious doubts about the extent to which the schools could deliver all the improvements they were aiming for simultaneously. Despite the production of an AP proforma, there was in practice no common method of completing this by the schools.

The analysis of schools' APs showed considerable variation and little consistency in how an AP could be approached in the first year. The evaluation mechanisms in particular showed wide variations and targets were often non-specific. It is clear that at the start of the MBW RSS no consistent approach was adopted for Action Planning. This had important consequences for the later evaluation of the impact of RSS (for details see Sammons, Taggart and Thomas 1998).

Where new techniques are introduced into schools it can be helpful to develop exemplification material for training purposes (to be used *before* implementation). The use of such material would encourage consistency and specific key areas i.e., evaluation mechanisms, could be highlighted.

Implementing the action plan
A number of factors helped Co-ordinators implement their Action Plans. The most significant factors were:

- having the co-operation of staff
- linking the school's staff development programme to the Plan
- the monitoring and evaluation of target setting
- being given the opportunity to keep staff informed of developments through regular meetings work at the very beginning of the project which had an immediate impact on the school e.g., the redecoration and re-equipping of rooms for library and reading club, the introduction of new books and resources etc.
- the courses run by BELB

Other factors mentioned as important in some schools were: the arrival of a new Principal, having a non-class teaching Co-ordinator, the re-deployment of staff and focusing on existing expertise on the staff.

Two groups of factors mitigated against Co-ordinators implementing their Action Plans. The first were concerned with staffing and personnel issues and the second with the central organisation of the initiative.

Staffing issues included:

- staff with entrenched attitudes
- poor departmental leadership
- not having the right people in post at the right time
- the Co-ordinator feeling isolated.

Central organisations issues included:

- the time scale for developments being too rushed at the outset
- the time scale between the submission of AP and the arrival of funding being too long
- complications in the system for ordering resources
- not enough support for staff development.

Improving the quality of action plans
It was recognised by BELB that the quality of the first APs were often poor, given the short time scale and lack of expertise and experience in school. The poor quality of the initial AP coupled with the emphasis on resources and the budget cycle, sent out mixed messages to schools. Schools were advised not to focus on financial matters but then felt pressured to spend large amounts of money very quickly without time for adequate reflection and planning in order to fit in with the demands of the financial year (1.4.95) for first year's APs. This meant that the initiative was perceived, in the early stages (possibly the first six to nine months) as being 'resources driven.'

In response to these difficulties BELB personnel began to look at other school improvement initiatives, other school improvement networks and to recognise that new strategies needed to be put in place if the initiative was to be refocused on its underlying aims. This led to a conference (9 and 10 May 1995) in which a leading school effectiveness researcher was a key note speaker. The conference was attended by Principals, Co-ordinators, DENI and BELB gave a general background to the 'school effectiveness movement' in both America and England, and focused on questions of urban disadvantage and school effectiveness and improvement. There was a particular stress on: academic achievement; reward based control systems with reference to areas of consistency, cohesion and constancy; and the role of the head teacher. In relation to School Development in RSS the emphasis was on: the means of achieving outcomes; organisational change and the change culture; altering behaviour; the focus on teaching and learning; and the development cycle.

After this keynote conference delegates were asked to get involved in workshop sessions which explore practical steps towards improvement. They also had the opportunity to analyse a practical example of an Action Plan teasing out from this example the key factors which could assist school improvement. The July conference was reported by many of the interviewees to be a *major turning point* in the development of the MBW RSS initiative. The subsequent evaluations of the day (by BELB) suggest that the key note speech and the input on Action Planning were the most valuable. After the conference, Principals and Co-ordinators assisted by BELB personnel felt they were able to apply much of what they had heard (and done) on that day to the situation in their own school. They were able to look at their APs in a new light and to set this into the overall context of school improvement.

> The involvement of outside consultants can be crucial in the school development process, whether as 'key-note speakers' or 'critical friends'. The ability to identify the most appropriate outside consultant is dependent on opportunities for senior personnel to network with colleagues in the field.

This subsequent emphasis on developing schools' capabilities for Action Planning meant that schools were more involved in the monitoring and evaluative process which, 'brought a new dimension to their schools, which is significant.' (BELB Advisor) This affected planning in almost every school, giving it a much sharper edge. At the start of the initiative, where schools had thought about school improvement they appeared to have visions, but lacked any sense of the detail needed to make them achievable. The conference was reported to have helped Principals, Co-ordinators and Advisors identify the important small steps which would help lead towards improvement.

Using Action Planning as an 'effective management tool' was also seen to have enabled Advisors to work with Principals and Co-ordinators 'on monitoring and evaluation and how you tie in finances and resources to your objectives.' Whereas schools were thought to be very weak at this before the initiative, subsequent training and guidance was felt to have done much to help them to improve this area of their work significantly. The approach to Action Planning taken after the first year appeared to help Principals and Co-ordinators narrow the focus of their APs and make their improvement programmes more realistic and manageable. Advisors encouraged schools to focus on one or two of their key objectives.

The impact of making Belfast work: Raising school standards

The evaluation mechanisms used to determine the success or otherwise of the initiative were inextricably linked to the schools' APs. Evaluation does not exist in a vacuum. It should not be seen a bolt on appendage but part and parcel of the planning cycle (MacGilchrist et al., 1995; Barber & Dann 1995; Sammons et al., 1997). The Board and Schools' own capacity to evaluate of the initiative improved markedly after the first year as schools became more adept at producing annual APs which showed a greater focus on the aims of RSS.

Although schools were given suggestions as to the type of monitoring and evaluating procedures they could engage for the initiative, this was not made a requirement at the start of the project. Advisors reported that some baseline measures were in place for evaluation purposes but this was inconsistent across the fourteen schools and practice in this area showed a great deal of variation. The mechanisms for evaluation identified in the APs ranged from some quantifiable data such as monitoring statistics, to more subjective perceptions of 'how things were going'.

Clearly some areas of the initiative are much easier to evaluate than others. In the core curriculum subjects of English language and mathematics pupil progress can, and was, in some schools explored by analysing gains in pupils' standardised test scores (see Impact on teaching and learning) although not all schools adopted this strategy.

An overall lack of data on baseline measures appears to have seriously weakened most schools' ability to take meaningful measures. As noted by OFSTED, 'the absence of success criteria in some initiatives makes it difficult for participants to know and understand the impact of their work and means that they depend on subjective judgement or anecdotal evidence, neither of which are likely to be accurate about the past or a good guide to future planning' (p. 36). The first year of the MBW initiative clearly reveals this as a problem in many schools. Subsequently schools' abilities to APs improved markedly in most cases.

The introduction to this evaluation acknowledged the particular difficulties faced by schools in areas of civil unrest. The National Commission on Education (1996) reported that any school which attempts improvements is influenced by factors which operate from *outside* as well as inside the school. Whilst none of the case study schools in Success Against the Odds experienced the particular circumstances of life in Belfast, they were nevertheless, in difficult and challenging environments. The Commission reported that what marked these eleven schools out was their, 'will to succeed was strengthened, rather than weakened, by calamities.' (p. 315). These were schools who were able to 'strengthen and hone their vision in relation to whatever vicissitudes are thrown up by their environment.' Working in Belfast is undoubtedly very challenging and school improvement here, more than perhaps anywhere else in the United Kingdom, has to be seen in terms of the extent to which 'challenges have been turned into opportunities. . .' (ibid).

The primary purpose of the RSS initiative was to tackle under-achievement and by considering 'pupil outcomes,' to assist teachers to raise standards and improve the quality of pupils' educational experience. When considering 'impact' the external evaluation addressed not only 'pupil outcomes' but changes in other areas identified in the original proposal as important to the project (management, the quality of teaching and learning and educational standards in literacy and numeracy).

Baseline measures
Although baseline measures were advocated at the start of MBW RSS (in 1994), schools did not appear to have received clear guidance on which to use and there was little consistency in approach across the schools as a consequence. However, when the later DENI Raising School Standards Initiative (expanded) was introduced in 1995 considerable time was spent establishing baseline measures.

The establishment of baseline measures can assist in the monitoring, evaluation and management of any new development. They can be used as the basis for the development of 'success criteria' and enable issues of impact to be examined critically and objectively for example, baseline measures have been used to provide value-added feedback to primary schools (e.g., Sammons & Smees, 1998; Strand, 1997; Yang & Goldstein, 1997) and at secondary level (Thomas & Mortimore 1996).

The extent of the impact
The evaluation data used to examine the initiative's impact were collected during the second and third year of the initiative (October 1996–July 1997). Development continued in schools after the data collection period and schools were encouraged to engage in exit planning via special meetings with BELB personnel. For some areas of the initiative it has been possible to demonstrate

clearly, on standardised measures, the impact of the initiative in the short term. These tend to be curriculum areas which have established assessment practices such as reading/maths tests. For other areas the evidence is more subjective reflecting participants' perceptions and in some instances may not be calculable for several years to come. Given the length of the initiative and the scope of the external evaluation, this paper cannot ascertain any long terms effect RSS might have. To explore this a follow-up study would be needed 2–3 years after the end of the project.

The use of the questionnaire data
As part of the evaluation, questionnaires were administered to three key groups: parents, pupils and teachers. The questionnaires have been used to inform the external evaluation and to inform individual schools.

In order to do this the quantitative and qualitative responses were aggregated together across the fourteen schools to give a picture of the initiative as a whole. It is important to note that the sample for this survey was small, using only fourteen schools. The schools themselves were not randomly chosen but fitted a specific criteria for selection on the initiative. Therefore, these data cannot be used to make wider claims about the performance of Belfast schools as a whole.

Outside of the remit of the research proposal, each Principal was sent a report analysing their own school's information. This aggregated responses from all fourteen schools set beside the information from their own school. The report was sent during September 1997, after the official end of the initiative in order to avoid any research bias or interference.

The feedback questionnaire analysis was used to identify strengths and possible areas for further development. It was also able to reveal any significant differences between perceptions in individual schools and in the RSS sample as a whole. The approaches used, and the cut-off points for determining strengths, represent particular ways of looking at the data and it was suggested that schools could also use different approaches. Schools were given guidelines on interpreting their data and on the ways they might address some of the issues raised by the questionnaires.

Reading and using the questionnaire data
Before considering the following sections the methodology used to interpret the questionnaire data needs some explanation. In each of the questionnaires respondents were asked questions which related to the aims of the initiative. The pupil questionnaires were similar in that *secondary* pupils were asked whether certain things happened in their school: always, usually, hardly ever and never, whereas *primary* pupils were asked if these occurred: always, most of the time, sometimes and never. Parents were also asked to respond to a range of statements as to whether they: strongly agreed, agreed, disagreed, strongly disagreed with them.

The purpose of the pupils' and parents' questionnaires was to explore perceptions of their school in relation to the developmental areas of the initiative.

The teacher questionnaire was different from the others as it sought to explore specifically the impact of the RSS initiative. In this questionnaire teachers were asked to respond to questions measured on two scales:

Scale A: the extent to which the school had made improvements during the last three years. This was graded: significant improvement, slight improvement, the same (no change), slightly worse or significantly worse

Scale B: the extent to which RSS had affected the above improvements. This was graded: significant effect, some effect or no effect.

The impact on school management

The importance of leadership and management which can help to enhance a shared vision for improvement has been highlighted in much of the literature on school effectiveness and school improvement both in Britain and internationally (Gray, 1990; Mortimore, Davies & Portwa, 1996; Stoll & Fink, 1996; Teddlie & Stringfield, 1993; Hallinger, 1996; Sammons et al., 1997). The starting point for exploring the impact which RSS had on management was the analysis of the teachers' questionnaire which sought views on the extent to which teachers (in all fourteen schools), thought their school had improved over the period of the initiative for a range of management indicators. This was compared with the extent to which teachers thought RSS had influenced these improvements. Both the additional written responses on the questionnaires and information taken from interviews were used in interpreting the quantitative data.

The questionnaire data was divided into two parts: (1) areas in which teachers considered RSS had a significant effect and (2) areas in which teachers considered RSS had no effect.

Table 12.1 displays the results for items rated as the most and those rated as the least affected.

Table 12.1 Areas in Which Teachers Considered RSS Had a Significant Effect.

% significant effect	% no effect		% improvement
40.0	17.2	staff contributing to the school development plan	74.1
38.3	10.4	clear vision of what the school is trying to achieve	92.4
34.8	22.5	the staff development programme	72.8
Areas in Which Teachers Considered RSS Had Least Effect			
22.0	26.4	communications between SMT and staff	69.7
18.2	33.0	the quality of SMT decisions	57.4
17.9	33.5	staff cohesiveness	53.3
27.4	34.1	staff morale	50.8
23.7	36.2	the leadership of the school	60.1
9.0	54.5	staff having a say in financial decisions	36.6

Note. $n = 189$.

In interpreting the tables the extent to which RSS was perceived to have affected an area needs to be considered alongside the extent to which that area was thought to have improved over the last three years. It is notable that the areas in which school have reported the most improvement (SDP 74 per cent, Vision 92 per cent and Staff development 73 per cent) are also the areas in which RSS was reported to have had the most significant effect. Similarly relatively less improvement was reported in the areas in which RSS was said to have had the least effect (Morale 53.5 per cent and collegiate decision making 37 per cent). The exception to this last case is the leadership of the school in which 60 per cent of teachers reported on improvements but 36 per cent perceived no effect on leadership from RSS. For these teachers *the appointment of a new Principal was seen as fundamental* to improvement in the management of the school rather than the RSS initiative. The importance of headship is widely recognised as crucial to successful change in the school improvement literature. 'The right sort of leadership is at the heart of effective schooling, and no evidence of effectiveness in a school with weak leadership has emerged from any of the reviews of research' (National Commission, 1996, p. 335).

The impact of new Principals cannot be attributed directly to RSS, nonetheless, improvements in the overall management at senior level were reported in most schools. In four schools (1 secondary and 3 primary), in particular, RSS was said to have enabled various individuals to hone their leadership qualities. This included comments such as 'the Senior Management Team ....has been allowed to develop' and Vice Principals were cited in particular for showing 'strong management style and expertise.'

An analysis of written comments showed that in general terms teachers from eleven of the 14 schools thought RSS had helped in sharpening the focus of management as it had encouraged them to 'look more closely at the running of the school' which had implications for school improvement.

Areas in which RSS appears to have had the most impact

School development planning
Nearly three quarters (74 per cent) of teachers thought SDP had improved during the period of RSS with 40 per cent crediting RSS as strongly instrumental in this. Teachers reported that the emphasis on planning (APs) had 'given direction for overall school development' which had also contributed to the 'development of schemes of work.' This conclusion is in line with Advisors' perceptions on this improvement in capacity to plan, engendered by RSS reported earlier.

There was evidence of considerable variation between schools in teachers' perceptions of improvement in management. In nine schools more than two thirds of respondents thought improvement had occurred over the last three years, although in only five schools was RSS seen to have a significant effect (by over half the sample) all of which were primaries. The range was from zero to 80 per cent.

Developing a vision of what the school is trying to achieve
The vast majority (92 per cent) of teachers thought their school's vision had improved, by far the highest rating for any item. Over a third (38 per cent) thought RSS had a significant effect on this change. Teaches wrote that RSS had helped develop vision by making 'staff more aware of (the) direction the school is moving because the 'aims of the school' had been made clear. Teachers had been set clear 'goals' and had a firmer notion of 'what we want to achieve.'

Teachers in all schools reported improvement in the school's vision (for 12 schools this was noted by over 90 per cent of respondents). Again for five schools (all primaries) RSS was seen to have had a significant effect in promoting this improvement. The range was from only seven to 83 per cent of teachers attributing a significant effect to RSS in their school. The small numbers of respondents in some schools (in two primaries and one secondary less than ten questionnaires were returned) means that differences between individual schools should be treated with great caution.

Improvements in the staff development programme
As with the two categories above, improvements in this area were also seen as significant. When teachers chose to write extra comments on the questionnaires thirty-six teachers from eleven schools chose to comment on this aspect of RSS. In written comments the schools' staff development programme was perceived to have been the area most affected by RSS. Whilst RSS enabled teachers to update their skills on INSET courses outside of school, teachers thought the opportunities for intra-school staff development of greater importance. In conclusion RSS was said to have, 'enabled teachers to work together in the classroom and learn from each other in a non-judgemental situation' which 'enabled staff members to share experiences and support each other.'

This perception concurs with the observations made on the importance of staff development programmes for school improvement by Barber and Dann (1996) who concluded that: 'a successful urban school requires a learning staff' (p. 22).

In all, for seven schools improvement in the staff development programme was reported by more than two thirds of teachers (the average was 73 per cent). However in only five schools (all primaries) was RSS seen to have had a significant effect on this. The range was from zero to 80 per cent.

Improved communication
Overall, improvements in communication between the Senior Management Team and the staff was reported by over two thirds of teachers (69.7 per cent). However this was not always seen to have been as a direct result of RSS. The difference between those teachers who thought RSS had no effect on this (26.4 per cent) and those who thought it had a significant effect (22.0 per cent) was small. An example of where RSS had been influential was summed up by one teacher who said 'RSS has 'opened' the management of the school in that our RSS co-ordinator has adopted a policy of informing the staff about decisions

etc. Staff morale is much higher as we feel we are all working together.' In some schools not only was communication improved at an institutional level, where 'the process of planning for RSS has focused SMT's minds on the issue of planning, communicating and working *with* staff,' but also at a personal level. 'We are able to spend quality time on a one to one basis with all the SMT.' The general benefits reported in this area were 'improved communications, staff being kept informed and made to feel part of a school team.' By contrast, RSS was not perceived to have led to improvements in the area of communication in schools where the project was seen as a primarily senior management initiative, little shared with the rest of the staff.

Areas in which RSS appears to have had the least impact
The three areas of management in which RSS would appear to have had a relatively lower impact are staff morale (34.1 per cent no effect), leadership (36.2 per cent perceived as noted earlier) and staff involvement in financial decisions (54.5 per cent no effect).

Staff morale
Just over half the teachers in the survey (50.8 per cent) thought staff morale had improved over the period of RSS. In some cases this was accounted for by factors outside of the scope of RSS, (34.1 per cent thought RSS had no effect in this area). Where RSS was thought to have had no influence on morale, teachers reported difficulties in coping with the pace and nature of 'change' brought about by 'initiative' type developments. This was summed up by one teacher who wrote that she/he was experiencing 'too many changes in too short a time. Bombarded with information and no time to have a quiet think about the implications.'

Where RSS was thought to improve teacher morale this appeared to be as a direct result of additional resourcing. Some teachers reported being 'delighted' with the additional resources they had access to as a result of RSS and one teacher wrote that this had a beneficial effect on teachers' morale because there had been improvements in not only working conditions but 'things I have needed for years I have now got.'

The morale of staff working in disadvantaged urban areas is an extremely important issue which needs to be taken into consideration when discussing school improvement. For example, National Commission on Education (1996) state that 'schools serving disadvantaged and frequently troubled areas, an abundance of energy and commitment is needed just to tread water' (p. 335). Schools in such areas often find it hard to recruit and attract staff and the stigma of being associated with a 'failing school', in raw league table terms may exacerbate recruitment difficulties.

Staff having a say in financial decisions
The questionnaire asked teachers nine questions about management issues in their school. Teachers reported the least improvement (36.6 per cent) in this area

and over half (54.5 per cent) felt RSS had not affected their involvement in financial decisions. By contrast less than one in ten (9.0 per cent) of teachers believed RSS to have had a significant effect on this area. Whilst Principals and SMT may have been increasingly willing to consult with teachers on other matters there still appears to be a reluctance in most schools to include teachers in financial decision making. Only five teachers from four schools chose to comment on this issue. These were from schools where they felt they had been involved in some financial decisions reporting that certain 'staff involved had an increased say on how money is spent in their co-ordinating areas.'

Other respondents' perceptions of impact
The questionnaires provide a picture of teachers' views of the impact of the initiative on management. This question was also explored with other groups involved in the initiative. Two particular improvements were cited by CoGs as being relevant to management, as well as observing more structured staff development programmes, they were also of the opinion that 'attitudes in general' were more conducive to school improvement as a result of RSS.

For the majority of Principals and Co-ordinators, the opportunities RSS presented for senior management to improve their communication and leadership skills and focus on 'target setting' were seen as the key to improvements in management. Positive developments were reported such as a clearer focus on planning and an increased awareness of the importance of the School Development Plan coupled with a better approach to the development of *whole* school policies. Principals reported that RSS had made them much more aware of their teachers' staff development needs.

The BELB Field Officers who worked with schools as part of the MBW RSS initiative echoed the findings of the CoGs and the Co-ordinators and added to be debate about management. They thought that RSS had been influential in highlighting the value of senior management *monitoring what is happening in classrooms* in a more systematic way.

Crucial to the success of RSS in some schools was, as reported by BELB personnel, the extent to which 'the management structure within the school was sensitive to the need for improvement.' It is evident that when respondents referred to 'management', they often meant the vision and commitment of individual Principals. The importance of the headteacher has already been commented on and in many respects RSS is no different to many other school improvement initiatives in that Principals demonstrated varying degrees of commitment and involvement. The relationship between the leadership of both senior and middle managers (heads of department) for secondary schools and school culture has been noted in studies of inner city schools in other contexts (Sammons, Thomas & Mortimore, 1997). A comment from Success Against the Odds echoes the sentiments of the BELB interviewees who drew attention to this areas: 'at the heart of each school is the headteacher's vision and analysis of what constitutes good learning and teaching. The refusal to be deflected from this and its corollary – an understanding of the difference between *means* (intensive and sensitive work with parents, for example) and *ends* (high academic and social achievement) – marks out these headteachers and their colleagues' (p. 336).

The impact on teaching and learning

The MBW RSS set out to tackle under achievement and raise school standards. Any attempt to improve pupil outcomes needs to focus on the quality of teaching and learning (Creemers, 1994; Slavin, 1996; Reynolds, 1997). The teacher questionnaire contained a number of items which attempt to address the impact on classroom practice.

Table 12.2 follows the pattern shown in the previous section on management. In line with the figures reported on management it is striking that the areas in which schools have reported the most improvements (curriculum development 88.6 per cent, teachers collaborating to plan teaching 81.1 per cent and the planning of lesson materials 82.7 per cent) are also the areas in which RSS was reported to have had a significant effect. Again, rather fewer respondents reported improvement in the areas in which RSS was said to have had no significant effect (SMT discussions 31.5 per cent no effect, pupil feedback 38.5 per cent no effect and homework 48.3 per cent no effect) by a substantial minority.

Areas in which RSS appears to have had the most impact

Curriculum development
Curriculum development was seen to have improved by nearly 90 per cent of teachers and over half felt RSS had played a significant part in this. Teachers reported that RSS enabled 'curriculum development to have a greater time

Table 12.2 Areas in Which Teachers Considered RSS Had a Significant Effect.

% significant effect	% no effect		% improvement
51.4	11.0	curriculum development	88.6
33.1	25.4	teachers collaborating to plan teaching	81.1
31.5	14.4	planning of lesson materials	82.7
30.4	17.1	quality of teaching generally	80.0
29.7	22.5	the quality of school policies	75.2
27.4	16.2	teachers focus on learning	77.3
26.9	20.9	the implementation of school policies	78.5
Areas in Which Teachers Considered RSS Had Least Effect.			
21.3	31.5	the frequency of discussion with SMT	59.0
16.2	38.5	the feedback given to pupils about their work	59.2
12.8	48.3	consistency in regular setting & marking homework	46.2

Note. $n = 189$.

allocation' which resulted in 'greater opportunity for curriculum development' and the ability to 'target certain areas of the curriculum.' Developments were reported at both institutional level because RSS 'facilitated curriculum panels to function and allowed us to maximise the curriculum/subject expertise present on the staff' and at an individual level where 'teachers have been free . . . to follow up discussions and develop their ideas on curriculum delivery.' Curriculum development in this context for many teachers was regarded as synonymous with staff development. Where improvements in teaching and learning occurred, particularly in the areas of language and maths, teachers reported that INSET opportunities, updating policies and the introduction of new programmes had brought about changes, 'practical in-service training which included demonstration lessons by Field Officers has helped in the teaching of reading and practical mathematics.'

The majority of teachers in all schools (89.1) thought there had been improvements in curriculum development in their school over the last three years. Only in one primary school did less than two thirds of staff think this area had improved (60 per cent). Amongst the secondary schools the range was from 67 per cent of teachers reporting improvement to 82 per cent. Only in six schools (all primaries) was RSS seen to have had a significant effect on curriculum development by over half the staff who completed the questionnaire.

Teacher collaboration
Over 80 per cent of teachers reported being drawn together in schools to work on various aspects of RSS. The result of this increased collaboration appears to have been more collegiate planning and, in some cases, the development of new schemes of work. A common comment from teachers was 'many areas of teacher collaborating and planning greatly improved by meetings with RSS appointed staff' and 'better schemes of work and better forward planning have led to greater improvements in development of (the) curriculum and have contributed greatly to the overall teaching within the school.' This emphasis on planning has felt to have led to great improvements in the quality of teaching and learning within many of the schools.

As well as reporting greater collegiality in planning the content of lessons, RSS appears to have encouraged more thought to be given to resourcing activities. Where this happened teachers reported that the additional resources offered to the schools as a result of RSS made an important difference to the quality of teaching and learning in their classrooms. One teacher commented, 'the availability of resources in various curricular areas has enhanced planning and widened the possibilities for children's learning experiences.' On the practical delivery of lessons another noted, 'purchase of added resources for English and mathematics have made the planning of lessons much easier.' A number of schools chose to use RSS money to appoint media resources officers/technicians to assist teachers in preparing lesson materials.

Improvements in the planning of lesson materials was perceived to have occurred in all but one school by a significant majority of teachers (the overall average was 83 per cent). The range was from 40 to 100 per cent at the school level. Again, however, the percentage who indicated that RSS had a significant effect on this varied markedly between schools from zero (3 schools) to 74 per cent. Only in five schools did half or more of staff think RSS had a significant effect. Again these were all primary schools.

The quality of teaching generally and teachers' focus on learning
Teachers were asked about both the quality of teaching and about teachers' focus on learning. Teachers reported that there had been improvements in both areas (88 per cent and 77 per cent respectively). Also over a third indicated a significant effect of RSS (30 per cent and 27 per cent respectively) on the quality and on the focus of teaching and learning. The teachers from ten schools who chose to write about this areas thought the initiative had helped them to look more critically at classroom practice. The further comments teachers made split them into three distinct sub-categories. The largest group were teachers who thought the initiative had enabled them to consider the *learning* taking place in their own classrooms: 'emphasis now on learning rather than teaching' and 'teachers are encouraged to focus on pupils' learning and completion of tasks.' The second, slightly smaller group were teachers who considered both *teaching and learning*, 'RSS has allowed teachers to focus on their teaching and the children's learning and make significant changes.' These teachers indicated an added depth to their work as a result of RSS, 'the staff have come together to look at all aspects of teaching and learning more thoroughly.' The third group, equal in size to the second group, placed the emphasis on *teaching* which was being done more thoughtfully: 'teachers are putting more thought now into what they are teaching and standing back more and thinking about the best way to put over a topic' and 'teachers more aware of what they deliver.'

Where changes in teaching and learning were reported teachers often put this down to the development of their own expertise arising from extra time to examine and develop their classroom practice. As one teacher wrote, 'discussions and training sessions and visits from people with expertise, reading documents, watching videos and keeping records. All these have caused us to re-examine our teaching methods, sometimes for the better.'

Views about improvements in the quality of teaching over the last three years were generally positive (over two thirds indicating this) in eleven of the 14 schools. The range was from 30 per cent to 100 per cent (2 schools). Again however, only in five primaries was RSS seen to have had a significant effect. The range was from zero (2 schools) to 69 per cent. In connection with teachers' focus on learning, most teachers believed improvements had occurred. The range was from 50 per cent to 100 per cent. However in eleven schools over two thirds reported improvement. Nonetheless in only two schools did more than half the sample attribute a significant effect to RSS (range zero to 67 per cent).

Areas in which RSS appears to have had the least impact

The feedback given to pupils about their work

The importance of appropriate teacher feedback on work has been identified in reviews of school effectiveness research (Sammons, Hillman, Mortimore 1995). Research by Tunstall and Gipps (1995) has likewise highlighted this aspect. Although 'feedback' was not targeted specifically by RSS it is an important aspect of pedagogy. Fifty-nine per cent of teachers reported improvements in teacher feedback to pupils about their work, although relatively few attributed RSS as having a significant effect (16 per cent). It is possible that teachers did not regard this areas as important in determining the quality of teaching and learning and /or that the focus of RSS development work did not give a priority to this area.

It should be noted that five teachers from four schools said the teaching and learning in their school was good before RSS, with good planning and policies which had been implemented in some of the target areas prior to their inclusion in RSS.

Consistency in regular setting and marking homework

Just under half (46 per cent) of teachers thought that this area had improved in the last three years and only a small minority (12 per cent) thought RSS had a significant impact. There is a strong tradition in Northern Ireland schools (both primary and secondary) of homework being set on a regular basis. Whilst there has been little directly reported differences in the teachers' attitudes to homework, there is evidence to suggest that parental attitudes have improved, with parents being more involved in 'learning at home schemes', especially in reading (see impact on parental involvement). The APs do not suggest that schools saw homework as a focus of RSS and the teachers' responses support this conclusion.

The pace of change

Six teachers chose to comment specifically on how RSS had altered the pace of existing improvements in teaching and learning, 'factors were already improving before RSS but it quickened the process and kept the improvements going' and 'there was a noticeable improvement in standards before RSS, but RSS certainly acted as a catalyst for change and speeded things up.' Again it must be remembered that many schools had been involved in other development projects prior to RSS and indeed in reading in particular the paired reading strategy was a focus of earlier development work.

Other respondents' perceptions of impact

Whilst most of the above has focused on teachers' views of the impact of the initiative on teaching and learning this question was also explored in interviews with other groups involved in the initiative.

Over half of the Principals claimed the initiative had enabled them to emphasis the importance of teaching and learning with their staff. Some Principals also reported an improvement in their teachers' subject knowledge, in particular IT skills. The most important impact RSS made on teaching and learning reported by Principals, however was changes in teachers' attitudes ('they lost some of their cynicism', 'they began to look beyond their own classrooms', 'staff see themselves as learners') and practices (a focus on *how* to teach as opposed to *what* to teach) which went beyond 'cosmetic changes' and resulted in greater consistency in delivering the curriculum. The RSS experience also made Principals increasingly aware 'that teachers need to reflect on their own teaching and learning.' RSS was credited by some Principals for giving flexibility to allow teachers time in which to 'review their classroom practice.'

All Co-ordinators agreed with Principals on the positive impact RSS had on staff development and changes to teachers' attitudes and practices. Co-ordinators considered that in addition, RSS had helped them to raise teachers' expectations of pupil performance which had implications for teaching and learning.

Field Officers identified a number of schools where teaching and learning, in their opinion, had improved as a direct result of RSS. In these schools teachers had been given opportunities to discuss 'good practice' and to examine their 'core values' and how this contributed to 'effective' teaching and learning in their school.

The impact on educational standards

The MBW RSS initiative was specifically instituted to improve levels of pupil performance in reading and numeracy in schools identified as under-achieving. This section will consider changes in pupil performance and access to Higher Education and Further Education from questionnaire, interview and schools' performance data.

The impact of RSS on educational standards as measured by the teachers' questionnaire shows some marked similarities and differences to the two previous sections on management and teaching and learning. It is similar because the area seen as most improved (reading 86 per cent) is also the area in which RSS was reported to have had the most significant effect (48 per cent) whereas the area viewed as least improved (number of pupils going on to Higher or Further Education 15 per cent) was also that in which RSS was seen to have had no effect by two thirds of respondents (66 per cent).

In the light of the aims of the MBW RSS it is encouraging that the vast majority of teachers perceived that improvement in standards of literacy and numeracy and in the quality of learning in the school had occurred over the preceding three years. By contrast the two areas in which only a minority felt improvements had been made concerned vocational results and the number of pupils going on to higher and further education (secondary schools only) may indicate that these outcomes had received relatively less attention than others. Secondary schools on the whole targeted Year 8 pupils (in 1994/5) who will not enter public examinations until 1998/9 onwards. The effect of MBW RSS on examination results would not therefore be evident until 1998/9. Examination results and entry into FE/HE are both areas which could be more susceptible to outside

Table 12.3 Areas in Which Teachers Considered RSS Had a Significant Effect.

% signific-ant effect	% no effect		% improvement
47.5	5.6	Standards of reading	88.5
36.8	8.6	Standards of numeracy	81.5
30.9	21.3	Teachers' goal/target setting for pupils	77.1
29.2	12.9	Quality of learning in the school	82.4
26.1	21.6	Teachers' focus on pupil outcomes/results	75.9
Areas in Which Teachers Considered RSS Had Least Effect.			
21.9	23.8	Standard of academic achievement for at Y7 (prim only)	71.8
13.2	29.1	Academic results	66.6
23.7	29.4	Consistency in standards between teachers	63.4
13.7	34.2	End of key stage assessments	57.1
14.4	37.4	Preparation of students for exams end or endt of KS assessment	57.4
8.2	38.4	Standard of academic achievement at Y8 (sec only)	63.4
1.8	58.9	Vocational results (sec only)	32.7
3.4	66.1	Number of pupils going on to HE/FE	15.0

Note. n = 189.

school influences (including HE and FE opportunities, parental and community expectations and local employment prospects). There was some variation evident in teachers' views about improvement and the impact of RSS between the 14 schools. The small numbers of respondents in some schools (in two primaries and one secondary less than ten questionnaires were returned) means that differences between individual schools should be treated with great caution.

Secondary school examination results 1993–94 to 1996–97
The four secondary schools' examination results were analysed to establish whether any trends could be discerned over the four year period 1993–94 to 1996–97. As noted earlier in the report schools had focused their efforts on Year 8 entry cohorts giving priority to literacy and numeracy. It is therefore too early to trace any impact on these pupils' performance at GCSE. Nevertheless, it is disappointing that in terms of published examination results, there was limited evidence of any improvement. None of the secondary schools showed any noticeable increase in the percentage of pupils gaining 5 GCSE passes (Grade A–C). In terms of the measure 1–4 GCSE (Grade A–C) there was also no evidence of improvement. Using a broader measure (percentage 5 GCSE passes A–G) there was slight evidence of improvement for two schools for the percentage gaining five A–G passes. In terms of no passes at GCSE or other examinations there was some evidence of small improvements.

There was also evidence of improvement in the percentage of pupils obtaining one or more passes in non-GCSE examinations for three out of the four secondary schools.

Reading and numeracy across all schools
The majority of teachers (over three quarters) in 12 schools believed reading and educational standards had improved over the last three years. In two schools was the trend weaker (60 and 63 per cent respectively citing improvement). For numeracy a similar pattern was evident although in two secondaries and one primary less than two thirds thought standards had improved. An analysis of schools' APs indicated that, during the first year of the initiative, all schools focused on reading with numeracy receiving more attention in the second or third year.

The RSS initiative was seen to have had a significant effect in raising reading standards by over half of teachers in five schools (two secondary, three primaries). The range was from zero to 89 per cent attributing a significant impact on reading. For numeracy RSS was perceived to have had a significant impact in four schools (one secondary, three primaries). The range was from zero (two schools) to 89 per cent.

With respect to target setting for pupils, teachers' views again varied at the school level. Although the majority reported improvement (over two thirds in 12 schools) the range was from 11 to 100 per cent. Only in three schools, however, did over half the teachers attribute a significant effect to RSS (one secondary and two primaries). The range was from zero (3 schools) to 67 per cent (2 schools). For the quality of learning improvement was cited by 12 schools the range being from 11 per cent in one institution to 100 per cent (5 schools). Again RSS was perceived to have had a significant impact by half the staff or more in four schools (all primaries). The range was from zero (five schools) to 67 per cent. Improvements in the extent to which teachers focus on improving pupil outcomes was noted by over two thirds of staff in 12 schools. The range was from 65 per cent (two schools) to 100 per cent (two schools). Nonetheless, only in three schools (two primaries) did more than half the staff think RSS had a significant effect on this area. The range reporting a significant effect was from zero (three schools) to 68 per cent.

Areas in which RSS appears to have had the most impact

Improvements in standards of reading and numeracy
Teachers generally had very positive views about the extent of improvement over the last three years. Nonetheless, only a minority attributed a significant effect to RSS. The interview data, however, indicated unanimous agreement between Principals, Co-ordinators, CoGs and BELB personnel that RSS had improved standards in reading and general numeracy across all schools. Several reasons were given for the improvements in reading which included the introduction of:

- additional reading time on the timetable
- Reading Recovery programmes
- new approaches to reading such as 'paired/peer reading' and 'Reading Clubs'

- the purchase of new published commercial schemes
- new library facilities and reading/library clubs
- a special focus on one curriculum area by all staff leading to consistency
- new equipment and book provision (which was reported to have motivated children who were 'displaying enthusiasm for and enjoyment of reading' unseen before RSS).

Numeracy was seen to have benefited mainly because of the additional resources provided by RSS. This was said to have affected the work of teachers by extending their teaching repertoire, most notably in the areas of practical work: 'extra equipment has been acquired enabling classes to benefit from practical activities.' Test scores were also cited in support of the assertions of improvement in reading and numeracy.

Whilst improvements were noted in connection with both of these core curriculum areas the emphasis in the first year of RSS for many schools was literacy with the focus on numeracy coming later on and, as noted later, test score evidence for numeracy was less widely available.

The evidence from school effectiveness research that reading performance is a crucial component of later academic performance (Sammons et al., 1995a) also indicates that socio-economic disadvantage has a greater influence on attainment and progress in this areas than in mathematics (Mortimore et al., 1988a; Brandsma & Knuver, 1989; Sammons, Mortimore & Thomas, 1993a). Given its importance for access to the curriculum, the focus on reading during the project's first year can be seen as an appropriate top priority for schools.

Teachers' focus on pupil outcomes/results

Where improvements were reported in this area teachers said RSS had helped their school give greater consideration to measured standards because of better tracking systems for monitoring progress, 'focus on outcomes and target setting has significantly improved percentage of top grades in the 11+ exam over the last three years' and 'teachers are focusing on what the pupils should be able to do at the end of teaching unit, RSS has given time to think about this.' In one school teachers reported being more aware of outcomes because of the introduction of a computerised learning system (ILS). However in four schools teachers reported that new assessment arrangements introduced *before* RSS had affected teachers' approaches in this area and therefore did not see RSS as the main catalyst for change. Many Principals and Co-ordinators indicated that RSS had made them more aware of pupil outcomes and performance scores and the importance of monitoring these. One secondary school reported the introduction of a new programme of regular tests for specific age groups as a result of discussions about pupil outcomes and this was a consequence of their involvement in MBW RSS.

The MBW RSS schools varied in the approaches they used to monitor the impact of their initiatives on pupils' educational outcomes. As illustrated in the section on Action Planning, schools' varied in their specific objectives, the pupils or year group targeted and the kinds of evidence collected. Because of this it is not possible to make systematic comparisons across schools of, for example, rates of pupil progress over the period of the initiative. By contrast, the expanded RSS (RSSI 1995) has used a value added framework, with common baseline and post test measures to enable such analyses to occur.

Areas in which RSS appears to have had the least impact
Whilst teachers thought there had been general improvements in many areas they appeared reluctant to ascribe this as a significant effect of RSS because, 'the programme has not really been in effect long enough to calculate improvements to standards through academic results' or as another teacher put it, 'it is too early to see any significant change in end of Key Stage assessments, change in this area might be observed in two or three years time.' Principals were able to cite improvements in specific areas but were cautious in making claims for sustained improvements. All four secondary schools thought it too early to see any impact on rates of entry to HE and FE and for other vocational and academic results.

The legacy of RSS and challenges for the future

The legacy of RSS must be viewed in the context of the many challenges experienced by UK schools during the 1990s, e.g., introduction of the National Curriculum and national assessments, publication of league tables, LMS, as well as internal changes in individual schools (amalgamations, new Principals appointed etc.). In any developmental project which has a fixed time scale and is based on additional funding, the inevitable question of 'what happens when the project is over and the funding runs out?' is bound to be asked. A number of respondents reported on what they perceived as challenges for the future.

The BELB's perspective
In many ways, BELB personnel considered the initiative as an amalgam of 'strategies for future issues', in that it introduced both schools, Officers and Advisors to some of the elements important in effective long term school improvement planning. It asked schools to identify priorities, and set targets through the drawing up of APs. It introduced them to alternative methods of monitoring, evaluating, staff development and self-evaluation. Concerns for the legacy of the initiative were reported for those schools which were perceived not to have taken 'ownership' of the *aims* of the initiative. These were schools where staff were perceived to have failed to see the school improvement potential of the initiative. They continued to focus on RSS as a simple curriculum focused project and saw the additional resources being the key to success. It was feared that, for these schools, the withdrawal of additional resources will mean in effect that work on school improvement, in this particular format, comes to an end. There was also a concern expressed that it would have been helpful to schools

if *exit strategies* had been 'built into the project from the beginning', it wasn't good enough to flag these up during the last year when the end was in sight.

Challenges for the BELB itself, in continuing the work started on this initiative were also cited. If other schools are to benefit from the lessons learnt during RSS it was said that the support services must have the 'philosophy and framework in place for taking on school improvement.' Respondents felt that BELB needed to look towards not only how it would continue to 'skill' schools in school improvement, but also how it trains its own teams. It was argued that BELB itself needed to ensure that the focus for schools is not on resources but the way these may be best used to enhance the quality of teaching and learning.

The Field Officers' perspective
Across the range of 'successful' RSS schools FOs considered the following a legacy of the project:

- teachers having been given the opportunity to discuss 'good practice' and to examine their 'core values' were now able to think about factors which contribute to 'effectiveness' in schools
- greater collegiality, more whole school planning with teachers feeling less isolated
- more sophisticated Action Planning and target setting
- better primary/secondary links and following on from this improvements in transition arrangements
- better managed staff development programmes tied to development planning
- a recognition of what can be achieved when teachers are 'empowered'
- the importance of the senior management monitoring what is happening in classrooms
- an increased awareness of pupil outcome information

The chairs of governors' perspective
Seven CoGs thought RSS would leave a lasting legacy in their school. Five of these Governing Bodies were reported to be considering their LMS budgets with a view to supporting work started by RSS, i.e., Reading Recovery programmes, class release for Co-ordinator etc. They wanted to maintain the 'uplift to carry teachers forward' which RSS had provided. The two other schools, which had little flexibility in their LMS budget to support development directly were, nevertheless, keen to continue the work of RSS by creating opportunities for further staff development, especially in the core curriculum areas.

By contrast four CoGs thought there would be little future for RSS in their school beyond the official end of the initiative due to the lack of continuing financial assistance and staffing difficulties (both loss of teachers and senior management conflict). One felt that an amalgamation had overtaken the school and was not sure how RSS would be built on in future.

The principals' perspective

Losing additional staff was the biggest concern for Principals (eleven out of thirteen interviewed) in continuing the work of the initiative. All of these said they were going to look critically at their LMS budgets to see if they could retain some of their RSS staff but for many this would be extremely difficult. Other concerns reported by Principals were the funding staff development and getting staff through the implementation 'dip' once the 'spotlight' of the Inspections was removed.

The co-ordinators' perspective

In five schools the Co-ordinators considered that budgetary constraints and the loss of additional staffing would have serious implications for the continuation of the work started on RSS. Similarly, the return of the Co-ordinator to the classroom was said to mean a period of consolidation rather than progress. Six Co-ordinators who reported success in the area of language (reading) said they intend to promote the RSS model of development in other curriculum areas, especially numeracy.

Nine Co-ordinators said they intended to continue with the work of the initiative in the areas of:

- continuing to develop the schools' reading provision
- the management of difficult/challenging children, through withdrawal provision or merit systems
- redrafting policies and schemes of work
- new year group programmes
- improvements in curriculum planning
- staff development
- primary / secondary liaison
- parental involvement and
- developing monitoring and evaluation processes.

The teachers' perspective

Teachers' comments revealed five areas which could be considered the legacy of RSS in their school. All percentages are of those who chose to write specific comments on the questionnaire.

- Working together

For example one wrote, 'if teachers like pupils are motivated sufficiently their goals rise and they achieve more' and another commented 'I feel that we have learnt that if you have a good team of teachers and a clear vision of where you are going RSS can help you achieve your goals.'

- The issues of resources

Some teachers said that resources were not always essential to improvements, 'resources can help but the best resource is teachers'.

- The importance of teaching and learning

One of the strengths of RSS specifically reported by teachers was it re-focused them on their primary job of teaching and forced them to consider contexts for learning. One teacher wrote, 'we have learnt a considerable amount about our teaching and the attitudes to learning in the school' and 'however difficult we must stop and think about what we're doing and *why.*'

- The importance of planning

Some teachers drew attention to the importance of planning both for special projects, 'an initiative of this magnitude needs careful planning,' and for school development. When considering school development they highlighted the benefits of having a School Development Plan, 'RSS has shown new ways to tackle old problems . . . it has helped to concentrate ideas on forward planning, i.e., The School Development Plan' and thus enhanced the school's capacity to develop and implement such plans.

- Managing change

Teachers who wrote about change thought RSS had helped staff in understanding 'how to undertake major change for sound educational reasons,' recognising that 'teachers have the power to change things when given the funding and resources needed.' However, the view was also expressed that 'it is impossible to make sweeping changes overnight.'

> Funding needs to be fit for purpose. Long term gains cannot be gauged by short term measures. Funders need to be clear in their initial aims and project outline how much work can reasonably be accomplished in the time scale they present. Unless this is done there is a danger of being over-optimistic with resultant disappointment.
>
> Wherever possible financial allocations should be phased so that they fit in with schools' natural methods of working. Where this is not always possible, schools should to be given reasonable notice of the financial cycle and more importantly they need to have the implications of the cycle made very clear to them. If specific deadlines have to be met e.g., in spending allocations, staff need to know the importance of these deadlines at the outset of the project to avoid rushed attempts to spend allocations.

Pupil outcomes

In some schools it appears that RSS encouraged a greater focus on monitoring pupil outcomes and the adoption of specific programmes to raise achievement of either targeted pupils with very low prior attainment in reading and mathematics, or whole year groups. The extent to which gains in reading and mathematics ages are sustained over the longer term cannot be established as yet, however, indicators will be provided by KS3 and GCSE results for these cohorts as they progress through school over the next four years. Having said this, improvements in these basis skills should improve these pupils' ability to access the secondary school curriculum, and may help to improve self-esteem, motivation and attendance.

School culture
There was evidence in some schools that significant improvements in curriculum planning and the quality of teaching and learning had been fostered by involvement in the MBW RSS. Improvements in the quality of education are of course hard to measure in any straightforward way but the views of different groups of respondents provided valuable sources of evidence. The emphasis on staff development facilitated by extra staffing and work with Field Officers and Advisors also enhanced the 14 schools' capacity to plan and cope with change. The extent to which good practice can be transferred within schools and the stability of staffing who have gained new skills will of course influence the extent to which schools reap long term benefits from this extra investment in staff development.

Cultural change is often seen as a key to school improvement (e.g., Fullan, 1991; Stoll and Fink, 1992; Hopkins, Ainscow & West, 1994; Reynolds, 1995; Sammons et al., 1997). Change in the pupil and local community culture is also relevant. The need to foster high expectations and a belief that all pupils can learn has been seen to be important in the context of highly disadvantaged inner city schools. The experience of MBW RSS had a positive impact on teachers' views in many project schools. In some schools parental involvement had also been tackled although success in this was not always attributed directly to the project.

Leadership
The importance of the Principal and Senior Management Team's commitment to the MBW RSS was found to be crucial to its overall success in all schools. Where there were long standing management problems the capacity of schools and Co-ordinators, in particular, to implement RSS was constrained. In schools which experienced a change of Principal this was often viewed as an important catalyst for improvement, and generally more influential than RSS itself.

Conclusions

In evaluating the overall achievements of the MBW RSS initiative, it is crucial to recognise the nature of the particular socio-economic and political context in which the 14 project schools are set. If the challenges for education in the inner city are many, then the particular problems which face Belfast's schools can only multiply these challenges. The initiative sought to tackle under achievement in schools in a selective system whose communities have had a long history of socio-economic disadvantage combined with experience of several decades of civil unrest. This initiative recognised that improvements do not necessarily come cheap and the funders are to be congratulated for investing in this work over a significant three year period. Nonetheless, the true benefits of the MBW RSS may not as yet be fully apparent. It must be recognised that in 'school improvement' there are 'no quick fixes' (Gray & Wilcox, 1995; Stoll and Myers 1997) and that the seed sown may take some time to come to harvest. Given the rushed and problematic start to the project in 1994 much of the first year was taken up with planning and in some schools it was not until the second year that developments really took off. Moreover, the focus on younger age groups in the secondary schools means that any long term impact on public examination results and HE/FE participation will not be picked up until 1999 onwards.

Management
The initiative was conceived with clear and ambitious aims and addressed a real need to focus on the difficulties faced by under performing secondary schools in particular. Given the nature of educational funding it was right that the opportunity to use significant central funding for the development of schools was taken. However, the lack of a realistic planning period to consider the nature of school improvement and the question of the most appropriate methods for tackling under-achievement meant that the start of the project was problematic. In an Authority with a history of innovative developmental projects an audit of past strengths and weaknesses associated with both management and development would have greatly assisted the MBW RSS in the initial phases.

The new monitoring and evaluation strategies which MBW adopted during the period of the initiative meant they became increasingly familiar with developments taking place on RSS. Similarly, their direct involvement with BELB and school personnel, through the regular update meetings, helped to reassure them, as funders, that the initiative was being effectively and efficiently managed. It is important that funders are represented on relevant management bodies and that they ensure that all initiatives are auditing/evaluated so that strengths can be built on and any weaknesses addressed.

The significant workload issues associated with managing the initiative were overcome and by the middle of year two, as the initiative progressed, the management became firmer. Personnel changes which occurred during the period of the work were dealt with effectively. There were many benefits for non-school based personnel (Advisors and Field Officers) from working on the RSS initiative. Involvement in the initiative has resulted in a greater understanding of the issues associated with school development and improvement which should be of great benefit for future projects. Much of the learning which has occurred on this initiative is already assisting in planning and developing the on-going DENI expanded RSSI which began in 1995.

Launching the initiative and selection of schools
The public naming of schools by newspapers as 'under-achievers' soured the launch of the initiative in the secondary schools because schools appeared unprepared for the publicity this attracted. There is a real need to ensure that the approaches made to schools to become involved in such a wide ranging ini-

tiative are uniform. Many of the problems associated with the launch were overcome, but this delayed commitment to the project in some schools. The relative merits of so called 'naming and shaming' approach to school improvement remains of course controversial elsewhere in the UK (Mortimore and Whitty, 1997).

The selection of schools in both phases was controversial. Despite clear criteria there was misunderstanding and suspicion from both schools which were included as well as those excluded. These problems were also overcome but made for a difficult period at the beginning of the initiative and meant that progress in the first year was limited in many ways.

Action planning
The schools involved in MBW RSS were not familiar in the initial stages of the initiative with the process of Action Planning. Schools struggled with trying to produce an Action Plan without, in many cases, sufficient training and support to tackle this complex operation. Where insufficient expertise was coupled with a time-table, perceived as unrealistic, and what appeared to be arbitrary decisions made about funding, many Principals in particularly, became jaded with the Action Planning process. Within the first year it was recognised by the CMC that Action Planning was a difficult issue for schools and this was addressed directly. From this point onward the initiative gained more credibility both in and outside schools and greater progress was evident subsequently. Schools and BELB staff made great gains in planning, targeting and implementing APs during the second year and the enhanced capacity to plan and manage change in schools was perceived by many involved to be an enduring legacy of the initiative.

Monitoring and evaluation
Whilst the procedures put in place for monitoring the initiative were adequate these could have been much improved if those involved with making the system work had been involved in the planning phase. In this way a more realistic time-table could have been set and the duplication of information avoided. Although monitoring was consistent outside schools, opportunities for evaluating the progress made in schools was often hampered by a lack of comparable baseline measures. Where such measures were in place they tended to be in areas which could be tackled by standardised tests. The approach schools took to evaluation improved markedly as the initiative progressed, this appeared to be related to the ability to set clearer and more realistic targets in Action Plans. Because schools focused on different groups of pupils and did not adopt a common approach to baseline assessment of pupils and later follow up testing, it has not been possible to make direct comparisons of rates of progress across all schools. As noted in the original evaluation proposal, the nature and availability of any information on pupil attainments from individual schools affects the extent to which the evaluation can address the issue of pupil achievement gains over the three years.

In some schools evidence of considerable short term gains in pupils' reading ages was submitted. If sustained in future years these gains should have a very beneficial impact on these pupils' subsequent progress at school and therefore on their long term educational outcomes. The adverse impact of poor literacy skills on young people's future ability to function in various aspects of adult life has been well documented by bodies such as The Basic Skills Unit. Although the APs in most schools emphasised reading, numeracy received much less attention in the first year. However, by the second and third year many schools had started baseline screening in mathematics and there was some evidence of attempts to transfer the success and achievements in focusing on reading into this core curriculum area.

Support for MBW RSS
Schools generally much appreciated the advice, help and support given to them as part of the MBW RSS initiative. The strategic use of Field Officers, in some schools, was particularly helpful as they coupled classroom credibility with curriculum expertise. There is evidence which suggests that many of the significant achievements made on the project could not have happened without this. It could be argued however that further developments, particularly during the initial stages of the project, could have been achieved had there been initial training for Advisors, FOs and Co-ordinators in issues associated with school improvement and school effectiveness for those taxed with the job of leading developments with and in schools.

DENI had an important role in RSS being influential in setting up, planning, managing and evaluating. Compromises had to be made to accommodate each of these different functions. Some schools found the process of inspection stressful, nonetheless there was clear evidence that schools which had the benefit of inspection evidence at the start of the project they were able to develop more realistic and meaningful Action Plans. The whole notion of 'improvement through inspection' could have been strengthened, if the criteria for subsequent 'evaluation' inspection (conducted after the first year) as opposed to a more formal inspection could have been made apparent and if there had been greater consistency in the verbal and written feedback schools received. Where the issue of management proved a barrier to school improvement there was a perception that this was not always addressed explicitly.

The impact of MBW RSS
There is clear evidence that the majority of schools involved in this initiative made improvements related to the main aims of the initiative. Nonetheless, there was variation between schools and within schools in the extent to which the 'school improvement' message had become embedded in practice. In schools that used the initiative to foster a whole school approach to developments, significant improvements were made, often transforming areas of practice. In schools where the Principals and Co-ordinator were less successful in involving all of the staff, little other than the benefits of three years of additional resources

may be the legacy of the project. This evaluation was commissioned to look at a 'school improvement initiative' – not individual schools. The encouragement of principals and their staff to consider *critically* how the MBW RSS experience relates to their own institution and staff and their approach to school development may enable others to learn from the initiative's legacy.

As noted earlier, the lack of common baseline measures has made the evaluation of literacy and numeracy problematic. Pupil outcomes in the literacy and numeracy foci were more easily assessed for schools which had baseline data, but this was not the case for all schools. Few schools had rigorous methods for establishing baseline data outside of literacy and numeracy. Although schools were able to produce information on attendance, suspensions, exclusions and other statistical measures, there was only limited evidence in their paperwork to suggest that this information was interrogated or analysed to set targets and aid the development of consistency in policy and practice.

Management
One of the major success of MBW RSS was that it broadened the focus of many other initiatives that had gone before to include a 'whole school approach' to development. It provided a vehicle for staff to articulate a 'vision' for the school and to identify the steps that need to be taken if the 'vision' is to be turned into a reality. It appeared that many senior management teams were able to focus specifically on teaching and learning shifting the emphasis in some schools away from a management dominated by administration. There was evidence to suggest that the improvements in staff development programmes, linked to better long, medium and short term planning will foster further developments in the future. Clearly the stability of staff who received additional help and the extent to which they are able to share this with other colleagues will also have an impact on the extent of long term cultural change in individual schools.

Teaching and learning
The MBW RSS enabled schools to develop more systematic and relevant staff development programmes which have fed into curriculum improvements. It provided time for key personnel to consider both subject knowledge and approaches to pedagogy. Linked to these developments has been a greater emphasis on collegiate planning which has had practical classroom application. The quality of lesson material has been improved and the awareness of the importance of a more appropriate curriculum, particularly at secondary level. It was clear that improvements in this area have been made in most schools. Nonetheless, it must be stressed that some (particularly the secondary schools which were operating from very low baselines) still have considerable scope for further improvements. The focus RSS gave to teaching and its links with learning, coupled with an emphasis on measuring pupil outcomes could be extremely important to future developments in many of these schools.

The MBW project has provided a considerable attempt to develop the improvement capacity of schools. There is evidence (Bollen 1989, Hopkins,

Ainscow & West, 1994), that school improvement is related as much to the quality of the policy (raising standards) as the improvement capacity of schools. The school improvement literature suggests that, given greater school autonomy in many systems in recent years, the concept of the thinking (self-reflecting) school has to be promoted. The evaluation suggests that the MBW RSS initiative did indeed stimulate reflection for staff, both individuals, and at an institutional level. It sharpened the measurement, monitoring, target setting and evaluating capacity of schools through the Action Planning process. In several schools significant curriculum development appears to have occurred which should have benefits over the longer term.

Educational standards

Without standard baseline measures it is difficult to be precise about improvements made across the board on this initiative. There is evidence to suggest that for individual children in some schools there have been considerable gains from being involved in RSS, especially for poor readers. The evidence for gains in mathematics however, is weaker. How much these short term gains can be sustained is difficult to predict, but it appears that, particularly at secondary school level, pupils are now better able to access the N.I. National Curriculum as a result of their experiences on MBW RSS. It is too early to make assertions about the impact of RSS on public examination results (GCSE and Vocational) and access to HE and FE. Improvements in primary pupil outcomes should mean higher baselines for Year 8 on entry into secondary schools. Given knowledge of the links between reading performance at age 7 and public examination grades improvements in primary pupils' reading ages are likely to have a positive impact on their later examination results and employment prospects (Sammons 1994b). In the long term the improvement of primary pupils' skills (for example related to the targets set by Literacy and Numeracy Task Forces) would do much to reduce the significant challenges faced by certain secondary schools where (as illustrated by some of the schools in MBW RSS) over 60 per cent of children may be two years or more below their chronological age in reading.

Time scale

Bollen (1989) argued that as school improvement is a process and not an event, time is a major factor that cannot be manipulated without a strong influence on the quality of the process. He noted that innovation is often set on a three year scheme which can be too short for institutional change to be embedded or for the impact of change to be fully assessed. A five year programme may be more appropriate and it is likely that a follow up of pupil achievement, attendance and behaviour over the next two years by BELB would provide valuable evidence of any longer term gains in the 14 schools involved in the MBW RSS experience.

The considerable difficulties faced by so called 'under-achieving' schools especially at the secondary level have been highlighted by inspectors and the media during the 1990s. The research evidence suggests that turning round such schools is extremely difficult (Gray & Wilcox 1995). The culture of such schools may be fragmented and focus on teaching and learning weak (Reynolds 1996). The MBW RSS initiative provided a major opportunity for schools to re-evaluate their approaches and a strong impetus for curriculum development. In most schools the capacity for planning was much improved during the three years and clear strategies for regular monitoring of reading were adopted in most schools, although the impact on numeracy to date appears weaker. In this context it is relevant to note that the need to develop specific approaches to develop pupils' basic number skills has been highlighted by the National Numeracy Project. This has recently reported substantial gains in pupil performance through structured interactive whole class approaches and it may be appropriate for the MBW RSS schools to consider ways in which such approaches could be adopted. The results of the National Numeracy Task Force may also prove valuable (DfEE 1998). There was some evidence of greater staff cohesion and clearer goals setting, although problems in senior management remain in some schools. The positive impact of a change of Principal was a noticeable feature in some schools and attributed with a greater impact in effecting improvements than the RSS initiative by teachers in these schools. The adverse impact of amalgamation on a school's ability to focus on the RSS initiative is also a relevant factor.

There is no doubt that the four secondary schools at the heart of the MBW RSS initiative continue to face considerable challenges in low levels of literacy and numeracy of their Year 8 intake. It is likely that this applies to several other secondaries serving socio-economically disadvantaged communities in Belfast. Ways of effectively targeting resources at the pupil level (without rewarding failing schools) to provide specific programmes to promote rapid gains across Years 8 and 9 to enable these pupils to access the curriculum may be needed if standards are to be raised in the long term. There may even be a case for temporary disapplication of the full national curriculum requirements in such schools to ensure a focus on basic skills for those with measured performance two or more years behind their chronological age. Intensive holiday clubs prior to secondary transfer may also be appropriate for pupils below a certain level although it is clear that raising achievement in the basic skills at the primary level is crucial. A number of RSS primaries had some success in these areas and ways of maintaining this emphasis and spreading good practice should be explored. The pupil culture in all boy secondary schools which cater for those with lower attainments in a selective system is also an important issue for staff seeking to promote better behaviour and discipline. It is hoped that the RSS secondary schools will share good practice in this area and continue to monitor trends in these outcomes.

The MBW RSS initiative provides an important example of a school improvement initiative which had very clear and laudable aims focusing on promoting pupil outcomes and which combined considerable financial support with external advice and guidance in seeking to develop participating schools' capacity to improve. It thus attempted to integrate both a 'top down', external approach to improvement with the encouragement of 'bottom up' strategies developed within individual schools. The majority of those involved valued the opportunities the initiative provided and believed that much was learnt from the MBW RSS experience, both by individual schools and BELB personnel. As with any initiative, there were areas of success and aspects where less progress was made than anticipated. Schools developed their capacity to plan, monitor and evaluate school improvement, and the evidence suggests that the quality of teaching and learning was improved in many cases. Nonetheless, significant challenges remained in several schools, especially at the secondary level. Some positive effects on pupils' reading were identified, and attempts to transfer this to mathematics were being made. Behaviour, discipline and attendance remained areas of concern especially for the secondaries, although modest improvements in attendance were found in most schools. On the whole, secondary schools faced greater challenges and experienced more difficulties in implementing their improvement strategies than their feeder primaries.

As yet the long term benefits of the MBW RSS are hard to judge, given the three year time scale. There is evidence that the experience of the project's first year provided a helpful input into the planning process for the expanded RSS province- wide (Thomas & Sammons, 1996). A further follow up of pupil cohorts which were targeted in schools' Action Plans at Key Stages and in terms of public examinations and post-school destinations would be valuable to explore this issue in subsequent years.

In the UK current work under the new Labour Government is attempting to develop ways of contexualising school performance and setting results in context to facilitate its policy of target setting and national strategies to raise standards. Moreover, the recently announced policy relating to Education Action Zones bears strong similarities to the strategy adopted in the MBW RSS initiative by focusing on low attaining secondary schools serving highly disadvantaged intakes and selected feeder primaries. This evaluation may therefore have implications which extend beyond the particular context of the 14 Belfast schools involved.

Appendix 12.1

Details of evidence collected via interviews and questionnaire surveys
The following interviews were conducted with:
  6 BELB personnel including Officers and Advisors
  13 Principals
  13 Co-ordinators
  12 Chairs of Governors
  4 Field Officers
  Making Belfast Work – one group interview

Teachers' views were sought both quantitatively and qualitatively in the questionnaires, with a total of 1,650 written responses analysed.

Questionnaire returns
The breakdown of returned questionnaires on which this analysis is based is as follows:

Questionnaires	Number of returns	% rate
All teachers	189	51
Teachers (Primary)	124	59
Teachers (Secondary)	65	41
All parents	983	57
Parents (Primary)	764	76
Parents (Secondary)	219	31
All pupils	1268	74
Pupils (Primary)	918	91
Pupils (Secondary)	350	49
Total	2440	64

13

Beyond the Millennium: Current Achievements and Future Directions for School Effectiveness Research

In preparing this volume I have been stimulated to reflect both on my own research career and more widely on the expansion, achievements and limitations of the school effectiveness research (SER) movement. The route is not always clear, many interesting avenues, byways and diversions have emerged over the last 18 years. This last chapter seeks to set my own work in the wider context of the school effectiveness field as a whole, to summarise key themes, identify what appear to me to be important milestones in the intellectual journey and to explore possible future directions as we move into the new age of the 21st century.

Inevitably, the selection of articles and material I have included in this book gives perhaps an idiosyncratic and certainly only a partial indication of the contribution of SER. It is, of course, a personal account, though it has depended much on the joint work of many colleagues working in various teams on different projects over the years. In particular, my research had benefitted enormously from the intellectual stimulation provided by colleagues since moving to the Institute of Education in 1993. Here the creation of the International School Effectiveness and Improvement Centre (ISEIC) in 1994, ably co-ordinated by Louise Stoll, has provided a particularly rich environment for the debate of ideas, methodologies and findings. ISEIC provides a valuable forum which brings together

> 'a group of people who are engaged in four complementary activities – research, development work, teaching and dissemination – in

order to enhance learning and foster pupils' progress and achievement in the broadest sense. In collaboration with each other and a variety of partners, we intend to play a powerful part in ensuring that the educational experiences of pupils and adults who work with them are of the highest possible quality, to support them in their role as citizens in a changing world and to help them shape a positive future' (ISEIC Enhancing Education, 1998).

We have been fortunate that a number of leading UK academics in the school effectiveness and improvement fields have served on ISEIC's Advisory Committee (Professors John Gray, David Hopkins and David Reynolds) as well as a number of headteachers and representatives of both the DfEE and OFSTED.

In creating ISEIC we were particularly concerned to help bridge the 'gap' between the two traditionally distinctive fields of effectiveness and improvement (Reynolds, Hopkins & Stoll, 1993; Hillman & Stoll, 1994; Stoll & Mortimore, 1995), by bringing together academics and practitioners so that both might learn from the others' areas of expertise. Through active engagement in this way, we believe it most likely that the best of the legacy of both fields can be build on and new approaches forged. I believe ISEIC is a good example of the attempts to reconceptualise school effectiveness and school improvement increasingly advocated by those concerned to improve the quality of education (see West & Hopkins, 1996).

ISEIC's principles

The purpose of ISEIC is underpinned by the following principles which I believe will continue to remain fundamental to the task of enhancing the quality of education as we move into the changing world of the new millennium:

- all children can learn, albeit in different ways and at different rates
- individual schools can make a substantial difference to the development, progress and achievement of all pupils
- effective schools add value to pupils' lives
- effective schools focus on a range of learning outcomes, including academic, practical, creative, personal and social
- schools improve most by focusing on learning and teaching, while also addressing their culture and internal conditions
- partnership is a fundamental element of successful school improvement
- intervention work needs to be based on appropriate research findings.

ISEIC'S aims

To achieve its purpose, ISEIC aims to support schools and those who work with them in order to:

- ensure pupils achieve a broad range of learning outcomes
- enhance schools' internal capacity for self-evaluation and for taking charge of change

- improve the quality of learning, teaching, leadership and management at all levels
- promote the development of schools as learning organisations
- build and sustain effective working relationships within and between schools and their partners

ISEIC's work covers seven main areas:

- developing new knowledge and fostering its understanding and use
- bridging theory and practice, through critical reflection
- raising awareness of, and debating, current national and international issues and, where appropriate, challenging existing assumptions
- providing leadership, guidance and facilitation
- producing comprehensive, high quality data and supporting their interpretation and use
- developing and maintaining constructive working relationships with national and international partners
- endeavouring to anticipate future global, social, political and economic trends and their potential impact on education.

My own contribution to ISEIC has been primarily through the research side. As an Associate Director of the Centre, with an oversight of research activities, I have found that links with ISEIC colleagues have helped to sharpen my focus on the practical implications and applications of the research. As Reynolds (1995) observed, this is an issue which

> 'awaits our urgent attention. The introduction of our findings into the educational system in general and into ineffective schools in particular, a process which is important to us for reasons of ideology (we are an applied science) and for the potential increments in our knowledge that can be generated by the interaction between school effectiveness research and the needs for schools to improve' (Reynolds, 1995, p. 66).

In drawing up a programme of work for ISEIC we have endeavoured to extend the corpus of knowledge about the implications for school improvement of school effectiveness work through particular projects. For example, by conducting multilevel analyses of examination or assessment results for individual LEAs and the provision of value added feedback about performance (Thomas & Mortimore, 1996; Yang et al., 1997; Sammons & Smees, 1998). We have also undertaken studies funded by the Economic and Social Research Council on secondary school and departmental differences, on primary school development planning and on the innovative use of staff in City Technology Colleges which have sought to improve our theoretical understanding of school effectiveness and improvement processes. Our collaboration with colleagues at the Quality in Education Centre of the University of Strathclyde provided an exciting opportunity to undertake a large scale project, Improving School

Effectiveness, which was explicitly designed to link the school effectiveness and improvement traditions and to inform the work of policy makers and inspectors in the particular context of Scottish education (Robertson & Sammons, 1996, 1997a, b; Thomas et al., 1998; Smith et al., 1998).

Our second strand of work in ISEIC focuses on methods of supporting schools working on improvement programmes via our group of 20 Associates (consultants drawn from senior roles within the education service) who work with us on a variety of programmes involving LEAs and individual schools. We firmly believe that intervention work needs to be based on relevant research findings and have created a School Improvement Network (SIN) which brings together people engaged in school improvement to share ideas and further refine improvement strategies. In addition to a series of conferences and workshops, SIN provides regular newsletters and brief summaries of research findings and their practical implications for its membership of 50 LEAs and 730 individuals and schools.

It is important to acknowledge that the term school improvement does not imply that only weak or failing schools need engage in improvement activities. In his inaugural lecture as Director of the London Institute, Peter Mortimore (1995) observed

> 'we reject this view and use this term as a shorthand for an international body of research and associated developments concerned with raising the quality of education. For us the crucial characteristic of the work is that the initiative stems not from government diktat nor from any academic or inspectional orthodoxy but rather from a commitment to view the staff of schools themselves as agents for change' (Mortimore, 1995, p. 15).

In this connection I believe that the definition of improvement used by Hopkins and colleagues as part of the Improving Quality of Education for all (IQEA) project is relevant, namely a distinct approach to educational change that enhances student outcomes as well as strengthening the school's capacity for managing change (Hopkins, Ainscow & West, 1994).

The *Improving Schools Journal* is a recent innovation (launched in 1998) supported by ISEIC which is published three times a year. The journal seeks to provide a forum where practitioners, researchers and others in the educational community can read about research, educational practice and policy making in the field of school improvement. The journal aims to be accessible in style and format to a wide range of educators and to bring to this readership work of high quality. As we enter the 21st century I believe it will become increasingly important for educational researchers to actively engage with the teaching profession as well as with policy makers to ensure that research findings are presented in a form which encourages dialogue. All too often educational change has in the past been driven by ideology rather than benefitting from the knowledge derived from well designed studies and evaluations. As my colleagues and I argued in *Forging Links*, the belief that changing structures rather than focusing on teaching and learning processes would raise standards has proved damaging to the profession and wasteful of resources (Sammons, Thomas & Mortimore, 1997). There remains a pressing need for evidence-based evaluations of policy and practice innovations which explore practitioners' perspectives and experiences, as well as the impact on students' educational outcomes (academic and social/affective). It is important also to recognise that improvement cannot be dictated by governments, they play an important part in creating a vision and structures which may foster the process but the active engagement of the teaching profession is an essential pre-requisite (Mortimore, 1995). It is for this reason that I believe a 'naming and shaming' policy is likely to be counterproductive in the long run.

As noted earlier, this volume seeks to provide an overview of SER based on my personal experience of working in the field. As such it cannot provide a complete picture of the rapidly developing research base, although I hope it provides a flavour of some of the key themes and issues. Rather than adopt an historical perspective charting publications in chronological order, I have sought to group items into the three main foci which have characterised the work of many in the field, myself included. The first part of this book examined *measurement* issues and included several papers which attempted to explore the nature of intake influences on students' educational outcomes. These issues remain pertinent today with continued debates about the appropriate methodology for valid comparisons of school performance. As John Gray commented in his thoughtful analysis of the contribution of educational research to the cause of school improvement,

> 'By the millennium there is a very good chance that schools will be evaluated in terms of a rigorous and common framework ... With a fair wind, however, schools can be relatively confident that they will at last be judged in terms of what they have added to their pupils' progress ... A new era of more sophisticated, research-based evaluation is potentially without grasp' (Gray, 1998, p. 10).

School effectiveness research has played a major part in raising awareness and shaping these developments and I believe will continue to have an important role to play in the critique and refinement of value added approaches to the study of institutional performance which I hope will encourage their appropriate and positive use (Thomas, Sammons & Street, 1997).

The second part of this volume attempted to draw together material which addresses the important issue of *understanding* school effectiveness through an examination of the links between processes and outcomes. How can the SER field help to illuminate the context/input/output equation, and how can we relate the growing body of empirical findings to the development of conceptual and theoretical understandings? These questions, I think, are vital for the field's further development and intellectual credibility. They remain for me among the most interesting and potentially fruitful avenues for future exploration.

In his Director's inaugural address, Peter Mortimore drew on Scheeren's (1995) thinking to outline six aspects in which he felt theory has a bearing on SER activity:

- conceptual clarification (i.e., what do we mean by terms such as effectiveness)
- explanations and predictions of positive pupil outcomes
- mechanisms of improvement
- delineation of models
- contingency effects and unintended consequences of different models
- relationships with established social, scientific and educational theories (Mortimore, 1995, p. 18).

He suggested that, although progress has been made in the first four of these aspects, by comparison the last two require particular attention in future studies.

The third section of this book considers the practical messages of SER. It focuses on the issues surrounding the use the school effectiveness research – and their implications for raising standards and school improvement. Inevitably, there are dangers of over-simplification and mis-interpretation or misuse, inherent in all attempts to apply research. The recent spate of criticisms of the field in the UK and elsewhere have, in part, stemmed from its success in attracting political and policy attention and also from the expansion of practitioners' interest in the potential contribution SER can make in the search for school improvement. Hamilton (1996), for example, castigated the growth of an 'international industry' in SER and suggests that the research is 'technically and morally problematic' (Hamilton, 1996, p. 55).

Of course, we must not take all criticism at face value. Many of these denunciations are, as Reynolds (1997) has observed, ill-informed and 'come from people who appear to have read very little school effectiveness research' (Reynolds, 1997, p. 99). In seeking to explain some of the strength of the anti-SER movement, Reynolds points to the historical and philosophical origins of the field which originated outside the mainstream of educational research. He concludes that the technological, applied nature of the field and its empirical approach to the study of students' outcomes lies at the heart of the hostility to SER evident amongst mainstream educational researchers in the UK who, he suggest, value the 'pure' above the 'applied'. 'Critics see effectiveness research as vulgar, rather grubby empiricism and simply not British' (Reynolds, 1997, p. 99).

The material I have included in the third part of this book is an attempt to illustrate some of the applied aspects of SER. The examples chosen include an exploration of practitioners' views of school effectiveness, an example of a value added project involving feedback to primary schools in a large LEA, and an evaluation of a major school improvement project undertaken in a highly disadvantaged urban area in the particular context of Northern Ireland. It is right to be cautious about the uses of SER for school improvement, to avoid simplistic interpretations and the 'quick fix' mentality (Stoll & Myers, 1997, p. 8). Yet while some educational sociologists have emphasised the importance of the detached study and understanding of educational issues, rather than seeking to influence policy and practice there is surely room for both approaches. Whitty (1997) has denigrated what he terms the 'simple hope' of the school improvement lobby in comparison with the 'complex hope' which he suggests can emerge from policy scholarship. Whitty rightly draws attention to the potential benefits of linking the school effectiveness and sociology of education fields, and stresses the need to remain aware of the 'bigger picture' or as he terms it the 'vulture's eye view of the world' (Whitty, 1997, p. 19) which remains aware of the structural influences on individuals and institutions, and recognises the importance of contextualisation. Recent work which he has conducted with Peter Mortimore at the London Institute of Education provides an example of the potential benefits of such linkage (Mortimore & Whitty, 1997). This reviewed the relationship between social disadvantage and achievement using the perspectives of both SER and the sociology of education, and in my view it provides a balanced account of both the possibilities and limits of the extent to which schools can attempt to ameliorate structural influences on students' educational outcomes.

I hope that the material in this volume provides some indication of the weight of research evidence generated by SER studies, and makes clear the importance many in the field attach to the study of the impact of social disadvantage and equity issues. In Part 1 complexity issues in the analyses and measurement of school effects are addressed in some detail, especially in Chapter 3. In a recent professorial lecture entitled *Models of Reality: New approaches to the understanding of educational processes*, Harvey Goldstein (1998) also focused on the issue of complexity arguing that

> 'in order to describe the complex reality that constitutes educational systems we require modelling tools that involve a comparable level of complexity' (Goldstein, 1998, p. 2).

Commenting on the 'unfortunate gulf between the exponents of qualitative and non-quantitative educational undertakings' he suggests that one reason is that the adherents of the latter 'tend to view the former as simplistic and reductionist'. His lecture powerfully demonstrates that in fact

> 'quantitative models do not need to oversimplify reality in the way they often do [and] I want to suggest that they can begin to provide usefully detailed descriptions of the world, and thus perhaps prepare the ground for a reconciliation of research methodologies' (Goldstein, 1998, p. 2).

The growth and current impact of SER owes much to the development of multilevel models and Goldstein's (and his colleagues in the multilevel models research group) contribution to the methodological advancement of the field has been highly influential over the last decade. This work provides, I believe, a powerful antidote to charges of simplistic and decontextualised approaches to the study of the complex and hierarchical systems which are typical of the organisation of education systems in most parts of the world. As such, it provides, I think, a good example of the 'complex hope' which the best studies in the traditions of school effectiveness and improvement research can offer.

The achievements of SER

There is fairly strong agreement amongst the SER community that the last quarter of a century has seen some important advances in the field. A feature of recent years is the emergence of wide ranging overviews such as Reynolds et al.'s (1994a) *Advances in School Effectiveness Research*, which brought together an international team of contributors, and Scheerens & Bosker's (1997) *The Foundations of Educational Effectiveness* which provides a valuable combination of critical reflection on the links between theoretical and conceptual understanding and empirical research. The publication of an *International Handbook* this year by Teddlie & Reynolds (forthcoming) is eagerly awaited and promises to be a key text for many years. Major reflective reviews such as these are important for consideration and act as powerful stimuli to the further development of the field, pointing up areas of doubt as well as consensus and identifying new questions which need to be addressed.

The creation of the International Congress for School Effectiveness and Improvement and the inception of the international journal *School Effectiveness and School Improvement* in 1990 likewise proved significant milestones, and have been vital to the international development of the field providing important forums for debate. There is growing interest in the results of comparative studies in school effectiveness and recognition of the need for an international perspective (Reynolds et al., 1994a). Work by Creemers (1995), for example, has examined process indicators on school functioning; Scheerens & Bosker (1997) have reported on an international comparative school effectiveness study using reading literacy data, while in the UK Reynolds & Farrell (1996) have reviewed international studies of achievement in mathematics and science. Much can be learnt from SER studies conducted in developing countries (for a review see Riddell, 1995). Such studies will enhance our understanding of the impact of contextual factors on student achievement and on the potential influence of schools.

In terms of achievements of SER, Reynolds (1995, 1997) and Mortimore (1995) have both highlighted some similar areas. Reynolds' (1997) retrospect and prospect for example has argued that 'the 1990s have so far been without doubt the decade of school effectiveness and school improvement' (Reynolds, 1997, p. 97) and lists four specific positive achievements by 'a relatively small number of people' (p. 98).

- 'We have convincingly helped to destroy the belief that schools can do nothing to change the society around them, and have also helped to destroy the myth that the influence of the family background is so strong on children's development that they are unable to be affected by school.
- In addition to destroying assumptions about the impotence of education, and may be also helping to reduce the prevalence of family background being given as an excuse for educational failure by teachers, we have taken as our defining variables the key factors of school and pupil outcomes which we 'back map' to look at the processes which appear to be related to positive outcomes.

- We have continuously in our studies shown teachers to be important determinants of children's educational and social attainments and have therefore hopefully managed to enhance and build professional esteem.
- We have begun the creation of a known to be valid knowledge base which can act as a foundation for training.' (Reynolds, 1997, pp. 97–98)

Elsewhere, however, Reynolds rightly draws attention to an unfortunate 'downside' associated with the popularity of SER within our various British educational words, namely that

> 'we have been instrumental in creating a quite widespread, popular view that schools do not just make a difference, but that they make all the difference' (Reynolds, 1995, p. 59).

This is an important point and relates to the criticism by Whitty (1997) of the 'simple hope' of school improvement. As the material presented in Part 1 of this volume demonstrates, structural influences on students' educational outcomes remain strong, and the links between students' background (the 'dowry' they bring to school) and their later educational outcomes should not be ignored. Indeed the methodology of SER provides a very powerful means of demonstrating the nature and extent of such influences, and the relative importance of the school. In my view, comparing the school's influence on absolute attainment levels at any one point over a period of time, with its impact on student progress over several years provides valuable evidence of both the potential strength and inevitable limitations of using schooling as a vehicle to implement social change (see Chapters 2 and 4 for examples of this approach).

Peter Mortimore's (1995) analysis of the impact of school effectiveness work has many items in common with that of Reynolds. He makes seven comments on the usefulness of the field which he claimed had:

- moderated over-deterministic sociological theories about home background
- qualified an over-reliance on psychological individualistic theories about learning
- focused attention on the potential of institutional influences
- provided – as a result – a more optimistic view of teaching and renewed attention on learning concerns we well as on school management
- advanced the methodology of the study of complex social effects
- stimulated many experiments in school improvement
- contributed to a growing set of theoretical ideas about how pupils learn in particular school settings. (Mortimore, 1995, p. 12)

To these lists of important achievements my experience over the last 18 years suggests that it is also important to emphasise the following:

- *a growing awareness of the importance of complexity in the study of institutional effects*
  As noted earlier, the methodological work of Goldstein (1995, 1998) and others has drawn attention to and facilitated the study of complexity in the

'real world' of educational settings. It is now increasingly recognised, as colleagues and I discussed in our secondary school study *Forging Links* (Sammons, Thomas & Mortimore, 1997), that effectiveness is a retrospective and relative construct which is both outcome and time dependent (see Chapter 9). Internal variations by subject (or type of outcome) and for different student groups can be highly significant. Over simplistic categorisations of 'good' or by implication 'bad' schools are therefore inadequate (Silver, 1994; Gray & Wilcox, 1995; Wilcox & Gray, 1996). Further work is needed to examine schools with mixed effectiveness, as well as those which are improving more or less rapidly than the average, or indeed are declining (Gray, Reynolds & Hopkins, 1998).

- *a recognition that fine distinctions (e.g., rank orders) between institutions are statistically as well as educationally invalid*
  For the majority of schools it is not possible to say much about relative effectiveness from one year's data. Rather, only statistical outliers (significantly positive or negative in a particular outcome and time period) can be distinguished. There is a need to be aware of the level of statistical uncertainty in results and to attach appropriate 'confidence limits' to estimates of school effects. The relative and retrospective nature of all measures of school effectiveness should be remembered. The past is not a perfect guide to the future, and SER is becoming increasingly interested in the problems and possibilities of mapping change over time both quantitatively (say in improvement in value added measures of effectiveness) and qualitatively using the perspectives and accounts of those involved in the process in schools.

- *an increased level of confidence amongst schools in the use and interpretation of performance data has resulted from the high profile attached to so called 'value added' studies*
  A variety of value added projects have been undertaken with LEAs to feed back results to schools to stimulate the process of school self-evaluation and review (an example of one conducted in Surrey is described in Chapter 11). These innovations are likely to provide an important vehicle for the creation of schools which are information rich and prove catalysts for change. Information about schools' relative performance (as provided by value added approaches) cannot of itself improve the quality of teaching and learning or influence students' results. It is only by using such information to evaluate current practice and to encourage staff in monitoring the progress of individuals and groups of students that a positive contribution to improvement efforts can be made. The case for the use of performance data by schools has been well argued by Fitz-Gibbon (1996). Further research exploring the ways practitioners actually use performance information is urgently needed, and a number of ISEIC colleagues are working on the ways of providing advice and guidance on the use of value added results (see Elliot, Smees & Thomas, 1998; Wikeley, 1998).

- *a greater focus on student outcomes in school improvement initiatives and a realisation that, in order to study improvement, it is necessary to investigate changes in school effectiveness over several years*

  The SER movement has provided, I think, an important impetus for reflection upon the aims and goals of schooling and a renewed recognition of the importance of students' educational outcomes and their implications for later life chances (Mortimore & Sammons, 1997; Reynolds, 1997). Longitudinal work by Gray, Goldstein & Jesson (1996) of change over time in academic effectiveness has addressed the important issue and revealed differences in schools' improvement rates over five examination cohorts. The follow-up work by Gray, Reynolds & Hopkins (1998) reporting the 'Improving Schools' research indicates that the effectiveness dimension accounted for around 80 per cent of the differences in pupils' performance potentially attributable to schools, with the remaining 20 per cent associated with the improvement dimension. Their evidence from case studies of 'slow' versus rapid 'improvers' is providing pointers to the correlates of improvement and ways of enhancing schools' capacity to improve.

- *an appropriate conceptual framework for the evaluation of school improvement initiatives*

  It is becoming increasingly recognised that the SER tradition provides a valuable methodological and conceptual framework for the evaluation of school improvement initiatives. A review of over 60 such initiatives by Barber & Dann (1996) in the UK pointed to the weakness of design in many of these in terms of lack of measurement of student outcomes, a point also raised by West & Hopkins (1996) and discussed in the Introduction to this book.

It is notable that school effectiveness approaches are being used in a number of areas to evaluate the impact of specific educational programmes. For example, Lapan, Gysbers & Sun (1997) adopted multilevel analyses to examine the impact of more fully implemented guidance programmes on the school experiences of high school students in Missouri. Such studies provide innovative examples of the analysis of social and affective outcomes of schooling using measures based on student self-reported views. The example of the Making Belfast Work Raising School Standards project (MBW RSS) described in Chapter 12 was included in this book as an illustration of the potential SER has to inform improvement initiatives and to provide an appropriate evaluative framework. This recognises of the need to include suitable baseline measures at the outset and the importance of exploring the multiple perspectives of different participants in the educational process.

Likewise, in the Scottish Improving School Effectiveness Project, discussed earlier, a value added dimension was adopted as an integral part of the research strategy to facilitate comparisons of the impact of improvement activities in 24 case study schools out of the 80 schools included in the project (see Thomas et al., 1998). This study also sought to increase understanding of the processes of change in schools (Robertson & Sammons, 1997a, b) and ways of evaluating evidence for school improvement (Robertson et al., 1998) through the linking of qualitative and quantitative data. I believe that the linking of school effectiveness methodology and school improvement approaches will become an increasingly common feature of research design in the next two decades and will foster the theoretical development of the field.

Challenges for the school effectiveness and improvement communities
As with the main stream of educational research, there is often an impatience from practitioners and policy makers who want preferably simple and quick 'answers' to the complex 'problems' they face and feel frustrated by the, as they would perceive it, over-cautious, seemingly common-sense and frequently non-prescriptive findings of researchers who emphasise the complexity of the social world. McGaw and colleagues (1992) have rightly argued that in any field

> '... current practice is, for the most part, today's monument to yesterday's research. Practice seldom operates at the frontiers of knowledge' (p. 166).

Although they emphasise that practitioners cannot afford to ignore the findings of research any more than researchers can afford to ignore the context of practice, they stress that the primary role of research is 'to inform, not to reform' (p. 166). I believe that SER has an important role in the process of informing policy makers and also of empowering practitioners. Current findings are by no means exhaustive but are sufficiently robust to provide a useful basis for the development of appropriate frameworks for school improvement initiatives and to encourage practitioners to engage in reflection on current practice and institutional self-evaluation and review. Nonetheless, as this book I hope stresses, there are no magic solutions, no tried and tested recipes for success and considerably more is known about the characteristics of effectiveness than about the processes of improvement, the dynamics of initiating and managing institutional and cultural change. In particular, as a number of authors have noted (Gray & Wilcox, 1995, Reynolds, 1996; Stoll & Fink, 1996), an over-emphasis on the effective end of the spectrum means that we know much less about ineffective and indeed average schools, although a number of important studies are currently underway which are addressing this question.

Gray (1998) has recently discussed the complaint made against educational research that it fails to provide clear cut answers and that the findings are obvious. He notes that 'the main reason, of course, for this is that it reflects reality' (p. 16). SER, in my experience, helps to illuminate the complexity of the interactions between the Context – Input – Process – Output equation (as reviewed in Chapter 9). Gray goes on to discuss the work by Gage (1991) on the 'obviousness' of social and educational research results, which he believes indicates 'that people tend to be more demanding when they assess research in education ... What they seem to be looking for is the educational equivalent of "wonder" drugs. Not surprisingly, they are frequently disappointed' (Gray, 1998, p. 16).

He further notes the pressures of 'short-termism'.

'Policy makers and practitioners want "solutions" and researchers want to publish. Yet many of the questions to which answers are sought turn out to be ones for which only relatively long-term perspectives will suffice. It is an unfortunate but unavoidable fact that the average child will complete their education under at least three governments' (Gray, 1998, p. 16).

This book has taken as its theme *School Effectiveness: Coming of Age in the 21st Century*. I believe that both the related fields of school effectiveness and school improvement are still in need of further research in order to promote their growing linkages, to clarify some conceptual and theoretical issues and to improve technical aspects of measurement and methodology. The practitioner-researcher interface also requires further development and study.

In terms of future directions of the field there are many possibilities and it is likely that different groups will take the lead in further advancement in particular topics. Scheerens and Bosker (1997) have argued the case for the redirection of SER, and point to the limitations of current knowledge and, in particular, to the need for more foundational work to settle important conceptual issues and improve research methods. As well as developments related to the statistical analysis of school effects, they also draw attention to the divergence in current instruments and the lack of standardised and validated instruments as an 'enduring set back to future school effectiveness studies'. They reiterate the need for more high quality, longitudinal studies and highlight the apparent

'discrepancy between more qualitative reviews on the one hand and meta-analyses on the other, with respect to school-level conditions that are expected to enhance effectiveness. The overall message of these analyses appears to be that classroom conditions have more impact than school organizational conditions in improving outcomes' (Scheerens & Bosker, 1997, p. 320).

Reynolds (1997) has highlighted some of the intellectual 'downsides' of British research which has tended to focus on schools in socio-economically disadvantaged areas, although more recent work has studied schools in a variety of contexts, (see Thomas & Mortimore, 1996; Gray, Goldstein & Jesson, 1996; Thomas et al., 1998). He specifically notes the lack of attention to classroom learning environments and teacher effects in the UK studies – in contrast to the attention these have received in American and Dutch research on instructional effectiveness (see, for example, Creemers, 1994). Reynolds (1997) further suggests weakness in the historic lack of 'interface' between the SER and improvement paradigms reflecting their different intellectual ancestries, although again work by David Hopkins and colleagues (1994), Gray, Reynolds and Hopkins (1998), Robertson and Sammons (1997) and Robertson et al. (1998) is helping to overcome this deficiency. Finally, he acknowledges that British research, by contrast to American and Dutch traditions, has only made very limited attempts at theory generation.

In my view current research conducted in the Netherlands which is attempting to build on meta analyses to provide quantitative research syntheses of empirically- based knowledge concerning effectiveness enhancing conditions (see Scheerens & Bosker, 1997 for example) or to test out particular theoretical models in empirical studies (see Creemers & Reezigt, 1996; Reezigt, Guldemond & Creemers, forthcoming) provide important pointers to the future methodological development of the field. Furthermore, the wide ranging data series currently being collected by leading Australian researchers such as Peter Hill, Ken Rowe and colleagues are providing important new evidence concerning the relative importance of classroom and school effects. The work being conducted at the University of Melbourne has already helped to advance SER methodology and has the potential to provide a growing body of evidence to address many of the key concerns of both SER researchers and school improvement practitioners particularly in relation to effective classroom practices. (Rowe & Hill, 1994, 1996; Hill, Rowe & Jones, 1995; Hill & Rowe, 1996).

In my personal judgement there are five areas where I feel important advancements can continue to be made in SER over the next decade or two:

- exploring the long term impact of schools across phases to establish the extent and nature of the long term and cumulative impact of schools;
- exploring the impact of schools on students with special educational needs (SEN) and investigating whether different classroom and school processes vary in their impact on different groups of students;
- informing and evaluating the impact of specific school improvement initiatives to enhance our understanding of educational and institutional change;
- exploring the impact of contextual variations as influences on school effectiveness, both at a regional and, through collaborative international studies, at a national level;
- explore the influence and nature of the relationships between home and school, particularly through the triad of relationships between student, parent and teachers (as described by Coleman, 1998) and the way these can contribute to school improvement initiatives, and
- further develop and link the theoretical bases of the school effectiveness and school improvement fields, particularly through the empirical testing and refinement of existing models.

In this last section I draw attention to some current work which colleagues and I are engaged in or plan to conduct in relation to the first and second of these five areas.

Long term impacts
The need for more detailed longitudinal studies which examine educational effectiveness over much longer time periods has been highlighted by a number of authors (Scheerens & Bosker, 1997; Goldstein, 1998; Gray, 1998) in recent years. Early work on the question of continuity of school effects from primary to secondary level is illustrated in this volume in Chapter 5. More detailed cross-classified analyses by Goldstein & Sammons (1997) have pointed to the particular importance of the primary phase. Further work is needed, however, using larger samples and conducted in a variety of contexts to establish whether it is indeed the case that primary schools continue to exert a long term impact on students' outcomes at the end of compulsory education and, if so, to explore the mechanisms by which such long term influences may operate, for example, through an impact on students' attitudes, self-esteem, motivation or study skills.

A number of studies have pointed to the long term benefits of early years' education (see Lazar & Darlington, 1982; Osborn & Millbank, 1987; Schweinhart et al., 1993). None, however, have sought to examine the separate influence of types of early years provision from the impact of specific pre-school centres on children's subsequent attainment and progress at school. In the US, McCartney and Jordan (1990) have drawn attention to the existence of parallels between research on child care and research on school effects and noted striking similarities in the methodological and conceptual advances in both areas. They argue that child-care researchers and school effects researchers would benefit by monitoring progress in one another's field as they assess what matters in terms of pre-school environments for which types of children.

The Effective Provision of Pre-School Education (EPPE) project with which I am currently engaged provides an example of a major longitudinal study which links the perspectives of researchers into early years' provision with an educational effectiveness approach derived from the SER tradition. This five year study (funded by the Department for Education and Employment) is assessing the attainment and development of children from ages three to seven years and began in January 1997. Multilevel approaches (including cross-classified models) are being used to explore both the effects of individual pre-school centres on children's attainment and social development at entry to primary schools and any continuing effects on such outcomes at the end of Key Stage 1 (age 7 plus years) in terms of national curriculum assessments in English, mathematics and science and standardised tests of reading and mathematics. Other non-academic outcomes affective and social behavioural measures are also being assessed. Currently the research is exploring four types of pre-school provision (nursery classes, playgroups, Local Authority day nurseries and private day nurseries) over a range of geographical areas in England. It was extended to include nursery schools in September 1998.

The EPPE study is investigating three issues which have important implications for policy and practice in early years education which is currently the subject of considerable Government interest in the UK and elsewhere as a means of improving the future educational prospects of disadvantaged children in particular and facilitating parents (mainly mothers) labour market participation. These issues are:

- the effects on children of different types of pre-school provision.
- the structural and process quality characteristics of more effective pre-school centres, and
- the interaction between child and family characteristics and the kind of pre- school provision a child experiences.

An educational effectiveness research design was adopted to explore these issues because it enables the project to investigate the progress and development of individual children (including the impact of personal, socio-economic and family characteristics and day care history), as well as the effects of individual pre-school centres on children's outcomes at entry to school (the start of reception which children can enter from just turned 4 to rising 5 year olds).

The EPPE study involves the collection of information about the process characteristics of pre-school centres and will include detailed case studies of particular centres identified as unusually effective or ineffective. A total of over 114 pre-school centres and over 2000 children are involved in the main study. Further details of the research design and the methodological challenges of studying pre-school provision (children exhibit very varied pre-school career patterns entering at a range of ages, attending for different numbers of sessions, and a substantial minority move between pre-school centres or are in dual provision) is given by Sammons et al., 1998a. It is hoped that the EPPE sample will be followed up over a longer period across the later primary years and into secondary education. In this way it will be possible to explore the extent of any continuity of pre-school effects and try andquantify the relative influence of different phases of children's educational careers.

My colleagues Kathy Sylva, Ted Melhuish and Iram Siraj Blatchford and I are the EPPE Principal Investigators and for me the project is providing another example of the benefits of working in multidisciplinary groups. Research such as the EPPE project represent, I believe, an example of the benefits of longitudinal designs with a cross-phase focus which will help to advance our understanding of children's progress and development and the nature of institutional influences on longer term educational outcomes. It will be many years, of course, before the full results of our research will be known. For example, the EPPE sample will not begin to enter secondary schools until 2005. Yet these early years are likely to be particularly influential on later life chances and experiences as long term cohort studies suggest (Parsons & Bynner, 1998). We need to know much more about the impact of early years' experiences and institutional influences in particular if we are to help to overcome the difficulties that certain groups experience and which increase the risk of later educational failure.

Special effectiveness and special educational needs (SEN)
The second of the four areas in which I see considerable scope for further development of the field concerns special educational needs. It is rather surpris-

ing, given the focus of early SER which, as discussed in Chapter 2, had a strong equity focus and involved the study of schools in socio-economically and often ethnically diverse areas, that school effectiveness studies have neglected SEN issues. Ainscow (1991) makes the case for an alternative approach to special needs education through 'a reconceptualisation of what we mean by educational difficulty' (Ainscow, 1991, p. 1). He suggests that the main problem is one of curriculum and argues that

> 'We can more usefully see pupils experiencing difficulty as indicators of the need for reform . . . I believe that such reforms would be to the benefit of all pupils. Consequently the aim is effective schools for all' (Ainscow, 1991, p. 3).

The need to further develop the SER base through specific studies of students with SEN has been noted by Reynolds (1995). Research on teacher effectiveness by Creemers (1994) and Slavin (1996) has suggested the importance of four factors, which Reynolds (1997) has summarised in connection with his arguments concerning the need to develop schools as High Reliability Organisations (HROs) and to foster a view of teaching as a craft or technology rather than an art:

- The quality of teaching
- The appropriateness of task
- The incentives used
- Time use
- Opportunity to learn.

In my view these factors provide a valuable starting point for the analysis of teaching practices. However, there remains a pressing need for more research on the detail of classroom processes and their relationship to student learning and outcomes if SER is to make a better contribution to teacher education and training and to the understanding and improvement of classroom practice (see Hill & Rowe, 1996).

Given increasing concern about the attainments of students who form the trailing edge of the UK education system, and current Government policy to raise standards of literacy and numeracy in the UK via a strategy of benchmarking and target setting, I believe the time is right for SER to begin to focus on the study of effectiveness in relation to SEN as well as regular students. Despite increasing recognition of the existence of differential effectiveness for different student groups (see Nuttall et al., 1989; Goldstein et al., 1993; Thomas et al., 1997b), the exploration of internal variations in effectiveness has tended to focus on questions of gender, ethnicity or socio-economic status but not on SEN. Recent research with which I was engaged on inner city infant schools conducted with colleagues at the London School of Economics examined primary teachers' definitions of and responses to SEN in relation to the implementation of national assessment and the impact of background factors including the category of SEN on attainment (West, Hailes & Sammons, 1995;

Sammons, West & Hailes, 1996). It did not, however, explore the question of internal variations in primary school effects on the progress and development of pupils with and without SEN. A study seeking to address this topic *School Effectiveness, Pedagogy and SEN in Mainstream Primary Schools* has recently been put forward by Dr Ann Lewis (a leading figure in the study of SEN) and myself (Lewis & Sammons, 1998). This study is designed to focus on students with and without learning difficulties and it is planned to examine a range of affective, social and cognitive outcomes for children aged eight to ten years. The aims are to:

- explore the range and characteristics of students identified as having SEN in mainstream primary schools for students in two age cohorts
- investigate teachers' understanding of, and pedagogical response to, students' perceived SEN
- analyse the progress and development of students with SEN, compared with other students in the same age cohort across a range of affective, cognitive and social outcomes of education
- identify and separate any systematic effects on student progress and development related to class, or school membership from those related to students' personal and family characteristics (e.g., age, gender, socio-economic, ethnic or language background)
- investigate any differential class/school effects for children with SEN compared with other students and to investigate the relationships between school and classroom process factors and student progress across a range of outcomes.

The concept of SEN and hence the provision that may follow is, of course, highly contentious. The identification of students in main stream classes as 'having SEN' will be influenced by a variety of factors including schools' and LEAs' SEN policies, resource availability, teachers' and parents' expectations and perceptions, and the curriculum as well as by the child's behaviour and prior and current attainments. Recent UK research suggests that individual approaches are more common place than ecological ones (Sugden, 1998) and that this is encouraged by recent legislation (1994 Code of Practice on SEN). Marsh's (1998) survey of LEA practice also reveals confusion about the purposes of additional resourcing and the links between social disadvantage and SEN.

In our proposed research we are deeply concerned with examining the equity implications of current primary school practices for different groups of students and teachers' responses to perceptions of SEN. In particular we believe there is a need to establish whether the correlates of effectiveness identified in SER (see Sammons, Hillman & Mortimore, 1995; or Bosker & Scheerens, 1997) apply equally to *all* students in schools. If this is so there

> 'is no support for a SEN-specific pedagogy. However, if certain school classroom characteristics are more or less effective for particular pupil groups, then we have a research base for claiming a group specific ped-

agogy. A key question is, does effective provision for pupils with SEN require "non-normal" features?' (Lewis & Sammons, 1998, p. 10).

In addition to policy and practitioner relevance concerning the influences of different school and especially classroom practices on primary students' educational outcomes, we believe that the proposed research will have implications for the theoretical and methodological development of the SER field by exploring the nature of school effectiveness for students with SEN and will contribute to current understanding of differential effectiveness.

Improving schools
The last four foci I have identified as areas in which SER can make significant advances over the next twenty years all relate in some way to the question of linkage with school improvement. This is a focus of concern within the field as the growing number of articles and papers on bridging the gap, or merging traditions illustrate. A useful analysis of the problem and current state of play is given by Gray, Jesson and Reynolds (1996). These authors point out the different ways of conceptualising schools' performance evident in the two paradigms. Their framework suggests a classification of schools by both effectiveness and improvement dimensions. It points out that the possible number of cells in such a classification can become very large if within school variations in terms of different outcomes are considered. Given this they emphasise the need for caution

'in recommending simple "treatments" for whole ranges of schools. A better grasp of each institution's strengths and weaknesses as well as their starting position will be required' (Gray, Jesson & Reynolds, 1996, p. 174).

They also draw attention to the challenge of how to *sustain* change (improvement) over *extended* periods.

The existence of sizeable gaps in the useability of SER, our understanding of how to turn knowledge about school effectiveness into enhanced strategies for school improvement, is all too evident. There is a need for greater attention to the exploration of the 'natural histories' of schools to learn more about how schools become more effective, or indeed how they decline. Further research tackling causal directions and chains and on the relative importance of different correlates of effectiveness is also advocated. In my view meta analyses such as those by Scheerens & Bosker (1997) and specific studies such as that by Creemers & Reezigt (1996; forthcoming) which seek to test empirically the applicability of currently accepted SER findings and models have much to offer in this respect.

The need for more research on the development and impact of relationships between parents, students and teachers (the triad drawn attention to in the work of Coleman and colleagues' Co-Production of Learning Project) may have important implications for the advancement of new approaches to school improvement (Coleman, 1998).

Case study research working with schools using SER findings has been suggested by Reynolds & Sammons (1997) as a fruitful avenue for testing the applicability of the methods and contrasted with the limitations of the action research tradition in effecting educational change. I anticipate that *collaborative school effectiveness and improvement research* – working with schools and teachers – will expand over the next decade. Increasing recognition of the importance of the classroom and pedagogy highlights the need for more research on the role of teachers in the improvement processes. 'By ignoring the classroom level there is a distinct risk that change effects will be only tinkering with the "variables which matters" ' (Gray, Jesson & Reynolds, 1996, p. 179). Of course, a greater focus on teacher effectiveness and on the role of teachers in the improvement process entails risks. There is a danger that the previous culture of 'blaming' student background and social disadvantage for educational failure evident in the 1970s, and to a lesser extent the 1980s, may be replaced by one of 'teacher blaming'. In the UK the so called 'naming and shaming' of schools has been a high profile feature in recent years. Research on effective pedagogy and the vision of teaching as a craft rather than an art – advocated by Reynolds, 1997 – may also be seen (wrongly in my view) to deprofessionalise teachers. At a time when there are difficulties, if not yet a crisis, in teacher recruitment in the UK, such research can be seen as a threat by a profession which feels beleaguered by the pace of educational change. This tendency may be exacerbated if policies to link teachers' pay and performance are adopted with the creation of different grades of classroom practitioner.

Recent initiatives in the UK which have built on the effective teaching research base, for example, citing the work of Slavin, (1996), are designed to ensure greater uniformity and higher standards in all primary schools (Literacy Taskforce, 1997; DfEE, 1988). The introduction of a literacy hour has been followed by a numeracy programme in England. It will be of considerable research interest to evaluate the impact of these initiatives and the associated strategy of target setting in the coming years to establish whether dramatic system-wide improvement is possible. It may be that standards rise overall, although it will be important to scrutinise the national assessments used and conduct independent standardised assessments of literacy and numeracy for random samples of students to check the extent of real change. It may also be the case that the differences between schools, and indeed between classes in schools, are magnified where a minority are unable or unwilling to adopt or accept the need for change. This could lead to inequities, at least in the short term, with disadvantaged groups less likely or able to exercise real choice in where they send their children to school and thus concentrated in less effective schools. Also, it should be remembered that the SER base has consistently demonstrated the impact of concentrations of disadvantaged groups which creates additional challenges for schools serving these communities.

The new UK Government has staked its reputation on its ability to improve education, seeing this as a route to reduce social exclusion and enhance economic progress. It will not, of course, be possible to evaluate the impact of its reforms properly in the lifetime of this parliament (before 2002). A true picture is only likely to emerge as the primary students involved move on to secondary school and beyond. The legacy of SER illustrates the need for increasingly long-time scales in studying educational change. For a school effectiveness researcher it is a fascinating social phenomenon to witness. Whether all aspects of the strategy will foster school improvement and a long term boost to standards remains hotly debated by academics and practitioners alike. Nonetheless, as Reynolds (1997) has noted the last decade has seen school effectiveness and improvement research moving to the centre stage in education in the UK. As a by-product of its educational experiment, the Government has provided the SER community in the UK with tremendous opportunities for the development of its research base, the results of which will be of considerable interest to the international SER community. The scope to evaluate and critique the impact of current Government initiatives is immense and, I believe, will act as a real catalyst to the further advancement of the field.

I am grateful to Bert Creemers and Dave Reynolds who asked me to contribute to the Contexts of Learning series for the opportunity and challenge of reflecting on my work over the last 18 years. I look forward to continuing the intellectual journey into the 21st century which I anticipate will see the 'coming of age' and perhaps even the flowering of SER as it becomes integrated with the school improvement tradition and begins to make a greater contribution to both educational policy and practice in an increasing number of countries, and thus benefit the education of all groups of students irrespective of their background.

References

ALBSU (1993) *The Basic Skills of Young Adults*. London: ALBSU.
ALBSU (1993) *Parents and Children: The Intergenerational Effect of Poor Basic Skills*. London: ALBSU.
APU (1981) *Language Performance in Schools*. Primary Survey Report No 1. London: HMSO.
APU (1982). *Language Performance in Schools*. Primary Survey Report No 2. London: HMSO.
Ainley, J. (1994) 'Curriculum Areas in Secondary Schools: Differences in student response', paper presented at the International Congress for School Effectiveness and School Improvement, Melbourne, Australia, 1994.
Ainscow, M. (ed) (1994) *Effective Schools for All*. London: David Fulton Publishers.
Ainscow, M. & West, M. (1991) *Managing School Development: A Practical Guide*. London: David Fulton Publishers.
Ainsworth, M. & Batten, E. (1974) *The Effects of Environmental Factors on Secondary Educational Attainment in Manchester: A Plowden Follow-Up*. London: Macmillan.
Aitkin, M., Anderson, D. & Hinde, J. (1981) 'Statistical modelling of data on teaching styles', *Journal of the Royal Statistical Society* A 144 (4): 419–461.
Aitkin, M., Bennett, N. & Hesketh, J. (1981) 'Teaching styles and pupil progress: A re-analysis', *British Journal of Educational Psychology* 51 (2): 170–186.
Aitkin, M. & Longford, N. (1986) 'Statistical modelling issues in school effectiveness studies', *Journal of the Royal Statistical Society* Series A 149: 1–43.
Alberman, E.D. & Goldstein, H. (1970) 'The "At Risk" register: A statistical evaluation', *British Journal Preventive and Social Medicine* 24: 123–125.
Alexander, R.J. (1992) *Policy and Practice in Primary Education*. London: Routledge.

Alexander, R., Rose, J. & Woodhead, C. (1992) *Curriculum Organisation and Classroom Practice in Primary Schools: A discussion paper.* London: DES.

Alston, C. & De Vaney, K. (1991) *Educational Priority Indices and School Resourcing in Hackney: LMS Supplementary Paper.* London Borough of Hackney, April 1991.

Angus, L. (1993) 'The sociology of school effectiveness', *British Journal of Sociology of Education* 14 (3): 333–345.

Armor, D., Conry-Oseguera, P., Cox, M., King, N., McDonnell, L., Pascal, A., Pauly, E. & Zellman, G. (1976) *Analysis of the Reading Program in Selected Los Angeles Minority Schools.* Santa Monica: Rand.

Ball, S.J. (1994) *Comprehensive Schooling Effectiveness and Control: An Analysis of Educational Discourses.* London: Centre for Educational Studies, King's College.

Bandura, A. (1992) *Perceived Self-Efficacy in Cognitive Development and Functioning,* invited address given at the annual meeting of the American Education Research Association, San Francisco, April 1992.

Barber, M. (1993) 'Raising standards in deprived urban areas', *National Commission on Education Briefing, No 16,* July 1993. London: NCE.

Barber, M. & Dann, R. (eds.) (1996) *Raising Educational Standards in the Inner Cities: Practical Initiatives in Action.* London: Cassell.

Barnes, J.H. & Lucas, H. (1974) 'Positive discrimination in education: Individuals, groups and institutions', in T. Leggatt, (ed.) *Sociological Theory and Survey Research.* London: Sage.

Bennett, N. (1976) *Teaching Styles and Pupil Progress.* London: Open Books.

Bennett, N. (1978) 'Recent research on teaching: A dream, a belief and a model', *British Journal of Educational Psychology* 48: 127–147.

Bennett, N. (1992) *Managing Learning in the Primary Classrooms,* ASPE Paper No 1. Stoke: Trentham Books.

Bennett, N., Summers, M. & Askew, M. (1994) 'Knowledge for teaching and teaching performance', in A. Pollard (ed.) *Look before you leap? Research evidence for the curriculum at Key Stage 2.* London: Tufnell Press.

Berman, P. & McLaughlin, M. (1977) *Federal Programs Supporting Change, Vol VII: Factors Affecting Implementation and Continuation.* Santa Monica: Rand.

Black, H., Hall, J. & Martin, S. (1990) *Learning, Teaching and Assessment: A Theoretical Overview.* Edinburgh: SCRE.

Blatchford, P., Burke, J., Farquhar, C., Plewis, I. & Tizard, B. (1987) 'Associations between pre-school reading and related skills and later reading achievement', *British Educational Research Journal* 13: 15–23.

Blatchford, P. & Cline, T. (1992) 'Baseline assessment for school entrants', *Research Papers in Education* 7 (3): 247–1269.

Bollen, R. (1989) *School Improvement: A Dutch Case in International Perspective. Foundation for International Collaboration on School Improvement (FICSI).* Acco Leuven Amersfoort, Belgium.

Bondi, L. (1991) 'Attainment in primary schools', *British Educational Research Journal* 17 (3): 203–217.

Bosker, R.J. & Scheerens, J. (1994) 'Alternative models of school effectiveness put to the test', in R.J. Bosker, B.P.M. Creemers & J. Scheerens (eds.) *Conceptual and Methodological Advances in Educational Effective Research: Special Issue of the International Journal of Educational Research* 21(2): 159–180.

Bosker, R.J. & Scheerens, J. (1989) 'Issues in the interpretation of the results of school effectiveness research', Chapter 4, in *International Journal of Educational Research* 13 (7): 741–751.

Bosker, R. & Witziers, B. (1995) 'School effects, problems, solutions and a meta-analysis', paper presented at the International Congress for School Effectiveness and School Improvement, Leeuwarden, The Netherlands, January 1995.

Bossert, S., Dwyer, D., Rowan, B. & Lee, G. (1982) 'The instructional management role of the principal', *Educational Administration Quarterly* 18: 34–64.

Boydell, D. (1974a) *The Teacher Record: A Technique for Observing the Activities of Junior School Teachers in Informal Classrooms.* Leicester: University of Leicester, School of Education.

Boydell, D. (1974b) *The Pupil Record: A Technique for Observing the Activities of Junior School Pupils in Informal Classrooms.* Leicester: University of Leicester, School of Education.

Boydell, D. (1975) 'Pupil behaviour in junior classrooms', *British Journal of Educational Psychology* 45 (2): 122–129.

Brandsma, H.P. & Knuver, J.W.M. (1989) 'Effects of school classroom characteristics on pupil progress in language and arithmetic', *International Journal of Educational Research* 13 (7): 777–788.

British Educational Management and Administration Society research cited in Earley, P., Fidler, B. & Ouston, J. (1996) *Improvement through Inspection? Complementary Approaches to School Development.* London: David Fulton.

Brookover, W. & Lezotte, L. (1979) *Changes in School Characteristics Coincident with Changes in School Achievement.* East Lansing: Michigan State University.

Brookover, W., Beady, C., Flood, P., Schweitzer, J. & Wisenbaker, J. (1979) *School Social Systems and Student Achievement: Schools Can Make a Difference.* New York: Praeger.

Brophy, J. & Good, T. (1986) 'Teacher behavior & student achievement', Ch 12, in M.C. Wittrock (ed.) *Handbook of Research on Teaching.* New York: Macmillan.

Brown, S. & Riddell, S. (1992) *Class, Race and Gender in Schools: A New Agenda for Policy and Practice in Scottish Education.* Glasgow: SCRE.

Brown, M. & Rutherford, D. (1995) *Successful Leadership for School Improvement in Areas of Urban Deprivation: A Framework for Development and Research.* Manchester: Schools of Education, University of Manchester/University of Birmingham.

Brown, S., Riddell, S. & Duffield, J. (1996) 'Possibilities and problems of small-scale studies to unpack the findings of large scale studies of school effectiveness', in J. Gray, D. Reynolds, C. Fitz-Gibbon & D. Jesson (eds.) *Merging Traditions: The Future of Research on School Effectiveness and School Improvement.* London: Cassell.

Bullock, A. & Thomas, A. (1993) 'Comparing school formula allocations', in G. Wallace (ed.) *Local Management and Central Control.* Bournemouth: Hyde Publications.

Byford, D. & Mortimore, P. (1981) 'School examination results in the ILEA 1979 & 1980', *Research and Statistics Report* 787/81. London: ILEA Research and Statistics.

California Assembly Office of Research (1984) *Overcoming the Odds: Making High Schools Work.* Sacramento: Author.

California State Department of Education (1980) *Report on the Special Studies of Selected ECE Schools with Increasing and Decreasing Reading Scores.* Sacramento: Office of Program Evaluation and Research.

Carley, M. (1981) *Social Measurement and Social Indicators.* London: Allen & Unwin.

Carroll, J. (1989) 'The Carroll Model: A 25 year retrospective and prospective view', *Educational Researcher* 18: 26–31.

Caul, L. (1994) *School Effectiveness in Northern Ireland: Illustration and Practice, Paper for the Standing Commission on Human Rights.* London: HMSO.

Central Advisory Council for Education (England) (1967) *Children and Their Primary Schools,* Vols. 1 & 2. London: HMSO (Plowden Report).

Chubb, J.E. (1988) 'Why the current wave of school reform will fail', *Public Interest* 90: 28–49.

Clegg, A. & Megson, B. (1968) *Children in Distress.* Harmondsworth: Penguin.

Cohen, M. (1983) 'Instructional management and social conditions in effective schools', in A.O. Webb & L.D. Webb (eds.) *School Finance and School Improvement: Linkages in the 1980s.* Cambridge, MA: Ballinger.

Coleman, P. (1998) *The Power of Three: Parent, Student and Teacher Collaboration.* London: Paul Chapman.

Coleman, J.S., Campbell, E., Hobson, C., McPartland, J., Mood, A., Weinfield, F. & York, R. (1966) *Equality of Educational Opportunity.* Washington: US Government Printing Office.

Coleman, J., Hoffer, T. & Kilgore, S. (1981) *Public and Private Schools.* Chicago: National Opinion Research Center.

Coleman, J., Hoffer, T. & Kilgore, S. (1982) 'Cognitive outcomes in public and private schools', *Sociology of Education* 55 (2/3): 65–76.

Coleman, P. (1994) *Learning About School: What Parents Need to Know and How They Can Find Out.* Montreal, Quebec: Institute for Research on Public Policy.

Coleman, P. & Larocque, L. (1990) *Struggling to be 'Good Enough': Administrative Practices and School District Ethos.* Basingstoke: Falmer Press.

Coleman, P., Collinge, J. & Seifert, T. (1993) 'Seeking the levers of change: Participant attitudes and school improvement', *School Effectiveness and School Improvement* 4 (1): 59–83.

Coleman, P., Collinge, J. & Tabin, Y. (1994) *Improving Schools From the Inside Out: A progress report on the coproduction of learning projects in British Columbia, Canada.* Burnaby, BC, Canada: Faculty of Education, Simon Fraser University.

Creemers, B.P.M. (1992) 'School effectiveness, effective instruction and school improvement in the Netherlands', in D. Reynolds and P. Cuttance (eds.) *School Effectiveness Research, Policy and Practice.* London: Cassell.

Creemers, B.P.M. (1994a) 'The history, value and purpose of school effectiveness studies', in D. Reynolds et al. (eds.) *Advances in School Effectiveness Research and Practice.* Oxford: Pergamon.

Creemers, B.P.M. (1994b) *The Effective Classroom.* London: Cassell.

Creemers, B.P.M. (1994c) 'Effective instruction: An empirical basis for a theory of educational effectiveness', in D. Reynolds et al. (eds.) *Advances in School Effectiveness Research and Practice.* Oxford: Pergamon.

Creemers, B., Peters, J. & Reynolds, D. (1989) *School Effectiveness and Improvement: Proceedings of the Second International Congress, Rotterdam, 1989.* Lisse: Swets & Zeitlinger.

Creemers, B. & Scheerens, J. (eds.) (1989) 'Development in school effectiveness research', special issue of *International Journal of Educational Research* 13: 685–825.

Creemers, B.P.M., Reynolds, D. & Swint, F.E. (1994) 'The International School Effectiveness Research Programme ISERP First Results of the Quantitative Study', paper presented at the British Education Research Association conference, Oxford, September 1994.

Creemers, B. & Scheerens, J. (1994) 'Developments in the educational effectiveness research programme', *International Journal of Educational Research* 21 (2): 125–139.

Creemers, B., Reynolds, D., Stringfield, S. & Teddlie, C. (1996) 'World Class Schools: Some further findings', paper presented at the AERA Conference, New York: April 1996.

Creemers, B. & Reezigt, G. (1996) 'School level conditions affecting the effectiveness of instruction', *School Effectiveness & Improvement* 7 (3): 197–228.

Creemers, B.P.M. & Reezigt, G. J. (1997) 'School effectiveness and school improvement: Sustaining links', *School Effectiveness & School Improvement* 8 (4): 396–429.

Crone, L.J., Lang, M.H. & Franklin, B.J. (1994) 'Achievement Measures of School Effectiveness: Comparison of consistency across years', paper presented at the annual meeting of the American Educational Research Association, 4–8 April, New Orleans.

Croninger, B. (1993) *Effects of Poverty on School Achievement and Behaviours* (3 August 1993) Programs for Educational Opportunity, Michigan: School of Education, University of Michigan.

Cuttance, P. (1986) *Effective Schooling: Report to the Scottish Education Department.* Edinburgh: Centre for Educational Sociology, University of Edinburgh.

Cuttance, P. (1987) *Modelling Variation in the Effectiveness of Schooling.* Edinburgh: Centre for Educational Sociology. University of Edinburgh.

Cuttance, P. (1988) 'Intra-system variation in the effectiveness of schooling', *Research Papers in Education* 3: 183–219.

DENI (1992) *Inspection Process in Schools.* Belfast: HMSO.

DENI (1992) *Evaluating Schools.* Belfast: HMSO.

DES (1988) Circular 7188 *Education Reform Act: Local Management of Schools.* London: HMSO.

DES (1989a) *Planning for School Development:Advice for Governors, Headteachers and Teachers.* London: HMSO.

DES (1989b) *School Indicators for Internal Management: An Aide Memoir.* London: HMSO.

DES (1990) *Developing School management: The Way Forward.* London: HMSO.

DES (1991) *Development Planning: A Practical Guide.* London: HMSO.

DfE Circular (7/93) *Inspecting Schools: A Guide to the Inspection Provisions of the Education (Schools) Act 1992 in England.* London: HMSO.

DfEE/OFSTED (1995) *Governing Bodies and Effective Schools.* London: OFSTED (HMI 071).

DfEE (1998) *Numeracy Matters: The Preliminary Report of the Numeracy Task Force.* (published for consultation) London: DfEE.

Daly, P. (1991) 'How large are secondary school effects in Northern Ireland?' *School Effectiveness and School Improvement* 2 (4): 305–323.

Davie, R., Butler, N. & Goldstein, H. (1972) *From Birth to Seven.* London: Longman.

Davies, H, Joslin, H & Clarke, L. (1993) 'Is it cash the deprived are short of?' Paper presented at a conference on research on the 1991 Census, University of Newcastle-upon-Tyne, 13–15 September 1993.

Davies, J. & Brember, I. (1997) 'The effects of pre-school experience on reading attainment: A four year cross-sectional study', *Educational Psychology* 17 (3): 255–265.

Dearing, R. (1993) *The National Curriculum and its Assessment: Interim report, July 1993*. London: National Curriculum Council & School Examination Assessment Council.

de Jong, M. (1988) 'Educational climate and achievement in Dutch schools', paper presented at the International Conference for Effective Schools, London: Institute of Education, University of London.

Department for Education (1992) *Choice and Diversity: A New Framework for Schools*. London: DFE.

Department for Education (1993) *Statistical Bulletin on Women in Post-Compulsory Education*, Issue No 26/93, December 1993. London: DFEE.

Department of Education and Science Inspectorate of Schools (1977) *Ten Good Schools: A Secondary School Enquiry*. London: HMSO.

Department of the Environment (1993) *An Urban Deprivation Index 1991*, Centre for Urban Policy Studies, Department of Geography, University of Manchester, March 1993—draft consultation document.

Douglas, J.W.B. (1964) *The Home and the School: The Study of Ability and Attainment in the Primary School*. London: MacGibbon & Kee.

Doyle, W. (1985) 'Effective secondary classroom practices', in M Kyle (ed.) *Reaching for Excellence: An Effective Schools Sourcebook*. Washington DC: US Government Printing Office.

Doyle, W. (1986) 'Classroom organization and management', in M. C. Wittrock (ed.) *Handbook of Research on Teaching*, 3rd edition. New York: MacMillan.

Drew, D. & Gray, J. (1991) 'The black-white gap in examination results: A statistical critique of a decade's research', *New Community* 17 (2): 159–172.

Earley, P., Fidler, B. & Ouston, J. (1996) *Improvement Through Inspection?* London: David Fulton.

Edmonds, R.R. (1979a) 'Effective Schools for the Urban Poor', *Educational Leadership* 37 (1): 15–27.

Edmonds, R.R. (1979b) 'Some schools work and more can', *Social Policy* 9: 28–32.

Edmonds, R.R. (1981) 'Making public schools effective', *Social Policy* 12: 56–60.

Edmonds, R.R. (1982) 'Programs of school improvement, an overview', *Educational Leadership* 40 (3): 4–11 (and interview, p14).

Elliott, J. (1996) 'School effectiveness research and its critics: Alternative visions of schooling', *Cambridge Journal of Education* 26 (2): 199–223.

Elliot, K., Smees, R. & Thomas, S. (1998) 'Making the most of your data: School self-evaluation using value added measures', *Improving Schools* 1 (3): 59–67.

Entwistle, D.R. & Hayduk, L.A. (1988) 'Lasting effects of elementary school', *Sociology of Education* 61: 147–159.

Epstein, J.L. (1987) 'Effects of teacher practices and parent involvement on student achievement in reading and maths' in S. Silver (ed.) *Literacy through Family Community and School Interaction*. Greenich, CT: JAI Press.

Essen, J. & Wedge, P. (1982) *Continuities in Childhood Disadvantage*. London: Heinemann.'

Evertson, C., Emmer, E. & Brophy, J. (1980) 'Predictors of effective teaching in junior high mathematics classrooms', *Journal for Research in Mathematics Education* 11 (3): 167–178.

Finn, C. (1984) 'Towards Strategic Independence: Nine commandments for enhancing school effectiveness', *Phi Delta Kappa* 65 (8): 518–524.

Firestone, W.A. (1991) 'Introduction: Chapter 1' in J.R. Bliss, W.A. Firestone & C.E. Richards (eds.) *Rethinking Effective Schools: Research and Practice*. Englewood Cliffs, New Jersey: Prentice Hall.
Fitz-Gibbon, C.T. (1985) 'A-level results in comprehensive schools', *Oxford Review of Education* 11 (1): 43–58.
Fitz-Gibbon, C.T. (ed.) (1990) 'Performance indicators', in *BERA Dialogues*, No. 2. Clevedon: Multilingual Matters.
Fitz-Gibbon, C.T. (1991) 'Multilevel modelling in an indicator system', Ch 6 in S.W. Raudenbush & J.D. Willms (eds.) *Schools, Classrooms and Pupils International Studies of Schooling from a Multilevel Perspective*. San Diego: Academic Press.
Fitz-Gibbon, C. (1992) 'School effects at A level: Genesis of an information system', in D. Reynolds & P. Cuttance (eds.) *School Effectiveness Research, Policy and Practice*. London: Cassell.
Forrest, R. & Gordon, D. (1993) *People and Places: A 1991 Census Atlas of England*. Bristol, School for Advanced Urban Studies, University of Bristol.
Fraser, B. J., Walberg, H. J., Welch, W. W. & Hattie, J. A. (1987) 'Syntheses of educational productivity research', *International Journal of Educational Research* 11 (2): 145–252.
Fullan, M. (1988) *Secondary Schools: Towards a More Fundamental Agenda*. The University of Toronto (mimeo).
Fullan, M. (1991) *The New Meaning of Educational Change*. London: Cassell.
Fullan, M. & Hargreaves, A. (1991) *What's Worth Fighting For? Working Together for Your School*. Toronto: Ontario Public School Teachers' Federation.
Fuller, B. & Clarke, P. (1994) 'Raising school effects while ignoring culture? Local conditions and the influence of classroom tools, rules and pedagogy', *Review of Educational Research* 64: 119–157.
Galton, M. & Simon, B. (1980) *Progress and Performance in the Primary Classroom*. London: Routledge & Kegan Paul.
Galton, M., Simon, B. & Croll, P. (1980) *Inside the Primary Classroom*. London: Routledge & Kegan Paul.
Gage, N. (1991) 'The obviousness of social and educational research results', *Educational Researcher* 20 (1): 10–16.
Gerrard, B. (1989) *A Longitudinal Study of a Primary Cohort With Special Reference to Truancy Behaviour*. (unpublished) PhD thesis CNAA.
Gillborn, D. (1992) 'Racism and education: Issues for research and practice', in S. Brown & S. Riddell (eds.) *Class, Race and Gender in Schools: A New Agenda for Policy and Practice in Scottish Education*. Glasgow: Scottish Council for Research in Education.
Gipps, C. (1992) *What We Know About Effective Primary Teaching*. London: Tufnell Press.
Glenn, B. (1981) *What Works? An Examination of Effective Schools for Poor Black Children*. Cambridge, MA: Harvard University Center for Law and Education.
Glennerster, H. & Hatch, S. (1974) *Positive Discrimination and Inequality*. Fabian Society Research Series, No. 314 London: Fabian Society.
Glover, D. (1992) 'An investigation of criteria used by parents and community in judgement of school quality', *Educational Research* 34 (1): 35–44.
Goldstein, H. (1987) *Multilevel Models in Educational and Social Research*. London: Charles Griffin & Co.
Goldstein, H. (1993) quoted in 'Taxed again by value added', *Times Education Supplement*. 13 August 1993.

Goldstein, H. (unpublished communication) *Results of a cross classified analysis of the extended JSP data set*. August 1994.

Goldstein, H. (1995) *Multilevel Statistical Models* (2nd Edition). London: Edward Arnold & New York: Halsted Press.

Goldstein, H. (1997) 'Methods in school effectiveness research', *School Effectiveness & School Improvement* 8 (4): 369–395.

Goldstein, H. (1998) '*Models for Reality: New approaches to the understanding of educational processes*', a professorial lecture. London: Institute of Education, University of London.

Goldstein, H., Rasbash, J., Yang, M., Woodhouse, G., Pan, H., Nuttall, D. & Thomas, S. (1992) 'Multilevel models for comparing schools', *Multilevel Modelling Newsletter* 4 (2): 5–6. London: Department of Mathematics, Statistics & Computing, Institute of Education, University of London.

Goldstein, H., Rashbash, J., Yang, M., Woodhouse, G., Pan, H., Nuttall, D. & Thomas, S. (1993) 'A multilevel analysis of school examination results', *Oxford Review of Education* 19 (4): 425–433.

Goldstein, H. & Sammons, P. (1995) *The Influence of Secondary and Junior Schools on Sixteen Year Examination Performance: A Cross-Classified Multilevel Analysis*. London: ISEIC, Institute of Education.

Goldstein, H. & Thomas, S. (1996) 'Using examination results as indicators of school and college performance', *Journal of the Royal Statistical Society A.* 159 (1): 149–163.

Goldstein, H. & Sammons, P. (1997) 'The influence of secondary and junior schools on sixteen year examination performance: A cross-classified multilevel analysis', *School Effectiveness & School Improvement* 8 (2): 219–230.

Good, T. (1984) 'Teacher effects' in *Making Our Schools More Effective:Proceedings of Three State Conferences*. Columbia, MO: University of Missouri.

Goodlad, J.I. et al. (1979) *A Study of Schooling*. Indiana: Phil Delta Kappa Inc.

Goodlad, J.I. (1984) *A Place Called School*. New York: McGraw-Hill.

Governors' Convention Report (1995) *Investing in Ethos*. London: Institute for School and College Governors.

Gray, J. (1981) 'A competitive edge, examination results and the probable limits of secondary school effectiveness', *Educational Review* 33 (1): 25–35.

Gray, J. (1990) 'The quality of schooling: Frameworks for judgements', *British Journal of Educational Studies* 38 (3): 204–233.

Gray, J. (1993) Review of J. Scheerens (1992) 'Effective schooling: Research, theory and practice', *School Effectiveness and School Improvement* 4 (3): 230–235.

Gray, J. (1995) 'The quality of schooling frameworks for judgement', in J. Gray & B. Wilcox (Eds.) *Good School, Bad School*. Buckingham: Open University Press.

Gray, J. (1998) *The Contribution of Educational Research to the Cause of School Improvement*, a professorial lecture. London: Institute of Education, University of London.

Gray, J., McPherson, A. & Raffe, D. (1983) *Reconstructions of Secondary Education*. London: Routledge & Kegan Paul.

Gray, J. & Hannon, V. (1986) 'HMI's interpretation of school examination results', *Journal of Education Policy* 1 (1): 23–33.

Gray, J., Jesson, D. & Jones, B. (1986) 'The search for a fairer way of comparing schools' examination results, *Research Papers in Education* 1 (2): 91–122.

Gray, J. & Jesson D. (1987) 'Exam results and local authority league tables', in *Education & Training UK: An Economic, Social and Policy Audit.* Newbury: Policy Journals.'

Gray, J., Jesson, J. & Sime, N. (1990) 'Estimating differences in the examination performances of secondary schools in six LEAs: A multi-level approach to school effectiveness', *Oxford Review of Education* 16 (2): 137–158.

Gray, J. & Simes, N. (1991) 'The stability of school effects over time' (summary tables), paper presented at the annual conference of the BERA, Nottingham Polytechnic, August 1991.

Gray, J., Jesson, D., Goldstein, H., Hedger, K. & Rasbash, J. (1993) 'A multi-level analysis of school improvement: Changes in schools' performance over time', paper presented at the 5th European Conference of the European Association for Research on Learning and Instruction, 3 September, Aix-en-Provence, France.

Gray, J. & Wilcox, B. (1994) 'The challenge of turning round ineffective schools', paper presented to the ESRC Seminar Series on School Effectiveness and School Improvement, Newcastle University, October 1994.

Gray, J. & Wilcox, B. (1995) 'The challenge of turning round ineffective schools', Chap 12 in J. Gray & B. Wilcox (1995) *Good School, Bad School.* Buckingham: OUP.

Gray, J., Jesson, D., Goldstein, H., Hedger, K. & Rasbash, J. (1995) 'A multi-level analysis of school improvement: Changes in schools performance over time,' *School Effectiveness and School Improvement* 6 (2): 97–114.

Gray, J., Goldstein, H. & Jesson, D. (1996) 'Changes and improvements in schools' effectiveness: Trends over five years', *Research Papers in Education* 11 (1): 35–51.

Gray, J., Reynolds, D. & Hopkins, D. (1998) *A Longitudinal Study of School Change and Improvement (The Improving Schools Research Project),* ESRC End of Award Report R000 235864. Cambridge: Homerton College, University of Cambridge.

Greenhill, S. & Chumun, S. (1990) *Examination Results in ILEA Schools—1988* RS 1275/90. London: Research and Statistics Branch, ILEA.

Grift, W. van der (1987) 'Self-perceptions of educational leadership and average achievement', in J. Scheerens and W. Stoel (eds.) *Effectiveness of School Organisations.* Lisse: Swets & Zeitlinger.

HMI (1990) *Standards in Education 1988–89.* London: HMSO.

Halsey, A.H. (1972) *Educational Priority, Vol 1, Problems and Policies.* London: HMSO.

Hallinger, P. (1996) 'The principal's role in school effectiveness: An assessment of substantive findings, 1980–1995', paper presented at the Annual Meeting of the Research Association, New York.

Hallinger, P. & Murphy, J. (1985) 'Instructional leadership and school socio-economic status: A preliminary investigation', *Administrator's Notebook* 31 (5): 1–4.

Hallinger, P. & Murphy, J. (1986) 'The social context of effective schools', *American Journal of Education* 94 (3): 328–355.

Hallinger, P. & Leithwood, K. (1994) 'Introduction: Exploring the impact of principal leadership', *School Effectiveness and School Improvement* 5 (3): 206–218.

Hamilton, D. (1996) 'Peddling feel-good fictions: Reflections on key characteristics of effective schools', *Forum* 38 (2): 54–56.

Hannon, P. & Jackson, A. (1987) *The Belfield Reading Project.* London: National Childrens' Bureau.

Hanushek, E. (1979) 'Conceptual and empirical issues in the estimation of educational production functions', *Journal of Human Resources* 14: 351–388.

Hanushek, E. (1986) 'The economics of schooling: Production and efficiency in public schools', *Journal of Economic Literature* 24: 1141–1177.

Hanushek, E. (1989) 'The impact of differential expenditures on school performance', *Educational Researcher* 18 (4): 45–65.

Hargreaves, D. (1995) 'School effectiveness, school change and school improvement: The relevance of the concept of culture', *School Effectiveness & School Improvement* 6 (1): 23–46.

Hargreaves, D.H. & Hopkins, D. (1991) *The Empowered School: The Management and Practice of Development Planning*. London: Cassell.

Hargreaves, D. & Hopkins, D. (eds.) (1994) *School Development; Development Planning for School Improvement*. London: Cassell.

Harris, A. (1995) 'Effective teaching. research matter', *SIN Network Institute of Education*. Summer 1995 No. 3.

Harris, A., Jamieson, I. & Russ, J. (1995) 'A study of 'effective' departments in secondary schools', *School Organisation* 15 (3): 283–299.

Heal, K. (1978) 'Misbehaviour among school children: The role of the school in strategies for prevention', *Policy and Politics* 6: 321–333.

Hedges, L.V., Laine, R.D. & Greenwald, R. (1994) 'Does money matter? A meta-analysis of studies of the effects of differential school inputs on student outcomes' (An exchange: Part 1), *Educational Researcher* 23 (3): 5–14.

Helmreich, R. (1972) 'Stress, self-esteem and attitudes', in B. King and E. McGinnies (eds.) *Attitudes, Conflict and Social Change*. London: Academic Press.

Hersh, R., Carnine, D., Gall, M., Stockard, J., Carmack, M. & Gannon, P. (1981) *The Management of Education Professionals in Instructionally Effective Schools: Towards a Research Agenda*. Eugene: Center for Educational Policy and Management, University of Oregon.

Hill, P., Rowe, K. & Holmes-Smith, P. (1995) 'Factors affecting students' educational progress: Multilevel modelling of educational effectiveness', paper presented at the International Congress for School Effectiveness, Leeuwarden, the Netherlands, 3–6 January 1995.

Hill, P. W., Rowe, K. J. & Jones, T. (1995) *SIIS: School Improvement Information Service, Version 1.1*. October 1995, Melbourne: University of Melbourne, Centre for Applied Educational Research.

Hill, P.W. & Rowe, K.J. (1996) 'Multilevel modelling in school effectiveness research', *School Effectiveness and School Improvement* 7 (1): 1–34.

Hillman, J. & Stoll, L. (1994) 'Understanding school improvement', SIN *Research Matters No 1*. London: Institute of Education, University of London.

Holmes, M. (1989) 'From research to implementation to improvement' in M. Holmes, K.A. Leithwood & D.F. Musella (eds.) *Educational Policy For Effective Schools*. New York: Teachers College Press.

Hopkins, D. (1994) 'Towards a theory for school improvement', paper presented to the ESRC Seminar series on School Effectiveness and School Improvement, Newcastle University, October 1994.

Holtermann, S. (1975) 'Areas of urban deprivation in Great Britain: An analysis of 1971 Census data', *Social Trends*, Vol 6. London: HMSO.

Holtermann, S. (1977) 'The welfare economics of priority area policies', *Journal of Social Policy* 7 (1): 23–40.

Hopkins, D. (1994) 'School improvement in an era of change', Chap 6, in P. Ribbens & E. Burridge (eds.) *Improving Education Promoting Quality in Schools*. London: Cassell.

Hopkins, D., Ainscow, M. & West, M. (1994) *School Improvement in an Era of Change*. London: Cassell.
Hutchison, D., Prosser, H. & Wedge, P. (1979) 'The prediction of educational failure', *Educational Studies* 5 (1): 73–82.
Hutson, N. (1998) *Making Belfast Work. Education Paper*. Urban Institute Belfast.
ISEIC (1998) *Enhancing Education*. London: Institute of Education, University of London.
Jencks, C., Smith, M., Acland, H., Bane, M.J., Cohen, D., Gintis, H., Heyns, B. & Michelson, S. (1972) *Inequality: A Reassessment of the Effects of Family and Schooling in America*. New York: Basic Books.
Jesson, D. & Gray, J. (1991) 'Slants on slopes: Using multi-level models to investigate differential school effectiveness and its impact on pupils' examination results', *School Effectiveness and School Improvement* 2 (3): 230–271.
Jesson, D., Bartlett, D. & Machon, C. (1997) 'Baseline assessment and school improvement—the use of data from the assessment of children at entry to school to support the raising of standards', paper presented at the Annual British Educational Research Association Conference, University of York, September 1997.
Jowett, S., Baginsky, M. & MacDonald, M. (1991) *Building Bridges: Parental Involvement in Schools*. Windsor: NFER-Nelson.
Joyce, B. & Showers, B. (1988) *Student Achievement Through Staff Development*. New York: Longman.
Kogan, M. (1986) *Education Accountability: An Analytic Overview*. London: Hutchinson.
Knuver, J.W.M. & Brandsma, H.P. (1993) 'Cognitive and affective outcomes in school effectiveness research', *School Effectiveness and School Improvement* 4 (3): 189–204.
Lacey, C. & Blane, D. (1979) 'Geographic mobility and school attainment: The confounding variables', *Educational Research* 21 (3): 200–206.
Lagerweij, N. A. J. & Voogt, J. C. (1990) 'Policy at school level', *School Effectiveness and School Improvement* 1 (2) 98–120.
Lapan, R., Gysbers, N. & Sun, Y. (1997) 'The impact of more fully implemented guidance programs on the school experiences of High School students: A statewide evaluation study', *Journal of Counseling & Development* 75: 292–302.
Lawton, D. (1994) 'Defining quality', in P. Ribbins and E. Burridge (eds.) *Improving Education: Promoting Quality in Schools*. London: Cassell.
Lazar, I. & Darlington, R. (1982) 'The lasting effects of early education: A report from the consortium for longitudinal studies', *Monographs of the Society for Research in Child Development* 47 (195): 2–3.
Lee, T. (1991) *Additional Educational Needs and LMS: Methods and Money 1991–2*. Bath Centre for the Analysis of Social Policy, School of Social Studies, University of Bath, July, 1991.
Lee, T. (1992) 'Finding simple answers to complex questions: Funding special needs under LMS', Ch 7 in G. Wallace (ed.) *Local Management of Schools: Research and Experience*. BERA Dialogue 6, Clevedon: Multilingual Matters.
Lee, V., Bryk, A., & Smith, J. (1993) 'The organisation of effective secondary schools', Chapter 5, in L. Darling-Hammond (ed.) *Research in Education* 19: 171–226, Washington DC: American Educational Research Association.
Levacic, R. (1990) 'Evaluating local management of schools: Establishing a methodological framework', in R. Saran and V. Trafford (eds.) *Research in Education Management and Policy. Retrospect and Prospect*. Lewes: Falmer Press.

Levine, D. (1992). 'An interpretive review of US research and practice dealing with unusually effective schools', in D. Reynolds & P. Cuttance (eds.) *School Effectiveness Research, Policy and Practice*. London: Cassell.

Levine, D. & Stark, J. (1981) *Instructional and Organisational Arrangements and Processes for Improving Academic Achievement at Inner City Elementary Schools*. Kansas City: University of Missouri.

Levine, D. & Lezotte, L. (1990) *Unusually Effective Schools: A Review and Analysis of Research and Practice*. Madison, WI: National Center for Effective Schools Research and Development.

Lewis, A. & Sammons, P. (1998) *School Effectiveness, Pedagogy and SEN in Mainstream Primary Schools*. ESRC Proposal R000 237981. Warwick: School of Education, University of Warwick.

Lezotte, L. (1989) 'School improvement based on the effective schools research', *International Journal of Educational Research* 13 (7): 815–825.

Lightfoot, S. (1983) *The Good High School: Portraits of Character and Culture*. New York: Basic Books.

Lipsitz, J. (1984) *Successful Schools for Young Adolescents*. New Brunswick: Transaction Books.

Literacy Taskforce (1997) *A Reading Revolution: Preliminary Report of the Literary Taskforce* chaired by Professor Michael Barbe. London: Institute of Education, University of London.

Little, A. & Mabey, C. (1972) 'An index for designation of educational priority areas', in A. Shonfield & S. Shaw (eds.) *Social Indicators and Social Policy*, Chap 5. London: Heinemann.

Little, A. & Mabey, C. (1973) 'Reading attainment and social and ethnic mix of London primary schools', in Donnison, D. & Eversely, D. (eds.) *London: Urban Patterns, Problems and Policies*, Ch. 9. London: Heinemann.

Louis, K. S. & Miles, M. B. (1991) 'Toward effective urban high schools: The importance of planning and coping', Chapter 7, in J. R. Bliss, W. A. Firestone & C. E. Richards (eds.) *Rethinking Effective Schools: Research and Practice*. Englewood Cliffs, New Jersey: Prentice Hall.

Louis, K. & Miles, M. (1992) *Improving the Urban High School: What Works and Why*. London: Cassell.

Luyten, H. (1994) 'Stability of school effects in secondary education: The impact of variance across subjects and years', paper presented at the annual meeting of the American Educational Research Association, 4–8 April, New Orleans.

Luyten, H. (1995) 'Teacher change and instability across grades', *School Effectiveness & School Improvement* 1 (1): 67–89.

Luyten, H. & Snijders, T. (1996) 'School effects and teacher effects in elementary education', paper presented to the Op de Onderwijs Research Dagen, Groningen, June 1995 and to appear in *Educational Research and Evaluation* 2 (1): 1–24.

MacBeath, J. (1994) 'Making schools more effective: A role for parents in school self-evaluation and development planning', paper presented to annual conference of the American Educational Research Association, 4–8 April, New Orleans.

McCartney, K. & Jordan, E. (1990) 'Parallels between research on child care and research on school effects', *Educational Researcher* 19 (1): 21–27.

McCullum, I. (1993) 'Contextualising educational performance by means of census data', paper given at a joint seminar of the Operational Research Society and the Local Educational Authorities Research Group, University of Warwick, 29 June 1993.

McDill, E. & Rigsby, L. (1973) *Structure and Process in Secondary Schools*. Baltimore: John Hopkins University Press.

McGaw, B., Piper, K., Barks, O. & Evans, B. (1992) *Making Schools More Effective*. Victoria: Australian Council for Educational Research Ltd.

MacGilchrist, B. (1995) 'The use of qualitative data in an empirical study of school development planning', paper presented at ICSEI, 3–6 January, Leeuwarden, The Netherlands, London: ISEIC, Institute of Education.

MacGilchrist, B., Mortimore, P., Savage, J. & Beresford, C. (1995) *Planning Matters. The Impact of Development Planning in Primary Schools*. Paul Chapman. London.

MacGilchrist, B., Myers, R. & Reed, J. (1997) *The Intelligent School*. London: Paul Chapman.

McGill, T.P. (1996) *Paper for MBW RSS Schools*. 7th June 1996 (unpublished).

McLaughlin, M. (1977) 'The Rand Change Agent Study revisted: Macro perspectives, micro realities', *Educational Researcher* 19 (9): 11–16.

MacKenzie, D. (1983) 'Research for school improvement: An appraisal of some recent trends', *Educational Researcher* 12 (4): 5–16.

McPherson, A. (1992) 'Measuring added value in schools', *National Commission on Education Briefing, No 1*. February 1992, London.

Mabey, C. (1981) 'Black British literacy', *Education Research* 23 (2): 83–95.

Madaus, G.G., Kellagham, T., Rakow, E.A. & King, D. (1979) 'The sensitivity of measures of school effectiveness', *Harvard Educational Review* 4: 207–230.

Mandeville, G.K. (1987) *The Stability of School Effectiveness Indices Across Years*. Washington: NCME paper.

Mandeville, G.K. & Anderson, L. (1986) *A Study of the Stability of School Effectiveness Measures Across Grades and Subject Areas*. San Francisco: AERA paper.

Marks, J. (1991) *Standards in Schools Assessment, Accountability and the Purposes of Education*. Social Market Foundation Paper No 11. London: Social Market Foundation.

Marks, J., Cox, C. & Pomian-Srzednicki, M. (1983) *Standards in English Schools*. London: National Council for Educational Standards.

Marks, J. & Pomian-Srzednicki, M. (1985) *Standards in English Schools, Second Report*. London: National Council for Educational Standards.

Marks, J., Cox, C. & Pomian-Srzednicki, M. (1986) *Examination Performance of Secondary Schools in the Inner London Education Authority*. London: National Council for Educational Standards.

Marsh, A.J. (1998) *Formula Funding and Special Educational Needs*. PhD thesis, Milton Keynes: School of Education, Open University.

Mathieson, E., Mico, E. & Morton, J. (1974) 'Pupil mobility survey', *Research and Statistics Report* 605/74. London: ILEA Research and Statistics.

Matthews, P. & Smith, G. (1995) OFSTED: 'Inspecting schools and improvement through inspection', *Cambridge Journal of Education* 25 (1): 23–34.

Miles, M., Farrar, E. & Neufeld, E. (1983) *Review of Effective School Programs, Vol 2: The Extent of Effective School Programs*. Cambridge MA: Huron Institute (unpublished).

Miles, M.B. & Ekholm, M. (1985) 'School improvement at the school level', *Making School Improvement Work*. Leuven: ACCO.

Ming, T.W. & Cheong, C.Y. (1995) 'School environment and student performance: A multilevel analysis', paper presented at the International Congress of School Effectiveness and Improvement, Leeuwarden, The Netherlands: January 1995.

Morris, R. & Carstairs, V. (1991) 'Which deprivation? A comparison of selected deprivation indexes', *Journal of Public Health Medicine* 13: 318–326.
Mortimore, P. (1990) *The Nature of Findings of Research on School Effectiveness in the Primary Sector*. Edinburgh: Educational Resources Unit Scottish Department of Education.
Mortimore, P. (1991a) 'The nature and findings of school effectiveness research in the primary sector', in S. Riddell & S. Brown (eds.) *School Effectiveness Research: Its Messages for School Improvement*. London: HMSO.
Mortimore, P. (1991b) 'Effective schools from a british perspective', Chapter 7, in J. R. Bliss, W. A. Firestone & C. E. Richards (1991) *Rethinking Effective Schools: Research and Practice*. Englewood Cliffs, New Jersey: Prentice Hall.
Mortimore, P. (1992) 'Issues in school effectiveness', in D. Reynolds & P. Cuttance (eds.) *School Effectiveness Research, Policy and Practice*. London: Cassell.
Mortimore, P. (1993a) 'School effectiveness and the management of effective learning and teaching', *School Effectiveness and School Improvement* 4 (4): 290–310.
Mortimore, P. (1993b) 'The positive effects of schooling', in M. Rutter (ed.) *Youth in the Year 2000: Psycho-Social Issues and Interventions*. Cambridge: Cambridge University Press.
Mortimore, P. (1995a) 'The positive effects of schooling, in M. Rutter (ed.) *Psycho-Social Disturbances in Young People: Challenges for Prevention*. Cambridge: Cambridge University Press.
Mortimore, P. (1995b) *Effective Schools: Current Impact and Future Possibilities*. The Director's Inaugural Lecture, 7 February 1995, London: Institute of Education, University of London.
Mortimore, P. (1998) *The Road to Improvement: Reflections on School Effectiveness*. Lisse: Swets & Zeitlinger.
Mortimore, J. & Blackstone, T. (1982) *Disadvantage and Education*. London: Heinemann.
Mortimore, P., Sammons, P., Stoll, L., Lewis, L. & Ecob, R. (1986) *The Junior School Project* (4 vols). London: Research & Statistics Branch, ILEA.
Mortimore, P., Sammons, P., Stoll, L., Lewis, D. & Ecob, R. (1987a) 'Towards more effective junior schooling', summer, *Forum* 29 (3): 70–73.
Mortimore, P., Sammons, P., Stoll, L., Lewis, D. & Ecob, R. (1987b) 'For effective classroom practices', autumn *Forum* 30 (1): 8–11.
Mortimore, P., Sammons, P., Stoll, L., Lewis, D. & Ecob, R. (1987c) 'The ILEA junior school project: A study of school effectiveness', spring *Forum* 29 (2): 47–49.
Mortimore, P., Sammons, P. & Ecob, R. (1988) 'Expressing the magnitude of school effects—a reply to Peter Preece', *Research Papers in Education* 3 (2): 99–101.
Mortimore, P., Sammons, P., Stoll, L., Lewis, D. & Ecob, R. (1988a) *School Matters: The Junior Years*. Wells: Open Books.
Mortimore, P., Sammons, P., Stoll, L., Lewis, D. & Ecob, R. (1988b) 'The effects of school membership on pupils' educational outcomes', *Research Papers in Education* 3 (1): 3–26.
Mortimore, P. & Stone, C. (1990) 'Measuring educational quality', *British Journal of Educational Studies* 39 (1) 69–82.
Mortimore, P., Mortimore, J. & Sammons, P. (1991) *The Use of Indicators, Effectiveness of Schooling and of Educational Resource Management*. OECD July 1991, London: Institute of Education University of London.

Mortimore, P., Sammons, P. & Thomas, S. (1994) 'School effectiveness and value added measures', *Assessment in Education: Principles, Policy & Practice* 1 (3): 315–332.

Mortimore, P., Sammons, P. & Thomas, S. (1995) 'School effectiveness and value added measures', paper presented at the Desmond Nuttall Memorial Conference 10.6.94, *Assessment in Education: Principle Policy & Practice* 1 (3): 315–332.

Mortimore, P., Davies, H. & Portway, S. (1996) 'Burntwood secondary girls' school', Chap 6, in *Success Against the Odds*. National Commission on Education. London: Routledge.

Mortimore, P. & Whitty, G. (1997) *Can School Improvement Overcome the Effects of Disadvantage?* London: Institute of Education, University of London.

Mortimore, P. & Sammons, P. (1997) 'Endpiece: A welcome and a riposte to critics etc.', Chapter 10, in J. White & M. Barber (eds.). *Perspectives on School Effectiveness and Improvement*. London: Bedford Way Paper, Institute of Education, University of London.

Moser, C.A. & Scott, W. (1961) *British Towns*. Edinburgh: Oliver & Boyd.

Murphy, J. (1989) 'Principal instructional leadership', in P. Thuston & L. Lotto (eds.) *Advances in Educational Leadership*. Greenich: JAI Press.

Myers, K. (ed.) (1996) *School Improvement in Practice: Accounts from the School Make A Difference Project*. London: Falmer Press.

Myers, K. & Goldstein, H. (1996) 'Get it in context?' *Education*, 16 February, No 187/7.

National Commission on Education (1996) *Success Against the Odds: Effective Schooling in Disadvantaged Areas*. (eds.) M. Maden & J. Hillman, London: Routledge.

National Institute of Education (1978) *Violent Schools—Safe Schools: The Safe School Study Report to the Congress*. Washington DC: Department of Health, Education and Welfare.

Northern Ireland Affairs Committee (Session 1996/7) 'Underachievement in Northern Ireland Secondary Schools', *House of Commons Report, together with the Proceedings of the Committee, Minutes and Evidence and Appendices*. London: The Stationery Office.

North West Regional Educational Laboratory (1990) *Onward to Excellence: Effective Schooling Practices: A Research Synthesis*. Portland, Oregon: North West Regional Educational Laboratory.

North West Regional Educational Laboratory (1995). *Effective Schooling Practices: A Research Synthesis – 1995 update*. Portland, Oregon: NREL School Improvement Research Series.

Nuttall, D. (1990a) 'The functions and limitations of international educational indicators', *International Journal of Educational Research* 14 (4): 329–333.

Nuttall, D. (1990b) *Differences in Examination Performance RS 1277/90*. London: Research and Statistics Branch, ILEA.

Nuttall, D. (1991) 'An instrument to be honed', *Times Educational Supplement* 13.9.91.

Nuttall, D., Goldstein, H., Prosser, R. & Rasbash, J. (1989) 'Differential school effectiveness', Chapter 6, in, special issue Developments in School Effectiveness Research, *International Journal of Educational Research* 13: 769–776.

Nuttall, D., Thomas, S. & Goldstein, H. (1992) *Report on Analysis of 1990 Examination Results*. Association of Metropolitan Authorities Project on Putting Examination Results in Context. London: Centre for Educational Research, LSE.

OECD (1989) *Schools and Quality. An International Report*. Paris: OECD.

OFSTED (1993) *Access and Achievement in Urban Education*. London: HMSO.

OFSTED (1994/5) *Inspection Quality*. London: HMSO.
Osborn, A.F. & Millbank, J.E. (1987) 'The effects of early education: A report from the child health (1992)', *Contemporary Issues in the Early Years*. London: Paul Chapman Educational Series.
Panton, K.J. (1980) 'Literacy in London', *Geography* 65 (1): 27–34.
Parsons, S. & Bynner, J. (1998) *Influences on Adult Basic Skills Factors Affecting the Development of Literacy and Numeracy from Birth to 37*. London: The Basic Skills Agency.
Paterson, L. (1992) 'Social Class in Scottish Education', in S. Brown & S. Riddell (eds.) *Class, Race and Gender in Schools: A New Agenda for Policy and Practice in Scottish Education*. Glasgow: Scottish Council for Research in Education.
Paterson, L. & Goldstein, H. (1991) 'New statistical methods of analysing social structures: An introduction to multilevel models', *British Educational Research Journal* 17 (4): 387–393.
Plowden Report (1967) See Central Advisory Council for Education.
Pollack, S., Watson, D. & Chrispeels, J. (1987) 'A description of factors and implementation strategies used by schools in becoming effective for all students', paper for American Educational Research Association.
Powell, M. (1980) 'The beginning teacher evaluation study: A brief history of a major research project', in C. Denham & A. Lieberman (eds.) *Time to Learn*. Washington DC: National Institute of Education.
Powell, J. & Scrimgeour, M. (1977) *System for Classroom Observation of Teaching Strategies*. Edinburgh: Scottish Council for Research in Education.
Preece, P. (1988) 'Misleading ways of expressing the magnitude of school effects', *Research Papers in Education* 3 (2): 97–98.
Preece, P. (1989) 'Pitfalls in research on school and teacher effectiveness', *Research Papers in Education* 4 (3): 47–69.
Pring, R. (1995) 'Educating persons: Putting education back into educational research' (the 1995 SERA lecture), *Scottish Educational Research Journal* 27: 101–12.
Prosser, R., Rasbash, J. & Goldstein, H. (1991) *ML3 Software for Three-Level Analysis Users' Guide for V2*. London: Institute of Education, University of London.
Purkey, S.C. & Smith, M.S. (1983) 'Effective schools, a review', *Elementary School Journal* 83 (4): 427–52.
Qualifications & Curriculum Authority (QCA) (1997) *The National Framework for Baseline Assessment*. London: QCA.
Ralph, J.H. & Fennessey, J. (1983) 'Science or Reform: Some Questions about the Effective Schools Model', *Phi Delta Kappan* 64 (10): 589–694.
Rampton Report (1981) *West Indian Children in Our Schools*. London: HMSO.
Raudenbush, S.W. (1989) 'The analysis of longitudinal multilevel data', Chapter 3, in special issue Developments in School Effectiveness Research, *International Journal of Educational Research* 13 (7): 721–740.
Reezigt, G., Guldemond, H. & Creemers, B. (forthcoming). 'Empirical validity for a comprehensive model on educational effectiveness', *School Effectiveness & Improvement*.
Reid, I. (1991) *The Development of Socio-Spatial Indices for Bradford's Schools*. Loughborough: Loughborough University Department of Education.
Reid, K., Hopkins, D. & Holly, P. (1987) *Towards the Effective School*. Oxford: Blackwell.
Renihan, P.J. et al. (1986) 'The common ingredients of successful school effectiveness projects', *Education Canada*, Fall, 16–21.

Reynolds, D. (1976) 'The delinquent school', in P. Woods (ed.) *The Process of Schooling*. London: Routledge & Kegan Paul.
Reynolds, D. (1982) 'The search for effective schools', *School Organization* 2 (3) 215–37.
Reynolds, D. (ed.) (1985) *Studying School Effectiveness*. Basingstoke: Falmer Press.
Reynolds, D. (1989) 'School effectiveness and school improvement: A review of the British literature', in D. Reynolds, B. P. M. Creemers and T. Peters (eds.) *School Effectiveness and Improvement. Proceedings of First International Congress*. Gronigen: RION Institute of Educational Research.
Reynolds, D. (1991) 'School effectiveness and school improvements in the 1990s', *Association for Child Psychology and Psychiatry Newsletter* 13 (2): 5–9.
Reynolds, D. (1992) 'School effectiveness and school improvement: An updated review of the British literature', in D. Reynolds & P. Cuttance (eds.) *School Effectiveness Research Policy and Practice*. London: Cassell.
Reynolds, D. (1994) 'School effectiveness research: A review of the international literature', in D. Reynolds, B.P.M. Creemers, P.S. Nesselradt, E.C. Schaffer, S. Stringfield & C. Teddlie (eds.) *Advances in School Effectiveness Research and Practice*. Oxford: Pergamon.
Reynolds, D. (1995) 'The effective school: An inaugural lecture', *Evaluation and Research in Education* 9 (2): 57–73.
Reynolds, D. (1996) 'Turning round ineffective schools: Some evidence and some speculation', Chap. 8, in J. Gray, D. Reynolds, C. Fitz-Gibbon & D. Jesson (eds.) *Merging Traditions: The Future of Research and School Effectiveness and School Improvement*. London: Cassell.
Reynolds, D. & Murgatroyd, S. (1977) 'The sociology of schooling and the absent pupil: The school as a factor in the generation of truancy', in H Carroll (ed.) *Absenteeism in South Wales: Studies of pupils, their homes and their secondary school*. University College of Swansea, Faculty of Education.
Reynolds, D., Sullivan, M. & Murgatroyd, S. (1987) *The Comprehensive Experiment: A Comparison of the Selective and Non-Selective System of School Organization*. Lewes: Falmer Press.
Reynolds, D., Creemers, B. & Peters, T. (1989) *School Effectiveness and Improvement: Proceedings of the First International Congress, London, 1988*. Groningen: University of Groningen, RION.
Reynolds, D. & Creemers, B. (1990) 'School effectiveness and school improvement: A mission statement', *School Effectiveness & School Improvement* 1 (1): 1–3.
Reynolds, D. & Cuttance, P. (1992) (eds.) *School Effectiveness Research, Policy and Practice*. London: Cassell.
Reynolds, D. & Packer, A. (1992) 'School effectiveness and school improvement in the 1990s', in D. Reynolds and P. Cuttance (eds.) *School Effectiveness Research, Policy and Practice*. London: Cassell.
Reynolds, D., Hopkins, D. & Stoll, L. (1993) 'Linking School Effectiveness Knowledge and School Improvement Practice: Towards a synergy', *School Effectiveness & School Improvement* 4 (1): 37–58.
Reynolds, D., Creemers, B., Nesselrodt, P.S., Schaffer, E.C., Stringfield, S. & Teddlie, C. (1994a) *Advances in School Effectiveness Research and Practice*. Oxford: Pergamon.
Reynolds, D., Sammons, P., Stoll, L. & Barber, M. (1994b) 'School effectiveness and school improvement in the United Kingdom', in C.P.M. Creemers & N. Osinga (Eds.) *ICSEI Country Reports*. Leeuwarden: GCO in Friesland.

Reynolds, D. et al. (1994) 'School effectiveness research: A review of the international literature', in D. Reynolds, B.P.M. Creemers, P.S. Nesselradt, E.C. Schaffer, S. Stringfield & C. Teddlie (eds.) *Advances in School Effectiveness Research and Practice*. Oxford: Pergamon.

Reynolds, D. & Farrell, S. (1996) 'Worlds apart? A review of international surveys of educational achievement involving England', *OFSTED Reviews of Research*. London: HMSO.

Reynolds, P., Bollen, R., Creemers, B., Hopkins, D., Stoll, L. & Lagerweij, N. (eds.) (1996) *Making Good Schools: Linking school effectiveness and school improvement*. London: Routledge.

Ribbins P. & Burridge E. (eds.) (1994) *Improving Education: Promoting Quality in Schools*. London: Cassell.

Riddell, A. R. (1995) *School Effectiveness and School Improvement in the Third World: A Stock-Taking and Implications for the Development of Indicators*. London: Institute of Education. University of London.

Riddell, S., Brown, S. & Duffield, J. (1994) 'The social and institutional context of effectiveness', paper presented at the annual conference of the British EducationalResearch Association, St Anne's College, Oxford, September 1994.

Robertson, P. & Sammons, P. (1997a) 'Improving school effectiveness: A project in progress', paper presented at the Tenth International Congress for School Effectiveness and Improvement, Memphis, Tennessee, 5–8 January 1997.

Robertson, P. & Sammons, P. (1997b) 'The improving school effectiveness project (ISEP): Understanding change in schools', paper presented at the British Educational Research Association annual conference, University of York, September 1997.

Robertson, P., Toal, D., MacGilchrist, B. & Stoll, L. (1998) 'Quality counts: Evaluating evidence for school improvement', paper presented at the International Congress for School Effectiveness & School Improvement, University of Manchester, January 1998.

Robinson, W.S. (1950) 'Ecological correlations and the behaviour of individuals', *American Sociological Review* 15 June 351–357.

Robinson, P. (1997) *Literacy, Numeracy and Economic Performance Centre for Economic Performance*. LSE, University of London.

Rosenshine, B. (1987) 'Direct instruction', in M. J. Dunkin (ed.) *The International Encyclopedia of Teaching and Teacher Education*. Oxford: Pergamon Press.

Rosenshine, B. & Berliner, D. (1978) 'Academic engaged time', *British Journal of Teacher Education* 4: 3–16.

Rosenshine, B. & Stevens, R. (1981) *Advances in Research on Teaching* (unpublished manuscript), University of Illinois.

Rowe, K.J. & Hill, P.W. (1994) 'Multilevel Modelling in School Effectiveness Research: How many levels?' paper presented at the International Congress for School Effectiveness and Improvement, The World Congress Centre, Melbourne, Australia, 3–6 January 1995.

Rowe, K.J., Hill, P.W. & Holmes-Smith, P. (1996) 'Assessing, recording and reporting students,' Educational Progress: The case for 'subject profiles', *Assessment in Education* 3 (3): 309–352.

Rutter, M. & Madge, N. (1976) *Cycles of Disadvantage*. London: Heinemann.

Rutter, M., Maughan, B., Mortimore, P. & Ouston, J. (1979) *Fifteen Thousand Hours: Secondary Schools and Their Effects on Children*. London: Open Books.

Sammons, P. (1985) *Participation in Vocational Further Education: A Study of Factors Influencing Entry into Commercial, Construction and Engineering Training in Inner London*. (unpublished PhD thesis) CNAA.

Sammons, P. (1987a) 'School climate, the key to fostering student progress and development?', keynote paper presented to the Annual Convention of the Prince Edward Island Teachers' Federation on *School Atmosphere, the Barometer of Success*. 29–30 October, Charlottetown, Prince Edward Island, Canada.

Sammons, P. (1987b) 'Findings from School Effectiveness Research: A framework for school improvement', keynote paper presented to the Annual Convention of the Prince Edward Island Teachers' Federation on *School Atmosphere: The Barometer of Success*. Charlottetown, Prince Edward Island, Canada, 29–30.10.87.

Sammons, P. (1989a) 'Measuring school effectiveness', in D. Reynolds, P. Bert, M. Creemers & T. Peters (eds.) *School Effectiveness and Improvement* 15: 169–188, Groningen, The Netherlands: School of Education University of Wales College of Cardiff & RION Institute for Educational Research.

Sammons, P. (1989b) 'School Effectiveness and School Organization', Chap 2, in L. Eldering & J. Kloprogge (eds.) *Different Cultures Same School: Ethnic minority children in Europe*. Amsterdam/Lisse: Swets & Zeitlinger.

Sammons, P. (1993) *Measuring and Resourcing Educational Needs: Variations in LEAs' LMS Policies in inner London*. Clare Market Paper No 6, London: Centre for Educational Research, LSE, University of London.

Sammons, P. (1994a) 'Findings from school effectiveness research: Some implications for improving the quality of schools', in P. Ribbins & E. Burridge (eds.) *Improving Education: The Issue in Quality*. London: Cassell.

Sammons, P. (1994b) 'Gender, socio-economic and ethnic differences in attainment and progress: A longitudinal analysis of student achievement over nine years', revised version of a paper presented to the Equity Issues in Performance Assessment Symposium of the annual meeting of the American Educational Research Association, New Orleans, 4–8 April. Later published in *British Educational Research Journal* 21 (4): 465–485, 1995.

Sammons, P. (1996) 'Complexities in the judgement of school effectiveness', *Educational Research and Evaluation*. 2 (2): 113–149.

Sammons, P. (1998) 'Diversity in classrooms: Effects on educational outcomes', in D. Shorrocks-Taylor (ed.) *Directions in Educational Psychology*. London: Whurr Publishers Ltd.

Sammons, P., Kysel, F. & Mortimore, P. (1983) 'Educational priority indices: A new perspective', *British Educational Research Journal* 9 (1): 27–40.

Sammons, P., Mortimore, P. & Varlaam, A. (1985) *Socio-economic Background, Parental Involvement and Attitudes, and Children's Achievements in Junior Schools*, RS 982/85. London: Research & Statistics Branch, ILEA.

Sammons, P., Mortimore, P. & Thomas, S. (1993a) 'Do schools perform consistently across outcomes and areas?' paper presented to the ESRC Seminar Series July 1993, University of Sheffield.

Sammons, P., Mortimore, P. & Thomas, S. (1993b) 'First weigh your ingredients', *The Independent*. 20 November 1993.

Sammons, P., Nuttall, D. & Cuttance, P. (1993) 'Differential school effectiveness: Results from a reanalysis of the inner London education authority's junior school project data', *British Educational Research Journal* 19 (4): 381–405.

Sammons, P., Cuttance, P., Nuttall, D. & Thomas, S. (1994a) 'Continuity of school effects: A longitudinal analysis of primary and secondary school effects on GCSE performance', originally presented at the Sixth International Congress for School Effectiveness and Improvement, Norrkoping, Sweden (revised version published in *School Effectiveness & School Improvement* 6 (4): 285–307.

Sammons, P., Thomas, S., Mortimore, P., Owen, C. & Pennell, H. (1994b) *Assessing School Effectiveness: Developing Measures to put School Performance in Context.* London: Office for Standards in Education [OFSTED].

Sammons, P., Thomas, S., Mortimore, P., Cairns, R. & Bausor, J. (1994c) 'Understanding the processes of school and departmental effectiveness', paper presented at the symposium *School Effectiveness and School Improvement: Bridging the Divide* at the annual conference of the British Educational Research Association, 9 September, St Anne's College, University of Oxford.

Sammons, P., Lewis, A., MacLure, M., Riley, J., Bennett, N. & Pollard, A. (1994d) 'Teaching and learning processes', in A. Pollard (ed.) *Look before you leap? Research evidence for the curriculum at Key Stage 2*, London: Tufnell Press.

Sammons, P. & Hillman, J. (1994) *Markets for Secondary Schools: The Interaction of LMS, Open Enrolment and Examination Results* proposal submitted to the Nuffield Foundation, London: ISEIC, Institute of Education.

Sammons, P., Thomas, S. & Mortimore, P. (1995) 'Accounting for Variations in Academic Effectiveness Between Schools and Departments: Results from the *Differential Secondary School Effectiveness Project* – a three-year study of GCSE performance', paper presented at the European Conference on Educational Research/BERA Annual Conference, Bath: 14–17 September.

Sammons, P., Hillman, J. & Mortimore, P. (1995) *Key Characteristics of Effective Schools: A Review of School Effectiveness Research.* London: Office for Standards in Education [OFSTED].

Sammons, P., West, A. & Hailes, J. (1995) *The Implementation of National Assessment at Key Stage 1: Practice in Five Inner City Schools.* London: ISEIC, Institute of Education, University of London.

Sammons, P., Thomas, S., Mortimore, P., Cairns, R. & Bausor, J. (1995a) 'Understanding School and Departmental Differences in Academic Effectiveness: Findings from case studies of selected outlier secondary schools in inner London', paper presented at the ICSEI, 3–6 January, Leeuwarden, The Netherlands.

Sammons, P., Cuttance, P., Nuttall, D. & Thomas, S. (1995b) 'Continuity of school effects: A longitudinal analysis of primary and secondary school effects on GCSE performance', *School Effectiveness & School Improvement* 6 (4): 285–307.

Sammons, P., Thomas, S. & Mortimore, P. (1995) 'Differential School Effectiveness: Departmental variations in GCSE Attainment', ESRC End of Award Report, Project R000 234130. ISEIC, Institute of Education, University of London.

Sammons, P., Mortimore, P. & Thomas, S. (1996) 'Do schools perform consistently across outcomes and areas', Chap 1, in J. Gray et al. (eds.) *Merging Traditions: The Future of Research on School Effectiveness and School Improvement.* London: Cassells.

Sammons, P., Thomas, S. & Mortimore, P. (1996) 'Differential school effectiveness: Departmental variations in GCSE attainment', paper presented at the School Effectiveness and Improvement Symposium of the annual conference of the American Educational Research Association, New York, 8 April 1996.

Sammons, P., Mortimore, P. & Hillman, J. (1996) 'A response to David Hamilton's reflections', *Forum* 38 (3): 88–90.
Sammons, P. & Reynolds, D. (1997) 'A Partisan Evaluation—John Elliott on school effectiveness', *Cambridge Journal of Education* 27 (1): 123–136.
Sammons, P. & Smees, R. (1997) *Developing Value Added Models at KS1 using Baseline Screening at Entry to School*. London: Institute of Education, University of London.
Sammons, P., Thomas, S. & Mortimore, P. (1997) *Forging Links: Effective Schools and Effective Departments*. London: Paul Chapman.
Sammons, P., West, A. & Hind, A. (1997) 'Accounting for variations in pupil attainment at the end of Key Stage One', *British Educational Research Journal* 23 (4): 489–512.
Sammons, P., Smees, R., Thomas, S., Robertson, P., McCall, J. & Mortimore, P. (1997) *The Impact of Background Factors on Pupil Attainment, Progress and Attitudes in Scottish Schools*. London: ISEIC, Institute of Education, University of London.
Sammons, P., Sylva, K., Melhuish, E., Siraj-Blatchford, I. & Dixon, M. (1998a) 'Effective provision of pre-school education [EPPE] project: Applying value added methods to investigate children's attainment and development over four years', paper presented at the International Congress for School Effectiveness & Improvement, University of Manchester, 4–7 January 1998.
Sammons, P., Taggart, B. & Thomas, S. (1998) *Making Belfast Work: Raising School Standards*. An Evaluation Report Prepared for the Belfast Education Library Board, London: ISEIC, Institute of Education, University of London.
Sammons, P., Thomas, S., Mortimore, P. & Walker, A. (1998b) 'Practitioners' views of effectiveness', *Improving Schools* 1 (1): 33–40.
Sammons, P., Thomas, S., Mortimore, P., Walker, A., Cairns, R. & Bausor, J. (1998) 'Understanding differences in academic effectiveness: Findings from case studies of selected outlier secondary schools in inner London', *School Effectiveness & School Improvement* 9 (3): 286–309.
Scheerens, J. (1992) *Effective Schooling: Research, Theory and Practice*. London: Cassell.
Scheerens, J. (1995). 'School Effectiveness as a Research Discipline', paper presented at the ICSEI Congress, 3–6 January, Leeuwarden, The Netherlands.
Scheerens, J. & Creemers, B.P.M. (1989) 'Conceptualizing school effectiveness', *International Journal of Educational Research* 13 (7): 691–706.
Scheerens, J. & Bosker, R. (1995) 'The generalizability of multilevel educational effectiveness models across countries', an EU-TSER proposal AREA II—Research on Education and Training, OCTO: University of Twente, the Netherlands.
Schweinhart, L.J., Barnes, H.V., Weikart, D.P., Barnett, W.M. & Epstein, A.S. (1993) 'Significant benefits: The High Scope Perry pre-school study through age 27', *Monograph of the High Scope Educational Research Foundation*, No 19. High Scope Press.
Schweitzer, J. (1984) 'Characteristics of effective schools', paper for American Educational Research Association.
Shipman, M. (1980) 'The limits of positive discrimination', in M. Marland (ed.) *Education for the Inner City*, Chap 5. London: Heinemann.
Shipman, M. & Cole, H. (1975) 'Educational indices in the allocation of resources', *Secondary Education* 5 (2): 37–38.
Sime, N. & Gray, J. (1991) 'The stability of school effects over time', paper presented to the British Educational Research Annual Conference, Nottingham Polytechnic, August 1991.

Sirotnik, K.A. (1985) 'School effectiveness: A bandwagon in search of a tune', *Education Administration Quarterly* 21 (2): 135–140.

Sizemore, B. (1985) 'Pitfalls and promises of effective schools research', *Journal of Negro Education* 54 (3): 269–288.

Sizemore, B. (1987) 'The effective African American elementary school', in G. Noblit and W. Pink (eds.) *Schooling in a Social Context: Qualitative Studies*. Norwood, NJ: Ablex.

Sizemore, B., Brossard, C. & Harrigan, B. (1983) *An Abashing Anomaly: The High Achieving Predominantly Black Elementary School*. University of Pittsburgh.

Slavin, R.E. (1987) 'A theory of school and classroom organisation', *Educational Psychologist* 22 (2): 89–108.

Slavin, R.E. (1996) *Education For All*. Lisse: Swets & Zeitlinger.

Smith, G. (1977) 'Positive discrimination by area in education: The EPA idea re-examined', *Oxford Review of Education* 3 (3): 269–281.

Smith, D.J. & Tomlinson, S. (1989) *The School Effect: A Study of Multi-Racial Comprehensives*. London: Policy Studies Institute.

Smith, I., Stoll, L., McCall, J. & MacGilchrist, B. (1998) *Improving School Effectiveness Project: Final Report April 1998—Teacher Questionnaires 1995/97*. Glasgow: Quality in Education Centre, University of Strathclyde.

Southworth, G. (1994) 'The learning schoo'l', Chapter 5, in P. Ribbens & E. Burridge (eds.) *Improving Education: Promoting Quality in Schools*. London: Cassell.

Stalling, J. (1975) 'Implementation and child effects of teaching practices in follow through classrooms', *Monographs of the Society for Research in Child Development*, 40 (7–8): Serial No 163.

Stalling, J. & Hentzell, S. (1978) *Effective Teaching and Learning in Urban High Schools*, National Conference on Urban Education, Urban Education Program, CEMREL St Louis, MO.

Stebbins, L., St Pierre, R., Proper, E., Anderson, R. & Cerva, T. (1977) *Education as Experimentation: A Planned Variation Model, Vol IV-A: An Evaluation of Follow Through*. Cambridge, MA: Abt Associates Inc.

Stedman, L. (1987) 'It's time we changed the effective schools formula', *Phi Delta Kappa*, 69 (3): 215–244.

Stoll, L. & Fink, D. (1989) 'Implementing an effective schools project: The Halton approach', paper presented at the International Congress for School Effectiveness, Rotterdam.

Stoll, L. & Fink, D. (1991) 'Effecting school change: The Halton approach', paper presented at the Fourth Annual Conference of the International Congress of School Effectiveness and School Improvement, Cardiff.

Stoll, L. & Fink, D. (1992) 'Effecting school change: The Halton approach', *School Effectiveness and School Improvement* 3 (1): 19–41.

Stoll, L. & Fink, D. (1994) 'Views from the field: Linking school effectiveness and school improvement', *School Effectiveness and School Improvement* 5 (2:, 149–177.

Stoll, L. & Mortimore, P. (1995) 'School effectiveness and school improvement', *Viewpoint 2*. London: Institute of Education, University of London.

Stoll, L., Myers, K. & Reynolds, D. (1996) 'Understanding ineffectiveness', paper presented as part of the symposium International Advances in School Effectiveness Research and Practice, at the Annual Conference of the American Educational Research Association, New York, 9 April 1996.

Stoll, L. & Fink, D. (1996) *Changing Our Schools: Linking school effectiveness and school improvement*. Buckingham: OUP.

Strand, S. (1997) 'Pupil progress during key stage 1: A value added analysis of school effects', *British Educational Research Journal* 23 (4): 471–487.

Stringfield, S., Teddlie, C. & Suarez, S. (1986) 'Classroom interaction in effective and ineffective schools: Preliminary results from Phase III of the Louisiana School Effectiveness Study', *Journal of Classroom Interaction* 20 (2): 31–37.

Stringfield, S. & Teddlie, C. (1987) 'A time to summarise: The Louisiana school effectiveness study', *Educational Leadership* 46 (2): 48–49.

Stringfield, S., Teddlie, C., Wimpleberg, R.K. & Kirby, P. (1992) 'A five year follow-up of schools in the Louisiana school effectiveness study', in J. Baslin & Z. Sass (eds.) *School Effectiveness and Improvement Proceedings of the Third International Congress for School Effectiveness*. Jerusalem: Magness Press.

Stringfield, S. (1994a) 'Outlier studies of school effectiveness', in D. Reynolds et al. (eds.) *Advances in School Effectiveness Research and Practice*. Oxford: Pergamon.

Stringfield, S. (1994b) 'A model of elementary school effects', in D. Reynolds et al. (eds.) *Advances in School Effectiveness Research and Practice*. Oxford: Pergamon.

Sugden, D. (1998) 'Helping children with learning difficulties', Chap 1, in D. Shorrocks-Taylor (ed.) *Directions in Educational Psychology*. London: Whurr Publishers Ltd.

Surrey Education Service (1994) *Surrey Screening Report 1993/4*. Kingston upon Thames: Surrey Education Service.

Swann Report (1985) *Education for All*, final report of the Committee of Inquiry into Education of Children from Ethnic Minority Groups, London: HMSO.

Tabberer, R. (1994) *School and Teacher Effectiveness*. Slough: NFER.

Taggart, B. & Sammons, P. (1988) 'Evaluating the impact of the raising school standards iniative in Belfast', paper presented at the American Educational Research Association Conference, San Diego, April 1998.

Teddlie, C. (1994a) 'The study of context in school effects research: History, methods, results and theoretical implications', in D. Reynolds et al. (eds.) *Advances in School Effectiveness Research and Practice*. Oxford: Pergamon.

Teddlie, C. (1994b) 'The integration of classroom and school process data in school effectiveness research', in D. Reynolds et al. (eds.) *Advances in School Effectiveness Research and Practice*. Oxford: Pergamon.

Teddlie, C. & Virgilio, I. (1988) 'School context differences across grades: A study of teacher behaviours', paper presented at the annual meeting of the American Educational Research Association, New Orleans, 1988.

Teddlie, C., Kirby, P. & Stringfield, S. (1989) 'Effective versus ineffective schools: Observable differences in the classroom', *American Journal of Education* 97 (3): 221–236.

Teddlie, C., Stringfield, S., Wimpelberg, R. & Kirby, P. (1989) 'Contextual differences in model for effective schooling in the USA', in B. Creemers, T. Peters & D. Reynolds (eds.) *School Effectiveness and School Improvement*. Lisse: Swets & Zeitlinger.

Teddlie, C. & Stringfield, S. (1993) *Schools Make a Difference: Lessons Learned From a 10 Year Study of School Effects*. New York: Teachers College Press.

Teddlie, C. & Reynolds, D. (1998) *The International Handbook of School Effectiveness Research*. Lewes: Falmer Press.

Thomas, S. & Nuttall, D.L. (1992) 'An analysis of 1991 Key Stage I results in Dorset: Multilevel analysis of English, mathematics and science subjects level scores', *British Journal of Curriculum and Assessment* 3 (1): 18–20.

Thomas, S. Nuttall, D. & Goldstein, H. (1992) 'The Guardian Survey (of A-level examination results)', October 1992, *The Guardian*. 20 October 1992.

Thomas, S. Nuttall, D. & Goldstein, H. (1993) The Guardian Survey (of A-level examination results)', November 1993, *The Guardian*. 30 November 1993.

Thomas, S. Nuttall, D. & Goldstein, H. (1993) *Report on Analysis of 1991 Examination Results prepared for the AMA*. London: University of London. Centre for Educational Research, LSE, London.

Thomas, S. Nuttall, D. & Goldstein, H. (1994) *Report on analysis of 1992 Examination Results*. Association of Metropolitan Authorities (in press).

Thomas, S. & Mortimore, P. (1994) *Report on Value Added Analysis of 1993 GCSE Examination Results in Lancashire*. London: Institute of Education.

Thomas, S. Sammons, P. & Mortimore, P. (1994) 'Stability and consistency in secondary schools' effects on students' GCSE outcomes', paper presented at the annual conference of the British Educational Research Association, 9 September, St Anne's College, University of Oxford.

Thomas, S. Sammons, P. & Mortimore, P. (1995) 'Stability and consistency in secondary school effects on students' outcomes over 3 years', paper presented at ICSEI, 3–6 January, Leeuwarden, The Netherlands.

Thomas, S., Sammons, P., Mortimore, P. & Smees, R. (1995) 'Differential secondary school effectiveness: Examining the size, extent and consistency of school and departmental effects on GCSE outcomes for different groups of students over three years', paper presented at the ECER/BERA annual conference, Bath, September 1995.

Thomas, S. & Mortimore, P. (1996) 'Comparison of value-added models for secondary-school effectiveness', *Research Papers in Education* 11 (1): 5–33.

Thomas, S. & Sammons, P. (1996) *Raising School Standards Initiative: The Development of Baseline and Value Added Measures for DENI*. London: ISEIC, Institute of Education, University of London.

Thomas, S. Sammons, P., Mortimore, P. & Smees, R. (1997a) 'Stability and consistency in secondary schools' effects on students' GCSE outcomes over 3 years,' *School Effectiveness & School Improvement* 8 (2): 169–197.

Thomas, S. Sammons, P., Mortimore, P. & Smees, R. (1997b) 'Differential secondary school effectiveness: Examining the size, extent and consistency of school and departmental effects on GCSE outcomes for different groups of students over three years', *British Educational Research Journal* 23 (4): 451–469.

Thomas, S. Sammons, P. & Street, H. (1997) 'Value added approaches: Fairer ways of comparing schools', *SIN Research Matters, No 7*. London: Institute of Education, University of London.

Thomas, S., Smees, R., MacBeath, J., Sammons, P. & Robertson, P. (1998) 'Creating a value added framework for scottish schools: A policy paper', paper presented at the International Congress for School Effectiveness & School Improvement, University of Manchester, January 1998.

Times Educational Supplement (1997) *Failing Schools*. Series of articles, 14th November 1997.

Tizard, & Hewison, J. (1982) 'Collaboration between teachers and parents in assisting children's reading', *British Journal of Education and Psychology* 52: 1–5.

Tizard, B., Blatchford, P., Burke, J., Farquhar, C. & Plewis, I. (1988) *Young Children at School in the Inner City*. Hove: Lawrence Erlbaum.

Tizard, J., Schofield, W. & Hewison, J. (1992) 'Symposium: Reading-collaboration between teachers and parents in assisting children's reading', *British Journal of Educational Psychology* 52 (1): 1–15.

Toews, J. & Barker, D.M. (1985) *The Baz Attack. A School Improvement Experience Utilizing Effective Schools Research 1981–1985*. Alberta: Ian Bazalgette Junior High School.

Tomlinson, J., Mortimore, P. & Sammons, P. (1988) *Freedom and Education: Ways of Increasing Openness and Accountability*. Sheffield Papers in Education Management 76, Sheffield: Sheffield City Polytechnic, Centre for Education Management & Administration.

Topping, K.J. (1992) 'Short and long-term follow-up of parental involvement in reading projects', *British Educational Research Journal* 18 (4): 369–379.

Trisman, D., Waller, M. & Wilder, C. (1976) *A Descriptive and Analytic Study of Compensatory Reading Programs*. Princetown: Educational Testing Service.

Tunley, P., Travers, T. & Pratt, J. (1979) *Depriving the Deprived*. London: Kogan Page.

Tunstall, P. & Gipps, C. (1995) 'How does your teacher help you to make your work better? Children's understanding of formative assessments', paper submitted to ECER/BERA Conference, Bath, September 1995.

Tymms, P. B. (1992) 'The relative effectiveness of post-school institutions in England and Wales', *British Educational Research Journal* 18 (2): 175–192.

Tymms, P., Merrell, C., & Henderson, B. (1997) 'The first year at school: A quantitative investigation of the attainment and progress of pupils', *Educational Research and Evaluation* 3 (2): 101–118.

United States Department of Education (1986) *What Works. Research about Teaching and Learning*. Washington, DC: United States Department of Education.

United States Department of Education (1987) *What Works Research about Teaching and Learning*. Washington: United States Department of Education (Revised edition).

United States General Accounting Office (1985) *Effective Schools Programs: Their Extent and Characteristics*. Washington DC: US General Accounting Office.

Van Der Werf, M.P.C. (1995) *The Educational Priority Policy in the Netherlands: Content, Implementation and Outcomes*. The Hague: CIP-gegevens Koninklijke Bibliotheek.

Van Velzen, W.G., Miles, M.B. & Ekholm, M. (1985) *Making School Improvement Work*. Leuven: ACCO.

Venezky, R. & Winfield, L. (1979) *Schools That Succeed Beyond Expectations in Teaching Reading*. Newark: University of Delaware.

Walberg, H.J. (1984) 'Improving the productivity of American schools', *Educational Leadership* 41 (1): 19–27.

Walberg, H.J. (1985) 'Homework's powerful effects on learning', *Educational Leadership* 42 (7): 76–79.

Walberg, H.J. (1986) 'Syntheses of research on teaching', Chapter 7, in M. C. Wittrock (ed.) *Handbook of Research on Teaching*. New York: Macmillan.

Wang, M.C. et al. (1993) 'Toward a knowledge base for school learning', *Review of Educational Research* 63: 249–294.

Watkins, C. (1996) 'Effective learning', *SIN Research Matter*. No.5, SIN Network Institute of Education, University of London.

Wayson, W. et al. (1988) 'Up from excellence', *Phi Delta Kappa*.

Weber, G. (1971) *Inter-City Children Can Be Taught to Read: Four Successful Schools*. Washington DC: Council for Basic Education.

Weinberger, J. et al. (1990). *Ways of Working with Parents Early Literacy Development.* Sheffield: University of Sheffield.

Weindling, D. (1989) 'The process of school improvement: Some practical messages from research', *School Organisation* 9: 1.

West, A. & Varlaam, A. (1991) 'Choosing a secondary school: Parents of junior school children', *Educational Research* 33 (1): 22–30.

West, A. Sammons, P. Thomas, S. & Nuttall, D. (1993) *Community Educational Indicators Final Report of Phase 1.* London: Centre for Educational Research, LSE, University of London.

West, A. West, R. & Pennell, H. (1993) *Additional Educational Needs Allowance: Examination of Options for Change.* Centre for Educational Research, LSE, London.

West, A. Hailes, J. & Sammons, P. (1995) 'Teaching and learning processes in inner city infant schools: Current policy and practice', *Clare Market Papers, No 10.* London: Centre for Educational Research, LSE, University of London.

West, A. Hailes, J. & Sammons, P (1995) 'Classroom organisation and teaching approaches at key stage one: Meeting the needs of children with and without additional educational needs in five inner city schools', *Educational Studies* 21 (1): 99–117.

West, A., Pennell, H., Thomas, S. & Sammons, P. (1995) 'The origins and development of the European Community educational indicators project', *European Educational Research Association Bulletin* 1 (3): 3–11.

West, M. & Hopkins, D. (1996) 'Reconceptualising school effectiveness and school improvement', paper presented at the School Effectiveness and Improvement Symposium of the Annual Conference of the American Educational Research Association, New York: 8 April.

White, J. (1997) 'Philosophical perspectives on school effectiveness research', chapter in J. White & M. Barber (eds.) *Perspectives on School Effectiveness and School Improvement.* London: Bedford Way Paper, Institute of Education, University of London.

White, J. & Barber, M. (eds.) (1997) *Perspectives on School Effectiveness Improvement.* London: Bedford Way Paper, Institute of Education, University of London.

Whitty, G. (1997) *Social Theory and Education Policy: The legacy of Karl Mannheim.* Karl Mannheim Memorial Lecture, 9 January 1997. London: Institute of Education, University of London.

Widlake, P. & Macleod, F. (1994) *Raising Standards. Parental Involvement Programmes and the Language Performance of Children.* Coventry: Community Education Development Centre.

Wikeley, F. (1998) 'Dissemination of research as a tool for school improvement?' *School Leadership and Management* 18 (1): 59–73.

Wilcox, B. & Gray, J. (1996) *Inspecting Schools—Holding schools to account and helping schools to improve.* Buckingham: OUP.

Willms, D. (1985) 'The balance thesis—contextual effects of ability on pupils 'O'grade examination results', *Oxford Review of Education* 11: 33–41.

Willms, J.D. (1986) 'Social class segregation and its relationship to pupils' examination results in Scotland', *American Sociological Review* 51: 224–241.

Willms, J.D. (1987) 'Differences between Scottish education authorities in their examination attainment', *Oxford Review of Education* 13 (2): 211–232.

Willms, J.D. (1992) *Monitoring School Performance: A Guide for Educators.* London: Falmer.

Willms, J.D. & Raudenbush, S.W. (1989) 'A longitudinal hierarchical linear model for estimating school effects and their stability', *Journal of Educational Measurement* 26 (3): 209–232.

Willms, J. & Raudenbush, S. (1994) 'Effective schools' research: Methodological issues', in T. Husen, & N. Postlethwaite (1994) (eds.) *The International Encyclopedia of Education*. 2nd edn, Vol. 4. Oxford: Pergamon.

Wilson, B. & Corcoran, T. (1988) *Successful Secondary Schools*. London: Falmer Press.

Wimpleberg, R. et al. (1989) 'Sensitivity to context: The past and future of effective schools research', *Educational Administration Quarterly* 25: 82–107.

Witziers, B. (1994) 'Coordination in secondary schools and its implications for student achievement', paper presented at the annual conference of the American Educational Research Association, 4–8 April, New Orleans.

Woodhouse, G., Rasbash, J., Goldstein, H. & Yang, M. (1992) *A Guide to ML3 for New Users*. London: Institute of Education, University of London.

Wynne, E. (1980) *Looking at Schools: Good, Bad and Indifferent*. Lexington: Heath.

Yang, M., Woodhouse, G., Goldstein, H., Pan, H. & Rasbash, J. (1992) 'Adjusting for measurement unreliability in multilevel modelling', *Multilevel Modelling Newsletter* 4 (2): 7–9. London: Department of Mathematics, Statistics & Computing, Institute of Education, University of London.

Yang, M. & Goldstein, H. (1997) *Report on Value Added Analysis for Primary Schools in Hampshire County*. London: Mathematical Sciences, Institute of Education, University of London.

Yelton, B.T., Miller, S. K. & Ruscoe, G.C. (1994) 'The stability of school effectiveness: Comparative path models', paper presented at the annual meeting of the American Educational Research Association, 4–8 April, New Orleans.

Young, P. (1990) 'The audit commission and accountability', in R. Saran & V. Trafford (eds.) *Research in Education Management and Policy. Retrospect and Prospect*. Lewes: Falmer Press.

Subject Index

academic attainment 80–94
  effectiveness 214
  emphasis 202–3, 245
  versus affective/social outcomes 83–4
Access and Achievement in Urban Education 51
accountability 72, 73, 94, 103, 153
achievement, sense of 268
achievements of the field 346
added value *see* value-added approach
adding value to pupils' lives 340
adjustment for intake 156
Advanced-level 82–4, 94
Advances in School Effectiveness Research 346
advantage, disadvantage and academic performance 55, 61
AEN (Additional Educational Needs) 32–4
age 43, 59, 63
aggregated analysis *see* OLS
aims
  of effectiveness research 75, 188–189
  of the school 314
AMA (Association of Metropolitan Authorities) 31, 49–54, 60
American Education Research Association Conference 294
American research 248
application of the research 253
APs (Action Plans) 303–9, 331–2, 340
area, effect of catchment ix, 10, 29, 39–40, 56
arithmetic 85; *see also* mathematics
articles, choice of xii, 339
Assessing School Effectiveness (1994) 22–70
assumptions of school effectiveness research 1
attainment
  predictors of 38
  prior *see* prior attainment

attendance 79
attitudes 79
Audit Commission's three E's 152–3
Australian study 231

background characteristics (factors) 5–20, 42, 45, 52, 55, 76–8, 88, 95, 98, 104–27, 135, 137, 141, 143–4, 158–62, 192, 279–80
  ethnic 11, 30, 38, 39, 42–5, 49–50, 59, 92–4, 120, 125
  occupational 11
  social 28
backward mapping 62
baseline (entry-to-school)
  assessments 271–90, 310
  measures, standard 341
basic skills (subjects) 73, 79, 87, 95, 126–7, 145, 161–2
behaviour 79, 244, 268
BELB (Belfast Education and Library Board) 293
  perspective 325–6
Belfast evaluation 257
bench-marking 23
BERA (British Educational Research Association) 103
BERA's Research Intelligence 226
Birmingham City Council 152
BLTs (Bradford Lifestyle Types) 37–9
BMT (National Foundation for Educational Research Basic Mathematics Test) 105–27
Bradford analysis 37
bridging theory and practice 341
British Births Cohort Study 129
buildings, school 12

car ownership 39
case studies 173, 194
catchments 295; *see also* area, effect of catchment

causality, views of 220
causes, underlying 229–51
caution 87, 127
census data 8, 10, 27, 29, 32, 34, 38–40, 42–3, 52, 56–8, 61
  surrogates 62
Centre for Educational Research 129
challenges
  intellectual 166, 177–9, 210
  for researchers 350–2
Changing Our Schools 256
Choice and Diversity (1992) 22, 154, 247
City Technology Colleges 341
class size 163–4
classroom and school processes 149–251
cluster analysis 47, 61
CMC (Central Management Committee) 297, 303–7
Co-ordinators' views 316, 321, 327
CoGs (Chairs of Governors) 298–9, 316, 326
collaborative work 358
collegiality 200
common sense 216
  versus research 170
communication
  between staff and Senior Management Teams 314
  between teachers and pupils 165–7, 169, 178–9, 180–2
comparison of 'like with like' 2, 25–70, 158, 223, 284
complex models 345
confidence limits 45, 54, 59–60, 62, 97
Conservative administration 21
consistency 82–84, 86, 94
consumers 103
context/input/output equation 343
context, performance more securely in 27, 63, 193–4
contextual *versus* differential effects 90
contextualisation of schools' results 24, 25, 345
continuity
  of school effects 87–9, 129–48
  of teachers 165
controlling for differences in intakes 230
controls, statistical 3, 74–6, 77, 84

correlation techniques 40, 175
criteria for adequate research 75
cross-classified models 131
cross-level influences 237–8
culture, school *see* ethos
cure-alls, no 222
curriculum development 318

definitions of effectiveness 75–80, 188–90
delinquents 155
DENI (Department of Education Northern Ireland) 293, 295, 298, 301
departmental differences 228, 235–7
deprivation index 34
DES (Department of Education and Science) 153
determinism, sociological x
DFE (Department for Education) 30, 58, 74, 134, 154, 218, 247
DfEE (Department for Education and Employment) 184, 340
dialogue, importance of 342
differences between schools *versus* between pupils 138
differential school effects (within-school differences) 90, 101–27
Differential School Effectiveness Project 101–27, 232–51
disadvantage
  educational 3, 7–9, 14, 21, 30, 161, 295
  social 345
discipline 156, 208
distribution of resources *see* extra resources, direction of
DoE (Department of the Environment) 32, 34–6
Dutch research 227, 231
Dutch School Improvement Project 299

EAL (English as an Additional Language) 274, 280–1
earnings 39; *see also* income
Economic and Social Research Council 341
economic-related component (E) 35
ED (enumeration district) 27–8, 34
Education (Schools) Act (1991) 103

Education Minister 298
educational
  goals 157–9
  quality 154
  standards, impact on 321–25
Educational Action Zones 7
effective
  classroom practice 176–81
  school, definition of 255
  schools 162–72
  teachers 238
  versus ineffective schools 134; and *passim*
effectiveness
  and 'quick fixes' xi
  and improvement, school 253–4, 269, 287, 292–336
  and improvement fields, gap between 340
  schools' relative 31–2, 53; *see also* residuals
efficiency, penalising 10
elementary schools 87
empirical
  approach 344
  findings and models, relations between 240
  studies outnumbered by reviews 221
  studies, policy and 229
English 85, 90
  fluency in 11, 19, 138, 272
Enhancing Education 340
EPA (educational priority areas) 5; *see also* area
EPI (Education Priority Indices) 6–20, 29, 33, 35
EPPE (Effective Provision of Pre-school Education) 353
equity 223, 271, 273, 280, 345
ERA (1988 Education Reform Act) 21, 84, 103, 152, 153, 156, 295
ERT (Edinburgh Reading Test) 105–27
ESRC (Economic and Social Research Council) 102, 129–48
ethnicity *see* background, ethnic
ethos (school culture) 2, 154, 168, 200, 244, 248, 250, 329
European Union 152

Evaluating Schools 304
examination performance 36–8, 40–1, 47–9, 95; and *passim*
expectations, high 206–8
explanatory power 244
expressive domain 248
external factors 71
extra resources, direction of 8, 29

factors to be taken into account
  secondary school 259–64
  departmental 264–68
family size 30, 39
FE (Further Education), entry to 325, 330
feedback
  to pupils 166, 320
  to schools 98, 126, 272–3
feel-good fictions 219–21
Field Officers' views 316, 321, 326
Fifteen Thousand Hours ix, x, 130, 151–2, 169–72, 229, 239
focusing on individual pupils 169
Forging Links: Effective Schools and Effective Departments 228, 256–69, 342
Forum 175
Foundations of Educational Effectiveness 346
Framework for the Inspection of Schools 184, 224
FSM (free school meals) 15, 17, 19, 24, 27, 29–30, 33, 37, 42–5, 49–53, 59, 63, 93, 137, 272, 274, 280
further research 63

GCSE (General Certificate of Secondary Education) 22, 25–70, 85, 94; and *passim*
gender *see* sex
glossary of terms 276
goals of the school 157–8, 198–200, 314
Governing Bodies 194
governors
  informing 184
  role of 218
grammar schools 41
grouping schools 47–55, 61–64, 70
guidelines, following 165, 200

Hackney 34
HE (Higher Education), entry to 39, 325, 330
Headstart anti-poverty programmes 130
high levels of noise 179
high quality data 341
HMCI (Her Majesty's Chief Inspector of Schools) 22
HMIs (Her Majesty's Inspectors) 32, 51, 71
HoDs (Heads of Department) 243, 245, 247, 251, 256
  views of 257–69
homework 320
hostility in the mainstream 344
housing conditions 30, 39
housing tenure 58
housing-related component (H) 35
HTs (Head Teachers) 195–8, 226, 244, 245, 247, 251, 313
  views of 257–69, 272, 316, 320–21, 327

ICSEI (International Congress for School Effectiveness and Improvement) 104, 346
ILEA (Inner London Education Authority) ix, 6–20, 33, 51, 104, 158–72
ILEA indices see EPI
implications for school improvement 341
improvement
  evaluating school 227–51, 291–336
  school, and effectiveness 25–6; and *passim*
  evidence of 332
Improving Education: Promoting Quality in Schools 151
Improving School Effectiveness Project 292, 341–2
Improving Schools Journal 342
income 30, 49, 121–2, 272
indicators, relevant 65–69
inequalities of opportunity xiii
influencing policy and practice 344
inner cities 60, 75, 83, 91, 249
input/output models 149
input-process-output equation 230, 233
Inspection Process in Schools 304

inspectors see HMIs
instrumental domain 248
intake factors 6, 28, 58–62, 82, 159, 223
  versus outcome measures 27
international studies 231
intervention based on research 340, 342
invidious comparisons, potentially 272
involvement
  of the teachers 165
  of the deputy headteacher 165
IQ (intelligence quotient), influence of ix, 232
IQEA (Improving Quality of Education for All) 256, 342
ISEIC (International School Effectiveness & Improvement Centre) xiv, 186, 272, 292, 297, 339
ISEP (Improving School Effectiveness Project) 257, 292–3
ISERP (International School Effectiveness Research Programme) 193

JSP (Junior School Project) 104–27, 133–4, 135; see also *School Matters*
judgement, use of 163

Key Characteristics of Effective Schools 183–218
Key Characteristics Review 292
KS (Key Stages, National Assessment at) 21, 28, 63, 153, 272–90, 295, 353

labelling children 9
Labour government 25, 152, 358–9
language factor 39; see also English, fluency in
language/reading, consistency of performance in 83
language-based measures 95
later
  attainment 125
  performance 147
leadership 164, 195–8, 247, 329
league tables see performance tables
learning, time spent on 202
LEAs (local education authorities) 32, 33, 152–4, 156, 184, 194, 254, 271–2, 301, 341

'like with like' *see* comparison of 'like with like'
limitations of the research xii
linking findings with theory 230–51
links between factors 240–1
literacy 272
Literacy and Numeracy Task Force 176
LMS (local management of schools) 7, 21, 33–34, 153, 171–2, 218, 258, 268, 295
Local Government Finance Act (1988) 153
London Institute of Education 228, 292, 339
London School of Economics 129
long-term
  benefits 334–6
  effect *see* continuity of school effects
longitudinal studies 30, 56, 85, 98, 104, 144, 236, 353
Louisiana School Effectiveness Study 87
LRT (London Reading Test) 134–6, 141–5

management 333
managers, informing 184
market forces 73, 238
mathematics/arithmetic 31, 83, 91, 105–27, 275, 278
MBW (Making Belfast Work) 292–335
means-related benefits 9
measurement issues 343
media coverage, impact of xiii, 23–24, 184, 298
methodology 28–40, 58, 98, 133, 189–92, 343
  substantial agreement on 104
millenium, beyond the 339–59
misleading results 161
mixed-activity sessions 166, 177
mobility 18
model
  an expanded 241–3
  'best fit' 58
Models of Effective Schools Project 102, 129–48
Models of Reality: New approaches to the understanding of educational processes 345
monitoring progress 209–10, 324, 331–2

multilevel modelling 26, 27, 32, 39–40, 46, 49, 53–4, 59, 61–3, 88, 96, 106, 124–7, 131–3, 135, 141, 146, 229–30, 272, 281, 345
Multilevel Models' Project 103–27

'naming and shaming' 71, 298, 343
national data 28–9, 46, 56–7
National Child Development Study 129
National Commission on Education 297
National Curriculum and its Assessment, The 26
natural justice 25, 74
NCA (National Curriculum and Assessment) 21, 28, 103, 121, 153, 156, 170–2, 218, 247, 258, 268, 272, 295
NCER (National Consortium of Examination Results) 38–70
neighbourhood *see* area
non-academic outcomes 64
non-cognitive outcomes 94, 176
non-manual workers 59, 93, 107, 162
non-school based personnel, benefits for 330
number on roll 59
numeracy 272, 323
NWREL (Northwest Regional Educational Laboratory) 96, 206

OECD (Organization for Economic Cooperation and Development) 155
OFSTED (Office for Standards in Education) 22–70, 78, 134, 184, 206, 220–4, 258, 291, 302, 340
OLS (ordinary least squares) regression 27, 34, 46–9, 52–4, 59–62
one-parent families 17, 30, 36, 39, 58
opportunities 5
oracy (speaking skills) 95, 162
outcomes
  educational xi, 26, 159, 328
  expected (predicted) 47
outlier schools and departments 80–2, 86, 230
over- or under-performing schools 55
over-emphasis on measurement of achievements 185

Pacific rim, educational systems in 231
parents'
　choice 21, 73, 156–7
　involvement 167, 211–13, 290
　on income support 32
　views xi, 1–2, 293
Parents' Charter 103
partnership 340
PCA (principal components analysis) 35
performance
　indicators 101–27
　school, in context 21–64
　tables ('league tables') 22, 24, 27, 28, 74, 97, 103, 121, 125, 192, 224, 225
personal account 339–59
PICSI (Pre-Inspection Context and School Indicator) 26, 39, 63
PIs (performance indicators) 154
pleasant environment 268
Plowden Report (1967) 5, 7, 8, 9, 11, 20, 29, 33
policies, school 163–5
policy
　government x, 7
　makers 94, 153–4, 280
　national 273
population
　density 39
　mobility 39
positive
　discrimination 5, 7, 8, 9, 20, 29
　reinforcement 208–9
post code classification 37, 39, 57
poverty 302; *see also* income
practical messages for school improvement 344
practitioners 154; *see also* teachers
　views of 253–69
praise, greater effect of 209
pre-school experience 280
predictors of later attainment 18
press *see* media coverage
primary schools 80, 142–6; *see also* 'School Matters'
Primary Yearly Record 167
Principals *see* Head Teachers
prior attainment 26, 28, 31, 42, 54–5, 61–4, 77, 82, 87–8, 90–2, 93, 98, 104–27, 135, 137, 143, 281

processes
　affecting school effectiveness 96
　and outcomes, understanding 343
　importance of linking outcomes and 185
　school 159, 162–4, 229
professional judgement 60
progress, relative 27
pseudo-science, accusations of and response to 222–6
public image 268
purpose, sense of 248

QCA (Qualifications and Curriculum Authority, *formerly* SCAA) 131, 283, 295
qualifications, parental 30, 44–46, 59
qualitative case studies 229–30, 244
quality
　in education 151
　of human relationships 248
　of teaching and learning 2, 173, 246, 319
Quality in Education Centre at the University of Strathclyde 292, 341
questionnaire data, using 311–12
quick fixes, no 194, 292, 344

race, influence of ix; *see also* background, ethnic
ranking, schools' 126; *see also* performance tables
raw *versus* 'value-added' results 122–3, 156
reading
　attainments 9, 31, 91, 105–27, 275, 277, 323
　versus maths and science 285
record keeping 167
register of children 'At Risk' 17
regression techniques 40, 175
relative *versus* absolute measure of disadvantage 35
reliability 2
researchers 154
residuals, schools' (i.e., school effectiveness) 45, 52, 54–6, 59, 61–2, 78, 87, 105, 124, 126, 192
resources 217, 304; *see also* extra resources

INDEX

RSS (raising school standards) 7, 257, 271, 292–336
sampling 96

SATS (standard assessment tasks) 153
SCAA (School Curriculum and Assessment Authority, *now* QCA) 131, 283, 295
school
    and classroom processes 149–251, 174, 187
    effectiveness and teacher effectiveness 186
    environment 132
    improvement and school effectiveness 170–2, 183, 188, 357–9; and *passim*
    leavers 73
    management, impact on 312
    practices 132
    versus classroom effects 236
School Development Planning 297
School Effectiveness and School Improvement 346
School Effectiveness, Pedagogy and SEN in Mainstream Primary Schools 356
School Matters (Junior School Project) ix, x, 77, 89, 130, 151–72, 224, 253
    follow-up 88–90
school-based staff development 213
school-level variation 59
schools
    advantaged 164
    influence of 1
    make a difference 190–1
    performance by 121–2
Schools and Quality 155
schools' effects *see* residuals
science 275, 278
Scottish research 257, 292
SDP (School Development Plan) 304, 313
secondary schools 25, 80
selective/non-selective schools 42–4
self-
    concept 79
    esteem, pupils' 210–11
    evaluation by schools xi, 63, 170–2, 340
    fulfilling prophecies 3, 204
SEN (Special Educational Needs) 29, 52, 59, 155, 354–57

senior management teams *see* SMTs
sense
    of achievement 268
    of purpose 248
SER (school effectiveness research) 185 and *passim*
SES (socio-economic status), influence of ix, 26, 42–5, 58, 90, 93–5, 193, 237, 293, 295
sex (gender) 26, 29, 30, 42–5, 49–50, 59, 63, 92, 115–8, 141, 143, 280
sick institutions, schools as 222
significant others 132
simplification, over, dangers of xii
SIN (School Improvement Network) 342
size of school effect (unexplained variance) 78, 79
SMTs (Senior Management Teams) 197, 247, 251, 256, 268, 313
social/affective outcomes 191, 214
social
    and cultural context 145, 231
    class 17, 30, 39, 58, 118–20, 223
    deprivation scores 37
    engineering 219–22
    indicators 12, 27
    (non-cognitive) outcomes 191, 214
socio/economic factors *see* SES
sociology
    educational 5
    linking with 345
SOEID (Scottish Office Education & Industry Department) 292
special
    circumstances 34
    help 29
    needs 42; *see also* SEN
SSAs (Standard Spending Assessments) 32–3
stability over time 84–8, 95, 285, 349; *see also* continuity of school effects
staff 152; *see also* teachers
    development 333
    experienced 51
    involvement 315
    morale 315
staffing difficulties 164
statistical
    analysis 40–7

significance 45, 52, 55, 63, 98, 224
uncertainty xii
statistically and educationally invalid 348
structured day 165, 176–7
structures *versus* processes 342
students' 'dowry' xi
  motivation 202
  progress 173
  views xi, 1–2, 293
subject (departmental) differences 81–3, 85, 93
subjective, too 27
Success Against the Odds 297, 316
Success for All 293
supervision of pupils' work 166
Surrey Code of Conduct 272, 274
Surrey Education Service 271–90
sustaining change 357

TAs (teacher assessments) 153, 275–6
teachers 226
  collaboration of 318–9
  effects of 95
  effective 250
  expectations of 51
  influence of 1, 148–251
  informing 184
  involvement of 165
  morale of 71
  perspective of 327–8
  qualifications and experience of 243
  views of xi, 1–2, 293, 316
  see also staff
teaching
  and learning, centrality of 201, 215
  styles 174–5, 204–6

Test/Task results 275–6, 283
test-reliability 125
theoretical basis of the field 227–51
theories and models 232–5
theory development 97
training of teachers 175
transformation of data 40
TSN (Targeting Social Need) 300–3

under-reporting 58
understanding academic effectiveness 239–47
unemployment 39, 58
University of Birmingham 152
urban deprivation 34–6
Urban Deprivation Index (1991) 36
US Department of Education 206

validity 2, 28, 54, 58
value-added approach 22, 24, 25, 26, 27, 39, 49, 53–5, 60, 62–4, 75, 88, 105, 145, 147, 156, 161, 171, 183, 223, 232, 268–9, 271–90, 348
values in education xi–xiv
variance, school-level 43
VR (verbal reasoning) 15, 17, 19, 49–52, 90, 135–6, 141–5

What Works 184, 185, 194
why things work 232
withering away 74
within-school variation 231
work-centred environment 166, 179–81
writing 275, 278

CONTEXTS OF LEARNING
Classrooms, Schools and Society

1. *Education for All.* Robert E. Slavin
    1996.   ISBN 90 265 1472 7 (hardback)
            ISBN 90 265 1473 5 (paperback)

2. *The Road to Improvement: Reflections on School Effectiveness.* Peter Mortimore
    1998.   ISBN 90 265 1525 1 (hardback)
            ISBN 90 265 1526 X (paperback)

3. *Organizational Learning in Schools.* Edited by Kenneth Leithwood and Karen Seashore Louis
    1999.   ISBN 90 265 1539 1 (hardback)
            ISBN 90 265 1540 5 (paperback)

4. *Teaching and Learning Thinking Skills.* Edited by J.M.H. Hamars, J.E.H. van Luit and B. Csapó
    1999.   ISBN 90 265 1545 6 (hardback)

5. *Managing Schools Towards High Performance: Linking School Management Theory to the School Effectiveness Knowledge Base.* Edited by Adrie J. Visscher
    1999.   ISBN 90 265 1546 4 (hardback)

6. *School Effectiveness: Coming of Age in the Twenty-First Century.* Pam Sammons
    1999.   ISBN 90 265 1549 9 (hardback)
            ISBN 90 265 1550 2 (paperback)